EXTENDING EDUCATION THROUGH TECHNOLOGY

SELECTED WRITINGS BY JAMES D. FINN ON INSTRUCTIONAL TECHNOLOGY

Edited by Ronald J. McBeath

Publication Steering Committee
Appointed by California Association
for Educational Media & Technology

William H. Allen
Diana Z. Caput
Robert C. Gerletti
Frank N. Magill
Paul V. Robinson

Association for Educational
Communication & Technology

INFORMATION AGE
PUBLISHING

80 Mason Street • Greenwich, Connecticut 06830 • www.infoagepub.com

Photo by Trojan Camera, Los Angeles
Compliments of Sue Meador

James D. Finn
March 7, 1915–April 2, 1969

ISBN: 1-59311-138-x

Printed in the United States of America

They're Back!

Thanks to the efforts of Information Age Publishing, seven classic publications in the field of instructional technology are once again available. These seven publications have been on my bookshelf for over 25 years, and are alongside Richard Clark's *Learning From Media*, also published by Information Age.

My copies were worn and torn, and often used. When asked for a list of publications that I considered "must" reading for those in the field, these were at the top of my list. While I do not use them every day, I use them often enough in classes, in meetings, and as references for newly written papers that I keep them close at hand. In my opinion, they are the basis for a professional library of any person in the fields of instructional technology, instructional media, instructional design, or distance education.

Extending Education Through Technology, a collection of writings by Jim Finn, long considered the "father of educational communications and technology," features articles written by Finn decades ago that are still widely quoted and directly relevant to the issues of the field today. One article alone, "Professionalizing the Audiovisual Field," is worth the price of this classic publication.

The history of the field, *The Evolution of American Educational Technology*, by Paul Saettler is *the* basic reference for how the field has grown and become the driving force in education and training that it is today. Every student of the field should read this book.

Three books on this list of classics, Ball and Barnes' *Research, Principles, and Practices in Visual Communications*, Chu and Schramm's *Learning from Television*, and Ofiesh and Meierhenry's *Trends in Programmed Instruction*, are the primary sources for research and design in our field. Some claim, and they are probably correct, that much of what are considered "best practices' today can be traced directly back to the conclusions provided by these three extremely important monographs. Change the terms visual communication, television, and programmed instruction to visual literacy, distance education, and e-learning, and you have classic publications with direct implications to the work of most of us today.

Robert Heinich's often quoted and rarely found classic, *Technology and the Management of Instruction*, is a masterpiece of writing and advice about the field that resonates strongly today. Heinich is still a leader in the field, and many consider this monograph to be his best work.

With little doubt, the 20 years of Okoboji conferences set the stage and provided a platform for leadership development and intellectual growth in the field. The Okoboji conferences have been often mimicked but never duplicated. This summary of the 20 years of conferences by Lee Cochran, the driving force behind them, is a must for every book shelf.

As Santayana noted, "Those who cannot remember the past are condemned to repeat it." These classic publications do not prevent us from making the same errors that these writers admonished us from committing. Once read however, they do help us remember, and remember well.

—**Michael Simonson**

Table of Contents

Acknowledgements

The editor wishes to acknowledge the professional interest and financial support provided by the Audio Visual Education Association of California (now known as the California Association for Educational Media and Technology) during the development of this publication. It was at the suggestion of Jerrold E. Kemp (President, 1970), that a steering committee was formed. One of the members of this committee, Robert C. Gerletti, as President of the Association for Educational Communications and Technology (1970) provided an important liaison with the national organization, the publishers of this book. The other members of the steering committee who assisted generously in supporting and guiding this publication through their comments, criticisms and expert knowledge were William H. Allen, Frank N. Magill, Paul V. Robinson, and Diana Caput, who also updated the bibliography, located copies of Finn papers, and provided them for the committee.

I would also like to thank Aleen Marquardt for her assistance in typing, O. B. Carleton for his valuable comments during the preliminary editing, and Gail Hipsher for her editorial assistance during the publication of this book.

One person I would like to thank in particular is Betty Finn who, through discussions with the editor and the provision of professional documents and recordings of her late husband, helped to establish a broader base of reference. Through her generosity, a portion of the income from this publication is being contributed to a memorial fund in the honor of James D. Finn for leadership activities in the field of Instructional Technology.

I will always be grateful to my wife, Marjorie, and my children for giving me the opportunity to spend so much time on this enterprise.

Ronald J. McBeath
1972

Notes About James D. Finn

Born in Great Falls, Montana, 1915, James Donald Finn attended the public schools there. He received his B.S. degree from Montana State College and his M.A. degree from Colorado State College of Education. He pursued further graduate study in educational radio and motion pictures at several institutions and research centers prior to World War II under a Rockefeller Fellowship, and after the war, completed his Ph.D. in Education at Ohio State University with Edgar Dale, one of the pioneers in the field. He was married, the father of two daughters, and resided in Whittier, California.

During World War II, Finn served as Chief, Instructional Aids Service, and Assistant Director for Training for the Command and General Staff School, Fort Leavenworth, Kansas. He was a captain in the Signal Corps at the time of his discharge.

As Professor of Education at the University of Southern California from 1949—1969, Finn's teaching brought him in close contact with students of education in general and instructional technology in particular. He taught summer sessions at Michigan State University and Syracuse University; delivered invited lectures in many school districts and at over 20 institutions for higher learning throughout the United States. In addition to being Chairman of the Instructional Technology Department in the School of Education, Dr. Finn also served for several years as Acting Head of the Department of Cinema in the College of Letters, Arts and Sciences at U.S.C.

Having served the Department of Audio-Visual Instruction of the National Education Association in many capacities, James D. Finn was elected president in 1960. He was also president of the Educational Media Council, 1961—1963, a council of 14 national organizations—educational and commercial—in the educational media field.

Finn's continuing involvement with both the theoretical and practical foundations of instructional technology is reflected in his bibliography of more than 100 publications. In addition to "The Audio Visual Equipment Manual" (1957), a major work for practitioners with immediate concerns about equipment operation, Finn made a long-range contribution by helping to found the *Audio-Visual Communication Review.* Finn served as senior consulting editor of the journal which is regarded as the leading theoretical organ for the field. He also edited *Teaching Tools* for many years.

Dr. Finn directed several national studies in the 1960s. The Technological Development Project of the National Education Association was designed to investigate and assess technological developments and trends in education, and was followed by the Instructional Technology and Media Project. Both of these were funded by the U. S. Office of Education, as were the Educational Media Specialist Institutes and the National Special Media Institutes, which were designed to provide media training programs for specialists. Going beyond the immediate field of public education, Finn directed the Medical Information Project for General Practitioners funded by the U. S. Public Health Service.

Other specialized studies include the development of the National Information Center in Educational Media, preparation of prototype instructional materials for use in Job Corps Training Centers, a study on newer media and education conducted with Robert O. Hall at the invitation of the Harvard Graduate School of Education, and a study on educational technology for the National Commission on Technology, Automation and Economic Progress.

As a consultant in instructional technology, Finn worked with state departments of education, teachers' associations, school districts, and universities. He was a special consultant to the U. S. Office of Education on educational media and also consulted with a wide range of industrial corporations during his career.

Introduction

Throughout his professional life, James D. Finn demonstrated in many ways his commitment to education and the necessity for extending education in our technological age. His extensive publications range through the why, what and how of education. Recognized nationally and internationally as an authority on instructional technology, he earned the respect of his field as a practitioner, researcher and leader.

The central purpose of this book is to bring together a selection of James Finn's writings for students of media and instructional technology and for decision makers who are concerned about education.

The first chapter, "Criticisms, Traditions and Challenges," provides us with a general perspective in which Finn examines the educational setting, asks why, and offers a challenge. He makes us aware of the importance of vigorously examining our ideas, practices and institutions. For Finn, philosophizing is an essential component of future planning if we are to go beyond the expedient.

In "Automation and Education," Finn focuses more specifically on the relationship between automation and education. He recognizes the realities of living in a technological age and considers the implications for education.

"From Audiovisual to Instructional Technology" is an historical review consisting of 11 papers developed over a period of 15 years in which Finn provided leadership as he fought to change a movement into a profession.

The last chapter, "Commitment to the Future," reveals Finn's ability to move with the times as he re-examines instructional technology and the problems of institutionalization. He assesses some of the changes in the social-political-economic scene and returns to the importance of keeping man in the center of our educational endeavors.

The selection of papers and the structuring of the book into these four major sections emerged after an examination of Finn's published and unpublished material, and discussions with others interested in this publication. The articles cover a period of 15 years. Co-authored articles and research reports have been excluded from the selection so that a concentration could be made on the distinctly individual statements which span Finn's career. A list of all Finn's writings is included in the bibliography at the end of the book.

Section I

Criticisms, Traditions and Challenges

Criticisms, Traditions and Challenges

Frontiersman in the field of education, James D. Finn as an instructional technologist ventured further than any other educator in exploring the impact, implications and consequences of the technological revolution in education. His scholarly concern for professionalizing the field of education through technology parallels the challenges he offered, the indictments he presented, and the penetrating attacks he made on the academic establishment.

Finn considered the need for trained intelligence and improved communication in this age of rapid change to be of paramount importance. His impatience with academicians and members of the "literary set," who loiter in their ivory towers while the great force of the scientific and technological revolution sweeps through Western culture, was also directed toward members of his own professional field who do not respond creatively to the humanistic challenges of this technological age.

Faith in the human potential and antagonism toward dictatorial elitists are characteristically portrayed throughout Finn's writings. He recognized the interdependence of men and machines and charged us to see technology as an organic process transforming our world. The positive approach he took toward this transformation is evidenced in this excerpt from "A Revolutionary Season" (1964):

> Organized education faces the time of revolution. It must seize on and keep (and change in form, if necessary) those values most necessary for ultimate survival; it must cast away much of what has been useful or comforting or status-giving in the past. It must take on new responsibilities, patterns of action, knowledge. We are presented, I believe, with the great opportunity of converting all the way from a subprofessional occupation into a true profession. This is what the revolutionary season has brought to us—opportunity. It is of this opportunity that it is time to speak.

Educational Criticism

Finn did speak out with pen, podium and presentations. "The Tradition in the Iron Mask" (1961) sets the scene as he uses Dumas' story to depict the "audiovisual tradition" as the victim of its dictatorial twin, the "literary tradition." After vindicating his thesis, Finn briefly reviews contributions of the audiovisual field at levels of operation, teaching, development, production, and theory. He then demands a bridging of the gap between the two traditions in order to conserve humanity in a technological age.

The difficulty in bridging the gap is brought into sharper focus in "The Good Guys and the Bad Guys" (1958) where Finn assembles critics and criticisms of education gathered over a period of years and dating back to 1932. While not attempting a comprehensive review, his "excursion along the edges of the critical literature" does probe deeply on occasion and provides us with a scholarly exploration of "contemporary critics back to their fountainheads." If the results seem to show a preponderance of Bad Guys, "that," says Finn, "is the way it is."

In "Some Notes for an Essay on Griswold and Reading" (1959), Finn rigorously examines, under the psuedonym of Peabody Nightingale, a statement written by President Griswold of Yale. He takes the opportunity to deal again with the general literary versus audiovisual position, but more specifically outlines his reasoning behind "some important answers to this recrudescence of anti-audiovisual criticism."

"The Sound and the Fury of Rudolf Flesch" (1955) is the concluding article in this section. Again Finn grasps the opportunity to place an issue in perspective. When Flesch takes the extreme position that the only way to teach reading is through phonics, his objectivity is questioned. Finn argues the importance of experience in a reading program and the relevance of "teaching tools" to build meaning so that words can be understood in context and thought about in relationships.

Academic Tradition

Finn indicts the spokesmen for the academic disciplines and their faithful followers in "Conventional Wisdom and Vocational Reality" (1965). "The great irony in this status-ridden situation is that, while the disciplinarians fiddle, a great social protest is burning through America demanding more, better and infinitely more functional vocational-technical-professional education."

The conventional wisdom of the vocational field is categorized and analyzed. Finn rejects the general viewpoint of Robert Hutchins because it lacks both time scales and degree scales. Finn proposes two approaches which could accommodate the problems of job obsolescence and personnel retraining.

The concept of leisure in automated society is put in a time-scale perspective and three fallacies are examined. We are admonished to consider the concepts of leisure and living in educational planning on both a short- and long-range basis. Finn does not envisage the academic disciplines contributing significantly to the planning needed and categorizes them as "the fountainhead of abstraction. Their goal is to leave life as far behind as possible."

In "An Uncomfortable Rejoinder" (1965), he extends the argument of the previous paper, this time directing it specifically toward the role of the junior college. Accepting abstraction as a characteristic of the age, Finn quotes Thomas Huxley: ". . . if encouragement is given to the mischievous delusion that brainwork is, in itself, and apart from its quality, a nobler or more respectable thing than handiwork—such education may . . . lead to the rapid ruin of the industries it is intended to serve."

Several arguments related to the academic-nonacademic controversy are dealt with under three headings: the problem of organization; the problem of teaching and general education; and the problem of philosophical position. Finn contends that the ultimate argument has to be philosophical since it is related to the concept of democracy.

"New Uses for Old Clichés" (1965) casts Finn in a traditional role of delivering a commencement address. Accepting his place among the over 30-year-olds, he addresses himself to the concerns and aspirations of his audience. Recognizing the problem of communication between the generations he attempts to find some new uses for old commencement address cliches. He considers several causes for the increasing generation gap such as accelerating change, instant communication, and abstract knowledge. In examining the metaphor of the relay race, Finn introduces some thoughtful suggestions for closing the generation gap.

Professional Challenge

"One thing that is new is the prevalence of newness, the changing scale and scope of change itself, so that the world alters as we walk in it. . . ." Finn opened "A Walk on the Altered Side" (1962) with this quote by J. Robert Oppenheimer and described the new world—the altered world—as the world of our children. The challenge for educators is to learn to walk in it.

As an instructional technologist, Finn outlined some of his philosophic concerns about the new world to his audience: "I come to indict. I feel that many educational philosophers have lost the way and that they have committed an even worse crime—they have failed to understand."

The organic process of technology is described as a legitimate object of study. Finn points out the need for continuing analysis and the need for direction and implications for the educators as midwives to the new era. He castigates intellectuals whose attitudes toward technology range from apathy to antagomism.

The challenge Finn offers in "A Revolutionary Season" (1964) is that the conservative majority in education develop appropriate responses to the revolution in education. "Things are not as they are (or, rather, were) and have not been for some time. Hence, the basis for educational conservatism no longer exists." He examines this proposition in detail and takes a brief look at the case for educational conservatism. The myth of the "little red schoolhouse" is shown as a basis for our "make-do" educational orientation.

Some aspects of irretrievable change, induced by the technological revolution, are exposed and educational implications are revealed by Finn, who states flatly that traditional responses no longer apply.

Originally titled "It Is Impossible to Be Chivalrous Without a Horse," the article "The Franks Had the Right Idea" (1964) was directed at educators in general. Finn translates this analogy and discusses the permeating nature of change. "In a broad sense, it has always been true that, in time of change, the world belongs to those who can grasp the nature of that change and fashion their life and culture to make the most of it."

The increasing problems of education are stated and their resolution held to be dependent upon an instructional technology which Finn challenges educators to understand and seize upon creatively. He sums up his analogy saying that "it will be impossible to be a professional teacher without instructional technology."

An editorial in *Teaching Tools*, "AV 864 vs A.D. 1970" (1959) takes issue with the narrow conceptual emphasis of Public Law 864 (the National Defense Education Act) toward science, math and modern foreign languages. In this article, Finn regards the law as a limited reaction to the Sputnik-induced fear that Russian education was superior to American education and that the Act could have a one-sided effect.

Finn saw U.S. national problems as being in the socio-cultural arena, not in the arena of science. He asks: "Will the problems surrounding China, Southeast Asia, Africa and the Middle East succumb only to rocket technology and tensor calculus? Or will they have to be approached with better ideas about man's relationships to man, about man's aspirations, about the way he organizes his lot and the lot of his children?"

The article and chapter conclude with Finn's challenge that national leadership depends upon having an educational system which can develop a superior culture for the future.

The Tradition in the Iron Mask

According to the romantic notions of Alexandre Dumas, the elder, Louis the XIV had a twin brother who was kept imprisoned in an iron mask lest he become king. The basic idea of the hidden and imprisoned twin appeals to me as an excellent starting point from which to examine the growth and place of the visual and audiovisual tradition in American education.

For it is my thesis that, even today at a time when the rapid rise of instructional technology in the form of language laboratories, teaching machines, mass film courses, experimental psychological research, television, 8mm sound film, and a host of other developments has stirred American education to the point where even the deans of school of education and the National Science Foundation have had to take notice, the very tradition from which these developments have sprung has been imprisoned in an iron mask. This tradition—the audiovisual tradition—is about 60 years old in its modern form and in its relationship to American education.

Changing the metaphor for a moment, it could be said that the visual or audiovisual tradition in education went through a gestation period of 60 years from conception to birth. This, however, does not really fit the case. The tradition, thought of as alive with potential but imprisoned in an iron mask by its twin, actually describes the situation much more accurately. And who, or what, is this twin that has contrived this imprisonment? I will explain in a moment. In the meantime, let me make the boundaries clear. Edgar Dale and other students back to William Johnson and Anna Verona Dorris have pointed out the debt our tradition owes to Erasmus, Pestalozzi, Herbart, Rousseau, Dewey, and many others. However, I am speaking of the modern visual-audiovisual tradition which began around the turn of the century. This beginning corresponds roughly with the development of certain photographic and sound techniques and with the advent of the motion picture.

Within these time boundaries, the audiovisual tradition has been the victim of a vengeful, anxious and dictatorial twin—*the literary tradition*. The literary tradition has exhibited a naked and psychopathic fear of the "rapid communication of intelligence" on which we put such a premium. One only has to look to find it in Wordsworth, Henry James, Clifton Fadiman, Joseph Wood Krutch, and Jacques Barzun; it has appeared again and again in the *Saturday Review* alongside pages dealing with film, record-

Reprinted from *Audiovisual Instruction*, Vol. 6, No. 6, June 1961, pp. 238-243. Presented as keynote address at DAVI Convention in Miami Beach, April 24, 1961.

ings and television; it was present in the reaction of newspapers to the introduction of radio, in the neurotic response of members of Congress during the days of the Roosevelt Administration when they discovered government agencies were making documentary films; it explains the ineptness of the vast majority of school administrators in handling the audiovisual problem in the past 20 years. It even explains the attempts by the American Legion to picket the great motion picture "Spartacus" and the pusillanimous cooperation of some educational administrators with these private censors of communication. It may even explain the genesis of the recent Supreme Court decision on motion picture censorship, as the law is essentially a literary achievement (Wald, 1961).

In any case, the basic reason is the same. The literary-oriented educational administrator has not been able to supply audiovisual materials and equipment for communication, but he can supply books. Our culture tends to fight censorship of the written word and to welcome it for the film. The literary tradition has remained supreme. The audiovisual tradition receives the second-class treatment of the unwanted twin.

The Literary Manifesto

It is not easy to understand this attitude of the literary tradition as it wages an eternal war on the audiovisual tradition. One would expect that all of the means of communication, no matter what their form or purpose—a sonata conveying a sense of nostalgia, a sonnet singing of love, a sonar beam carrying intelligence of an underwater obstacle, a sound wave riding with a television frequency—would be respected for their potential to communicate fact and feeling in a complex, essentially uncommunicative world. But it is not so with the literary tradition. The literary tradition, in fact, says: *There is but one God and it is the Word; there is but one human and he is the man with literary sensibility; there is but one world, and it has been printed on a press for all to see. Everything else is either a false god, an inhuman man, or a phantom world.* This is the literary manifesto.

All this has been documented before (Finn, 1955). However, it is important for our purposes and in these times to review and bring up to date the state of this documentation. For I am not speaking of some old-fashioned viewpoint but of a position that is in current exchange among men of literary sensibility. Consider this quotation from Joseph Wood Krutch in his 1959 volume, *Human Nature and the Human Condition:*

Are what our school principals grandly call "audio-visual aids" usually anything more than concessions to the pupils' unwillingness to make that effort of attention necessary to read a text or listen to a teacher's exposition? Can anything be said in favor of most of them except that they are, at best, a surrender to the delusion shared by children and adults alike that the mechanical techniques of communications are interesting in themselves, no matter what (even if it happens to be genuine information) is being communicated? Are they not, at worst, merely devices for "catching" an attention which can never be given freely or held for long? How often can it be said that any movie, film strip, or recording teaches the so-called student—who has dwindled into a mere listener or viewer—more than could be learned in the same time with a little effort, or that the mechanical method has any virtue other than the fact that such effort is not required? Is there anything a picture can teach the pupil which is worth as much as that ability to read which he stands in great danger of losing? (Krutch, 1959, p. 134)

This polemic is obviously a declaration of war and vindicates my thesis that, from the belles-lettres side of the fence, the literary tradition intends to keep the audiovisual tradition confined. The objection may be raised, however, that a man like Krutch is not very influential and that his work can have little, if any, effect upon the audiovisual state of things. To take this view is to underestimate the power of the verbalists that control our society—even if they are working for someone who sits behind the scenes. They control the magazines of opinion that, in turn, are read by the opinion-molders—the *Saturday Review*, *Harper's Magazine*, *The Reporter*—and the "think" sections of great newspapers like the *New York Times*.

Krutch is a perfect case in point. He is, of course, a great literary figure and critic. He writes with a style that is admired throughout the English-speaking world. His scholarship, however, is disgraceful. Mr. Krutch, in one case that I can document, uses Vance Packard as an authority for activities of the NEA. I doubt if even the *Chicago Tribune* ever went to Paul Bunyan for facts on the Minnesota lumbering industry. Krutch's work is shot full of this kind of scholarship. Yet he continuues to influence professional opinion-molders and is fully protected by a band of worshipping editors.

When, for example, Krutch's errors in the particular case mentioned—and it had something to do with audiovisual materials, by the way—was protested to the *Saturday Review* on the grounds that his story was completely false, Mr. Harrison Smith, associate editor, wrote me as follows: "A good many years ago, I agreed with John Mason Brown that Joseph Krutch was one of the few genuine critics we have, and since then I have found no reason for doubting it." The iron curtain fell shut with the swish of the silken drape. In the meantime, the fable concocted by Packard and Krutch has been immortalized in shortened form in the book I referred to (Krutch, 1959).

While Krutch might be called the psychopathic wing of the literary opposition to the audiovisual tradition, this same attitude, albeit a little less stringent, may be found in other great spokesmen for the technology of print. Speaking of the invention of communication devices in 1957, A. Whitney Griswold, president of Yale University, speculated satirically on whether or not Yale should consider the implications of these devices in its curriculum. Following a freshman through a list of courses relating to cinema, radio, television, etc., he ends up on this bitter note.

. . . The freshman reads on in despair. He is looking for a course in English. He can't find one. He goes to the Dean. "English?" says the Dean. "Oh we don't bother with that any more. We have developed more effective means of communication." (Griswold, 1957, p. 47)

Arthur Bestor, who is nothing more nor less than a literary critic of education, takes the position, as is well known, that audiovisual means of communication are for babies (Bestor, 1953). The entire Council for Basic Education, Professor Bestor's organization dedicated to wiping out the

NEA, follows the literary tradition with a vengeance, including, in the last few months a tongue-in-cheek attitude toward teaching machines.

As sort of a pinnacle of literary bias, I would like to cite the case of a Mr. Joseph Kerman, an associate professor of music at the University of California at Berkeley. Mr. Kerman's attitude epitomizes the never-never land in which the literary tradition resides. He argues that students will never learn to appreciate music by performing it (Kerman, 1959). I gather, too, from what he said, that they will not get much out of "long sleepy stretches of record playing" (Kerman, 1959, p. 222). Music, it seems, must be approached solely through symbols. Someone should tell Leonard Bernstein.

Finally, the literary manifesto against the audiovisual tradition is expressed in its best form by Jacques Barzun in his book, *The House of Intellect.* Since Barzun *is* a great scholar and possesses flexible and even, at times, poetic qualities of mind, he needs an ampler review than time and space permit. However, it is easy enough to make the main point. Barzun sees three great influences destroying what he calls "intellect." His definition of intellect leaves something to be desired; but, essentially, he is referring both to the accumulated knowledge in a rather narrowly defined liberal arts (the furnishing of the mind), and an attitude of mind which, boiled down, seems to consist principally of the ability to appreciate and handle language with precision.

The three influences destroying this intellect are Science, Philanthropy and Art. All of these are involved, one way or another, in the audiovisual tradition. Science—particularly the social science area from whence we get our research data—has been attacked by literary spokesmen for so long it needs no further comment. Philanthropy not only refers to foundations and gifts, but to the attitude of mercy toward other humans—the psychiatric or guidance approach, for instance. Thus Barzun condemns efforts at mass education which attempt to take care of *all* students through technological means such as the motion picture (Barzun, 1959). Art, in many ways Barzun's number one demon, is condemned for its nonverbal aspects and its emphasis on creativity without the disciplines of language and thought.

Permit me some illustrative quotations:

The artist turning toward ideas has been met halfway by a public turning aside from words and greedy for speechless art. . . . Everywhere picture and sound crowd out text. The Word is in disfavor, not to say in disrepute. . . . (Barzun, 1959, p. 16)

. . . art has put a premium on qualities of perception which are indeed of the mind, but which ultimately war against Intellect. (Barzun, 1959, p. 17)

. . . we cannot make intellectuals out of two million pupils—too many are incapable of the effort even a modestly bookish education requires. . . . (Barzun, 1959, p. 95)

The fundamental role that language, or rather the neglect of language, plays in all these losses and failures, including the scientific, is a characteristic of democratic schooling to which I shall return. I mention language here, as at every relevant point, not only to show

how frequently the reference is in order, but also to point out how the defect perpetuates itself: the unquenchable talk about education is useless because most of the speakers, having no regard for the implications of words, cannot criticize their own thought or discover its absence. (Barzun, 1959, p. 114)

[Speaking of illustrated textbooks] What buyers and sellers cannot see is the great gap in quality between the visual and the conceptual matter they thrust on the pupil. . . . they forget the two limitations of visual aid. . . . Pictures by themselves say little or are ambiguous. . . . Moreover it is evident that visual memory and the power to summon up ideas are not the same. Unless the student learns to turn a verbal account into the right vividness in his mind's eye, and conversely to frame his imaginings in words, he always remains something of an infant, a barbarian dependent on a diagram. (Barzun, 1959, p. 138)

To be fair, Barzun at times evidences an ambivalent attitude toward the language-audiovisual conflict. See, for example, his comments on the preparation of humanities professors (Barzun, 1959, p. 131).

A Summing Up

These are but examples, but they show, I think, the substance of the literary manifesto—the bill of particulars that has kept the audiovisual tradition shut up in the tower. These are powerful voices. Their influence extends far beyond the people who hear or read them directly.

And what, in sum, is the case of the literary tradition versus the audiovisual? It is the assumption, first of all, that all knowledge is developed as a result of a great Conversation (usually capitalized in order to personify it) carried on solely by words. Audiovisual communication, it is charged, destroys this Conversation.

Second, it is assumed that reading involves hard work and effort on the part of the reader and that this is so good in and of itself that it must be engaged in all the time. Audiovisual means of communication, it is claimed, do away with this hard work and effort and therefore destroy this alleged good, even if they communicate the same material. This view might be called the Puritan Theory of Linguistic Perception.

Third, books and language in printed form are mistakenly thought to be individual—as opposed to mass means of communication. This individual concept is somehow and in an unstated manner transformed into an elite concept. Books, reading and the Conversation are for the elite. Into this dream world have come means of communication that are supposed to be mass communication devices. They transmit to all and destroy the elite concept. This is nothing less than the deadliest of crimes against society. One might call this the Jacobin Theory of Communication.

Fourth, and last, it is assumed that the human heritage—writing, art products, and mathematical symbols—can be stored only in books. Any violation of this concept—the storage and use of information, ideas, and even such products as dramatic tragedy on film, magnetic tape, or in the memory drums of a computer—amounts to a violation of the literary herit age. It is thought to be, unconsciously perhaps, a form of book burning.

The Tradition in the Iron Mask

Causes and Consequences

I think the cause of this attitude toward the audiovisual tradition by men of literary sensibility is the same phenomenon that Sir Charles Snow has so well described in his now famous book, *Two Cultures and the Scientific Revolution*. I have just reread this little book for the third time. It is so insightful concerning our problem here that I am tempted to quote pages. Instead, let me give you Sir Charles' words in one of the ways he states his thesis. Because he is both a novelist and a scientist, he has lived all his life between these two groups. And he says:

. . . For constantly I felt I was moving among two groups—comparable in intelligence, identical in race, not grossly different in social origin, earning about the same incomes, who had almost ceased to communicate at all, who in intellectual, moral and psychological climate had so little in common that instead of going from Burlington House or South Kensington to Chelsea, one might have crossed an ocean.

In fact, one had travelled much further than across an ocean—because after a few thousand Atlantic miles, one found Greenwich Village talking precisely the same language as Chelsea, and both having about as much communication with M.I.T. as though the scientists spoke nothing but Tibetan.

. . . I believe the intellectual life of the whole of western society is increasingly being split into two polar groups. (Snow, 1959, pp. 2-4)

It seems to me that the audiovisual tradition is much closer to the scientific culture than to the literary culture. The anti-audiovisual bias, then, is easily explained as a special case of this general phenomenon.

Further, the technology that transacts with science is misunderstood even more by literary men. It is also, at times, as Sir Charles points out, not understood by scientists themselves. But it is the technological revolutions that have been important. As Snow says," . . . those two revolutions, the agricultural and industrial-scientific, *are the only qualitative changes in social living that men have ever known*" (Snow, 1959, p. 24).

I have spent the last five years of my professional life arguing that the cascade of events we are now experiencing in the audiovisual field is the extension of the technological revolution to education and can only be understood in those terms.

If I am right about the cause of this split—this confinement of the unwanted twin of the audiovisual tradition—what are the consequences? The consequences, themselves, I think are informative. They cannot all be spelled out here, but let us look at a few.

Most school districts, county units, states, and institutions of higher education in this country have never solved the problem of the logistics of instruction. School administrators, at any level, have never been able to deliver the funds, the organization, the equipment, or the materials for audiovisual instruction. This explains the lack of room darkening, acoustic treatment, and electrical power; it explains high schools of 2,500 students with a person euphemistically called an audiovisual coordinator who has one period a day at his disposal; it explains the poverty of nonverbal materials; it explains why a teacher never has enough maps, films or tapes.

The Tradition in the Iron Mask

School administrators and the entire school organizational apparatus have a literary orientation. The mail and the textbooks come through and that's it.

Another consequence of this allegiance to the literary side of the cultural split is a predictable resistance to innovation—especially educational innovation related to research, invention and development. Thus, many curriculum people are horrified today at the prospect of programed instruction—no matter in what form it may occur. The standard curriculum position, with its rather unscientific group dynamics orientation, is the opposite of the precision now being developed by Galanter, Skinner, Komoski, Calvin, and others.

The generalized fear of "machines" is a simple adoption of the literary position by many educational workers who have been brought up in the literary tradition—even though they may pay lip service to other approaches to communication. The literary humanist is, by the nature of his milieu, fearful of the "inhuman" machines. Thus, for example, the German Friedrich Georg Juenger, in his assault on the machine age called *The Failure of Technology*, has a chapter attacking photography as somehow sinful—shades of Barzun (Juenger, 1949, p. 138)!

(Nothing in this argument should be construed as taking sides, one way or another, on the so-called "instructional materials concept" or the role that librarians should or should not play in the logistics of instruction. This is a technical question, the answer to which must be determined by a variety of technological factors.)

The Contribution of the Audiovisual Tradition

Opposed to this situation stands the audiovisual tradition. Although kept masked in the tower, it has managed to develop (almost discontinuous with the educational enterprise) as a tradition. A slender thread, strong and fine, stretches back through the fabric of American education these last 60 years. To follow my second metaphor, strands are now rapidly being added to this thread as the coming technological revolution takes over. To return to the first, the Bastille has been stormed, and the audiovisual tradition is being freed.

It is a tradition of which we may well be proud. It has given much to American education. At the level of operation it has created the beginning of a logistics of instruction ready for introduction as the great wave of instructional technology sweeps in. The service concept (the right materials at the right time in the right place for the right teacher) and the technical concept (good presentation under proper technical conditions) are fundamental requirements of the new technology—and they are ours.

At the teaching level, we have contributed the theory of the abstract-concrete relationship and the variety-of-materials concept leading into the cross-media approach. These are well known. Not so well known is the concept of optimum synthesis. In production this means the proper combination of subject matter expert, communications expert, curriculum expert, and artist; in teaching this means what has suddenly been discovered as the teaching team, but which we were working with long ago. Also, the whole business of self-instruction stems in part from the audiovisual tradition. The listening-viewing corners we have put into progressive elementary schools are examples.

At the level of development, the audiovisual tradition has brought the engineering concept into educational reality. When W. W. Charters, one of the great unsung fathers of this movement, spoke of educational engineering he was at least partly referring to the planned use of the media of communication in instruction. Invention, too, is a part of our tradition. There is nothing intrinsically wrong in being gadgeteers; I think we apologize too much for this. We have not only invented many small but useful gadgets, but we have also developed the first man-machine systems of instruction.

Fundamentally, however, the great contribution of the audiovisual tradition to educational development is that it has always been founded upon an experimental base. The closest ties in all of education, with the possible exception of the testing movement, are between experimental psychology and audiovisual education. This relationship has held since it began in 1919 with Lashley and Watson, two of the great experimental psychologists of their time. It extended across the years to include Freeman, Brownell and others down to the present when the work of Carpenter, Skinner, Pressey, Glaser, Lumsdaine, Galanter, and many others receives its widest currency among us. This movement is so extensive that it is hard to separate our own researchers from the pure psychologists. Dale, Hoban, VanderMeer, Meierhenry, and many more fit this description.

A portion of our tradition can be identified at the level of production. And this is very important. We have had great pressure in recent years to return to "subject matter." The audiovisual tradition never left it. In fact, we have always of necessity carried this banner and in a sense represent a conservative trend in American education. For the audiovisual tradition knows, better perhaps than the literary tradition, that when you design a communication you have to say something—you have to have an objective, a defined audience, and a subject. Anyone who has committed anything to the screen knows the discipline this imposes. Contrary to Mr. Barzun, it is a discipline in many cases more exacting than that required by the literary tradition.

The concept of programming is not new to the audiovisual tradition. Perhaps this is why we have not become afraid of it. Any cross-media approach to instruction—the linking of films, direct experiences, filmstrips, etc. together to achieve an instructional objective—must be programmed. The sequencing of a film or television program is a kind of programming. Granted that the current concepts of programed instruction are somewhat different, they are, in a broad sense, of the same pattern.

Most important, however, at the production level, has been the contribution of the creative artist, the attempt to put, as Robert Wagner has said, design in exposition. This means that the audiovisual tradition recognizes the possibilities of a rhetoric of film and other forms of visual communication. The literary tradition denies this possibility. (I would speculate from this point of view that many of the current educational productions on film or television which consist almost entirely of a teacher in front of a blackboard are basically inept efforts of literary-oriented thinkers to design visual presentations.) For the rhetoric—the grammar—of audiovisual communication exists. I suspect we will find it before the literary men find the true grammar of the English language.

The fifth and last level where the audiovisual tradition has made a contribution is that of theory. We introduced the concepts of communication into education. All of the applied theoretical work on communication in education began with us and most of it is still published in our journals. Now, here is a very important point. The dichotomy I have been describing might imply that the audiovisual tradition had no use for language. Nothing could be farther from the truth. We have made an effort to understand language; we are, perhaps, as sensitive to its contributions and possibilities for the communication of facts and emotions as our literary brethren. There are those among us who have carefully studied language. For 30 years, for example, Edgar Dale has been carrying on vocabulary studies. And there are others. The work of Korzybski, Ogden and Richards, and other students of meaning, the work of linguists—all of it finds a home in our councils. I venture to say that this is not true of great segments of the literary world. It may be that their appreciation of language is intuitive and ours is technical, but the reverse concept—the literary tradition paying the same attention to the audiovisual tradition—does not exist.

A Disciplined Intellectualism

Fundamental to our theoretical formulations is a rigorous discipline. Again, I feel that, within all of education, the audiovisual field, with its theoretical formulations, is the most disciplined. To borrow a wonderful expression from the sports writers, our intellectual tradition is "hard nosed."

I have discussed briefly and much too hastily the contributions of the audiovisual tradition at the levels of operation, teaching, development, production, and theory. There are some others that defy categorization. The contribution our tradition made to and the growth it acquired during World War II has yet to be properly assessed. But we handled large-group instruction, self-instruction, and new patterns of instructional organizations; we created systems approaches to instruction and adapted new communication devices to learning—all in the space of four years. The Ford Foundation and others interested in these matters might have saved themselves much time and money if they had deigned to ask the proper questions in the proper places concerning this experience which is found in our tradition and nowhere else. As I have said before, it is no accident that during World War II, Hoban handled the Army program, Noel the Navy program, and Brooker the U.S. Office of Education program.

These great contributions are why I was so anxious tonight to initiate a tradition honoring the past presidents of DAVI. The men and women on this platform and those whose names were called have carried the audiovisual tradition through American education these last 38 years. They are symbolic, of course, for there have been many others as well. But each of these has played a major part in a very difficult effort. Our field stands where it does today because of them and their associates. Many of them will continue to carry the burden of leadership far into the future. All of us, but particularly those of us associating with the audiovisual movement for the first time, can find great inspiration here.

Rapprochement

Now that the iron mask has been removed, now that our tradition and its contributions have been made apparent to American educators and the public, do these two traditions—the literary and the audiovisual—have to remain as unbridgeable as the literary and the scientific cultures? I do not think so. An age of technology does need the softness of the humanists. And literature and criticism will become sterile, as they are rapidly on the way to doing, if they cannot come to grips with and understand this technological age. In a word, we need each other. A rapprochement is demanded; the gap must be bridged.

I do not think that the literary humanists will attempt to bridge this gap; I do not think that they could do it if they wanted to. There are, of course, many literary men who understand the audiovisual tradition. There are writers in several media, like Rod Serling; there are men like Orwell who never lost touch with concreteness. (There is even a small literary tradition that has attempted to come to grips with technology. The ambivalence of Samuel Butler toward machines and Henry Adams' embracing of technological concepts are examples.) These we can work with. The Barzuns, the Krutches, the Griswolds, and their close followers are, I suspect, lost causes.

But I submit, first, that words alone and the literary sensibility will never solve our serious educational problems. They can help, but our culture is turning—must turn—to technology for this job. Thus, the demands on the audiovisual tradition are great. Realizing this, we certainly do not wish to turn people or schools into automatons and factories. Yet we must wield the technology we are developing to help solve the problems of education. There are 2,000 years of human values in the literary tradition. I think it is up to the audiovisual tradition to save these values from dying at the hands of their hapless literary guardians in order to keep our own traditions, and theirs, human.

This is the paradox that is our greatest challenge today.

References

Barzun, Jacques. *The House of Intellect.* New York: Harper & Brothers, Publishers, 1959.

Bestor, Arthur. *Educational Wastelands.* Urbana: The University of Illinois Press, 1953.

Finn, James D. "A Look at the Future of AV Communication," *Audio-Visual Communication Review*, Vol 3, No. 4, Fall 1955, pp. 244-256.

Finn, James D. "Some Notes for an Essay on Griswold and Reading," *AV Communication Review*, Vol. 7, No. 2, Spring 1959, pp. 111-121.

Griswold, Alfred Whitney. "On Conversation, Chiefly Academic" in *In the University Tradition*. New Haven, Conn.: Yale University Press, 1957, pp. 34-48.

Juenger, Friedrich Georg. *The Failure of Technology: Perfection without Purpose.* Hinsdale, Ill.: Henry Regnery Company, 1949.

Kerman, Joseph. "Music" in *The Case for Basic Education* by James D. Koerner. Boston: Little, Brown and Company, 1959, pp. 217-224.

Krutch, Joseph Wood. *Human Nature and the Human Condition.* New York: Random House, 1959.

Snow, C. P. *The Two Cultures and the Scientific Revolution.* New York: Cambridge University Press, 1959.

Wald, Jerry. "Movie Censorship: The First Wedge," *Saturday Review,* April 8, 1961, pp. 53-54.

The Good Guys and the Bad Guys

We have a television-watching tradition in our family which is probably much the same as that observed in many other American families. If one of us enters the living room while a show is in progress, the first question asked is usually, "Who are the bad guys?" or "Is he (or she) a bad guy or a good guy?" Once these questions are answered, it is fairly easy to understand the show, television being what it is these days.

Unfortunately, the heap of criticism that has been piled on public education, school administrators, teacher education institutions, programs and faculties, John Dewey, and driver education in the last decade or so is not so simple to understand. You can't tell the bad guys from the good guys without a program similar to the annual yearbook of the *Daily Racing Form*. And the labels don't help much in understanding educational issues of the day.

Educational criticism has become a complicated, entangled and, oftentimes, profitable business for a mished-mashed assortment of pundits ranging from the Roosevelt-hating Raymond Moley through the educationist-hating Arthur Bestor and the write-for-pay boys like John Keats and Howard Whitman to clear-headed social critics like David Riesman and conscientious scientists like Glenn Seaborg. After collecting materials in this field for almost 10 years and now attempting to make some kind of sense out of the mass, I find that generalizations are hard to come by. It is absolutely unfair to Seaborg, for example, to compare him with Bestor; it is equally unfair to consider Albert Lynd in any category not including Baby Face Nelson, Dillinger, and the Clanton brothers—the killer instinct is equally present, the only difference being that between using words and bullets.

Over the years, of course, various analyses and refutations of criticism have been attempted. A good example of a refutation is an article by Harold Hand (1957). Robert Skaife has written several excellent analyses (1957); the pioneering work of Scott and Hill has been helpful as the first attempt to establish categories of criticisms (Scott & Hill, 1954). The Research Division of the NEA has also attempted to categorize criticisms (Research Division, 1957). That the work, for the most part, has been scattered and spotty is no discredit to the profession. The critical material comes in so many different guises, in such great volume, and is carried

Reprinted from *Phi Delta Kappan*, Vol. 40, No. 1, October 1958, pp. 2-5+.

by so many different channels that, taken as a whole, it is quite intractable.

One way to attempt to describe the educational criticism of the last decade is to look at it in the form of products of various media of communication. With the exception of poetry (and I could be wrong about that, although I try to keep up with the social poets like Sec), almost every form of expression known to man has, in the last 10 years, carried critical comment on education over every possible medium of communication. Sam Levenson, the television comic who makes derisive comments on modern report cards, is an outré example of how far out into left field you may wander and still play this new All-American game.

Criticism has appeared in books, pamphlets, speeches, essays, on television and radio, and in legislative reports, to mention only a few of the vehicles. Even the critical product appearing in books, pamphlets and magazines is too large a burden for this article; only a portion of it, namely some critical books and pamphlets, will be examined.

No attempt will be made to review this literature in the comprehensive style set by Professor William Brickman with his monumental reviews in *School and Society*. Rather, what follows will be a brief literary excursion along the edges of the critical literature, with some deeper probings here and there where it is thought they may be illustrative and helpful. Let us begin with books.

There have been, for example, novels. The most vicious criticism of professors of education in print is still to be found in Virgil Scott's novel *The Hickory Stick*. Someday one of the bright young English instructors so loved by the Council for Basic Education is going to remember that novel and, under the imprimatur of Arthur Bestor, we shall once again be able to read:

. . . every Professor of Education has been run through the same mold and has emerged with the same dry, expressionless voice (if you walked from class to class with a tuning fork and a metronome, you would find that they all maintained a perfect middle-C pitch and a constant speech tempo throughout every forty-five minute lecture), and these men also have the same scaly skin which looks sandpapery, and the same suits, brown tweed or gray pin-stripe, and the same brown-stained brief cases, sagging and limp-looking when they are laid clip-up on a desk, and the inevitable rubbers on the inevitable black oxfords, and the inevitable rimless glasses on the inevitable thin bridges of noses, and the inevitable smell of chalky dryness about their bodies. (Scott, 1948, p. 95)

Virgil Scott hated. No man could write like that if he didn't hate.

On the other hand, Evan Hunter obviously turned out *The Blackboard Jungle* to make green hay out of his short experience as a teacher in a New York vocational high school. He only incidentally took a swipe at education in the process. If Hunter hates, he hates a social order that produces a Blackboard Jungle. More likely, he just enjoys money.

Much more common, of course, are the nonfiction books that have been pouring into the bookstores each of the past several years. We are indebted to Bestor for two (1953, 1955), Smith for three (1949, 1954, 1956), Bell for two (1949, 1952), Woodring for two (1953, 1957), Flesch for two (1954,

1955), Hutchins for two recently (1953, 1956), Lynd (1950), Hildebrand (1957), Keats (1958), Riesman (1956), and others. All of these books are easily available to any citizen who wants to read.

The byroads of the book publishing business turn up even more fascinating items. There is, for instance, Hilda Neatby, the Canadian Bestor (1953); there is an absorbing account of a teacher's long life in California schools with a gratuitous (and *non sequitur*) chapter or two at the end which sound as if they had been appended and ghost-written by Albert Lynd (the publisher, as with so many of these books, is Henry Regnery of Chicago) (Nathan, 1956). There is the Pasadena-inspired volume by Mary Allen which proves beyond all doubt to members of the Pasadena School Development Council that Goslin and the progressives had a blueprint for a socialist-communist-collectivist America and had decided that Pasadena was the Concord Bridge (Allen, 1955).

Back even further on the byroads, the going gets stickier. A man by the name of John Howland Snow, a name not unknown in extreme reactionary circles, markets a remarkable book under the title, *The Turning of the Tides*. This book turns out to be an abridgment of an earlier study by the late Congressman Paul Shafer of Michigan, proving that all education is hell-bent for socialism; to this abridgment, Mr. Snow has added his own comments, including the fact that the "old debbil" John Dewey founded the American Association of University Professors and, hence, caused the California loyalty oath controversy.[1]

A retired doctor, Lewis Albert Alesen, M.D., is the author of a book, entitled *Mental Robots*, which attacks current efforts to do something about mental health. *Mental Robots* turns out to be another blast at modern education, John Dewey, and the progressives, as well as ammunition for the battle the right wing is now carrying on against mental health programs, rabies vaccination, and water fluoridation.

These books from the byroads of publishing are no doubt officially considered beyond the pale by such people as Bestor and Smith, although the unofficial relations are by no means clear. The importance of books of the type just described is that they are advertised and circulated to small groups that can cause trouble in any community in the name of "Basic Education Council," "Friends of the Public Schools," or whatnot.

Returning to the main highway, however, the serious student will soon begin to realize that most of the books by Smith, Lynd and company are, to use Mortimer Smith's own words, "derivative"—that is, they derive from sources buried in what turns out to be a vast literature. One thing which intrigued me from the beginning was the fact that Smith and Bell, for example, made several references to a man named Albert Jay Nock. Lynd makes references to Smith; Keats has obviously read Lynd. Way in the

---

[1] Actually, to be fair, Mr. Snow's pitch is not too different from Mortimer Smith's treatment of the same general subject in *The Diminished Mind* (Chapter IV: "Educational Brain-washing, Democratic Style"). In Chapter VI, Smith also spiritedly defends the Pasadena School Development Council, a group presently very enthusiastic about Mr. Snow. In this case the far byroads seem to get very close to the main highway.

background sits the crochety figure of Albert Jay Nock, a misanthropic essayist of two generations ago. Nock was a character so fascinating that he deserves study for his own sake; educationally, he ended up his life with a simple view, namely that, *outside of a few others like himself, all men are apes and, therefore, should be so educated.* It is to Nock's credit that he came up with a pristine theory of education for the elite so simple and yet so elegant (Nock, 1932, 1943).

The derivations from Nock in some of the literature (especially that emanating from the so-called Council for Basic Education, which will be discussed later) are fairly easily traced.[2] There are others. A good case can be made that the so-called "New Conservatives" (more specifically, Mr. Russell Kirk) are a sort of intellectual fountainhead to much of the Bestor-Smith-Fuller stream.[3]

The first item to notice is the ubiquitous presence of Henry Regnery as the publisher of much of this critical material. His authors include Mortimer Smith, Robert M. Hutchins, Russell Kirk, and Beatrice Nathan. Mortimer Smith's first attempt at batting down the professional educators, *And Madly Teach,* was Regnery's first successful book. Since then, Regnery has had his share of ideological successes, including Buckley and Bozell's *McCarthy and His Enemies.*[4]

With the exception of Mr. Hutchins, who obviously doesn't belong with this crowd, the general educational-ideological axis of Regnery and his authors is remarkably consistent with the Bestor-Smith axis. An examination of Russell Kirk soon reveals that this can be documented. Consider this:

> The turgid style and the unimaginative epistemology of Dewey seem to have exercised a peculiar fascination for the stubborn doctrinaires who, in any age, make up too large a part of the body of teachers; and those dull persons (true conservatives of stupidity, for they make the sentences of Dewey, Kilpatrick, and Counts into unalterable secular dogmas) have obtained, by virtue of the dogged and dreary lust for "administrative positions" which characterizes them, a mortal clutch upon our poor educational institutions. Professor Arthur Bestor recently described some aspects of this pedantic tyranny in his *Educational Wastelands.*
>
> . . . These young people have been robbed of their true natural right

---

[2]Woodring and others make much of the idea that what is needed is a new philosophy of education to meet the criticism of the times. This may be true, but what is needed first is a study of what I call the para-philosophers of education—men like Nock, Kirk, Eliot, and others. The most vocal critics, such as Bestor, criticize *from* a viewpoint. This viewpoint is principally derived from the para-philosophers, not from anything that could remotely be considered a philosophy of education; and until Woodring, Brubacher, et al. understand this fact and understand these men, they will continue to whistle in the dark.

[3]Some of the writers, for example, John Keats, might be horrified to find this out—I've no way of knowing—but that is the penalty Mr. Keats must pay for trying to make money off a sure thing like kicking professional educators in the teeth. You never know, in a case like that, what sort of company you will end up in.

[4]Not exactly an attack on McCarthy.

to genuine instruction in the works of the mind . . . . And this is only one of the several ways in which the Deweyites have converted our educational institutions into so many weapons for a concerted assault upon true Reason. Mr. Mortimer Smith's *And Madly Teach*, or Canon B. I. Bell's *Crisis In Education*, or Mr. Albert Lynd's writings . . . draws up an indictment of the failure of the Deweyites which no amount of platitudes from the masters of most of our schools can refute. (Kirk, 1954, pp. 63-65)

While professional educators are probably fairly familiar with the general run of critical literature mentioned in the first part of this article, it is not as likely that they will be familiar with Mr. Kirk and his associates, and with the content of magazines such as the *National Review*. All things considered, the *National Review* can be taken as the current vehicle for the Kirkian viewpoint. It is, therefore, extremely interesting to find that Professor Bestor's *CBE Bulletin* editor, Mr. Mortimer Smith, takes great pains to review favorably articles on education appearing in the *National Review*.[5]

Personally, I don't care whether Arthur Bestor is married to Russell Kirk's sister or if Henry Regnery has a mortgage on Mortimer Smith's house. For all I know, they may all belong to the same bridge club. However, having established the *intellectual* connection, it is pertinent to inquire what Russell Kirk's position is with respect to education. Again, the answer is simple. Although not quite as simple as in the case of Albert Jay Nock, it is much the same answer. Mr. Kirk believes in an elite, although he is not sure, apparently, precisely who should belong to his elite. At any rate, whatever else Kirk may think, he has a very low opinion of human nature. His guides for the salvation of mankind are two—prescription and prejudice. By prescription, he means rules (constitutional, etiquettal or ritual). It is not as clear as to what he means by prejudice, but it is apparent that men of various orders and classes are to be conditioned by education into prejudices which will, in turn, keep the orders, classes and rules intact.[6] The Basic Education group has another full-blown elite theory here. For my money, they would be better off with Albert Jay Nock. At least he was clear and to the point.

The elite concept from which much of a certain sort of modern educational criticism stems can be traced back many generations. There is a limit, however, to what can be done in this article. It would suffice to mention that Mr. Kirk, in turn, relies heavily on T. S. Eliot. Eliot is not known as a friend of the common man. He once defined education (and

---

[5]In the cute style of the *CBE Bulletin*, mention is made of the fact that "we read everything from the *Nation's Schools* to the *National Review*." No doubt they do; however, when it comes to reviewing, things are different. Kirk gets a real good press; Harold Hand or any professional educator you want to name does not.

[6]It is impossible to say how many people (outside of the Bestor group of educational critics) take Russell Kirk seriously. The great British political scientist Dennis Brogan said of him, "Mr. Russell Kirk is not a conservative. He is a reactionary. He preaches a return to an economy and a polity that has never existed in the United States and has not existed [anywhere] since the Reformation, perhaps since the impact of a money economy" (Brogan, 1956, p. 10).

Mr. Hutchins took him to task for it) as a process designed to preserve the class and select the elite.[7],[8]

The point to all this, of course, is that the professional scholar in education has a great task ahead of him merely to explore the literature of criticism contained in books. He can begin with the contemporary and well-known critics; he can slip off into the strange byways of rightwing ideology; he must search back of the contemporary critics to their fountainheads.

Professor Bestor and his friends, of course, do not by any means exhaust the sum total of modern educational criticism contained in books, although they account for the great bulk of it. Rudolf Flesch, for example, eats his alphabet soup alone. Other critics, such as Riesman (1956), Whyte (1956), and Philip Wylie (1954), insert their educational criticism into larger critical works and are, perhaps, best encountered in their articles and essays; and there are other books we must, for this time, just pass by.

In view of the volume of critical books, the question can be raised as to the number of replies or counterattacks which have been printed in book form. These have been very few and (it seems almost deliberately) little publicized. The only exception to this was Woodring's first book, *Let's Talk Sense About Our Schools*. For some reason, Woodring has since become a hero to Henry Luce's communication empire and, with *Life* and *Time* behind you (and now the Ford Foundation), how can you miss? Whatever Woodring did in *Let's Talk Sense*, he undid in his latest pronouncement, *A Fourth of a Nation*, in which, while mildly remonstrating with some of the more feral critics, particularly about education studies at the graduate level, he in effect went over to their side. As a gladiator, Woodring not only refused to fight the lion, he climbed up in the emperor's box and began drinking his wine.

There is no doubt that Charles H. Wilson's *A Teacher Is A Person* is the best book appearing in print to date which attempts to explain "the other side," to refute some of the calumny, and to make some real sense out of the problems of education. Autobiographical in nature and exceptionally well written (perhaps this is why it has been so largely ignored—an educationist who can write must be kept under wraps), it has been practically unheard of in all the clamor. Another interesting effort is Caley's *A Teacher's Answer*, a book which, while critical (to the point where Smith would approve) of many things that have happened in education since 1900, still deals with down-to-earth problems and down-to-earth answers. Caley's book, unfortunately, did not receive the impetus that a major publisher might have given it, and has been lost in the shuffle. There have been other books in this vein, but they seem also to have been passed by. The Melby-Puner (1953) volume is a collection of articles. The Dreiman

---

[7]See Eliot (1948, p. 95 & following) for Eliot's statement and Hutchins (1950) for Hutchins' criticism.

[8]In his *Quackery in the Public Schools*, Albert Lynd makes much of the idea that the ladies of the PTA would not vote for John Dewey if they "understood the philosophical ballot." I wonder what the good ladies would do with Kirk, Eliot and Nock under the same conditions.

book is an account of some successes of the National Citizens Commission for the Public Schools (1956)—an outfit that has been soundly criticized by the rightwing press. When these works are looked at together, the impression is gained that either publishers don't think it monetarily worthwhile to publish the professional answer to these attacks or the professionals have simply left the field to the critics.[9] The first explanation is the more likely, particularly when viewed in connection with the lack of publicity given such books as do exist.

Between the far boundary of books and the near edge of magazines lies the never-never land of pamphlets, organization pronouncements, and reprints. Here anything goes—inconsistency, anti-semitism, pro-segregationism, pro-report cards, pro-phonics, anti-phonics, and side comments about rabies vaccine and hidden persuaders. It is impossible to review this material, but two or three samples may be taken from the mass much as the geologist does when he drives a shaft to obtain a core.

The burden of the reactionary right wing is no longer carried by Allen Zoll. His pamphlets, such as *The Little Red Schoolhouse*, are old stuff these days. Today his same audience is treated to reprints (from the *Congressional Record*, of course) of comments by Congressman Ralph W. Gwinn (R.-N.Y.), or they subscribe to a nationally distributed bi-monthly sheet called *Facts in Education*, edited by Frances P. Bartlett of Pasadena and the famous but now somewhat moribund School Development Council.

Recent issues of *Facts* make spectacular reading. Right next to an article entitled "How Pink is the PTA?" is an attack on the distinguished anthropologist Ashley Montagu of Rutgers. Montagu is accused, among other things, of having prepared the UNESCO statement on race, of having lectured at the New School for Social Research, of having helped in the production of the film on the atomic bomb called "One World or None" (*Facts*, November-December 1957).

In the next issue, *Facts* readers are urged to join Arthur Bestor's Council for Basic Education, to read his statements in *U.S. News and World Report*, and to buy at least one of his two books. In an adjoining column, the praises of Admiral Rickover are sung and it is suggested that faithful followers write to the National Sojourners, Inc., P.O. Box 171, Worthington, Ohio, for an address by an Admiral Wallen, and to the Associates of Americanism, P.O. Box 6476, Pt. Loma Station, San Diego 6, California, for a reprint of a speech by Dr. Arnold Beckman entitled "Public Education . . . A Menace to Science." Recently, it might be added, *Facts* has extended itself somewhat beyond the field of education to belt President Eisenhower and such convenient targets as the Supreme Court.

Discussion of rightwing pamphleteering should stop there, but the temptation to refer to one more item is too great. An organization known as The Long House publishes, among many pamphlets, a chart on elemen-

---

[9]We are speaking here of trade books—books that are sold in bookstores. There is discussion of this decade of criticism in the professional literature, as, for example, in Lieberman (1956), but this material never reaches the public except in a distorted form through the eyes of someone like Howard Whitman.

tary education, which sells for 25 cents. The chart sets up a dichotomy between what it calls "Traditional Education, Up-to-Date," and "Education for a 'New Social Order.'" Under these two headings it lists the traditional subject matter divisions and other items such as "teaching technique," "play," "homework," and "audiovisual aids." These divisions are then commented upon from each viewpoint. The comments on audiovisual aids are revealing. The traditional program, it seems, will make careful, low-cost use of audiovisual materials; the new social order school makes too profuse use of these materials and—here comes the kicker—"There is abundant evidence of a breakdown of time-tested traditions and beliefs. Films slanted toward a vague 'internationalism' and a 'welfare state' are in wide use. Audiovisual aids tend to supplant mental discipline and to retard individual reasoning when used to excess"[10] (*The Elementary School*, 1954).

The back of the chart has been used by the publisher, in the interest of true economy, to advertise his other wares and also to sell a group of books of which he apparently approves so much that he is willing to assist in their distribution. Included in this list are Clarence Manion's *The Key to Peace*, Bestor's *Educational Wastelands*, and the previously mentioned *Turning of the Tides.*[11]

A lifetime could be spent in the twilight world of rightwing, anti-education pamphleteering—and without much profit. The great gold mine of critical pamphlet publishing today is located in Bestor's Council for Basic Education. Since most professionals are, or ought to be, familiar enough with the general outline of the CBE and its view of life, no general discussion will be attempted here. Readers who desire background information are referred to the excellent article by Skaife (1957).

The CBE's main instrument is its *Bulletin*, published monthly. The *Bulletin* was formerly edited by Harold Clapp and is now under the pen of Mortimer Smith.

At the moment, according to the last *Bulletin*, the CBE has a grant of about $35,000 to begin a program of completely revamping the curricula of American schools. As a first step, various scholars, such as the chemist-educational critic Joel Hildebrand,[12] are to prepare papers enunciating exactly what it is they wish learned in their fields at what levels. As an amusing sidelight, the *Bulletin* makes much of the fact that this will not be a

---

[10]I cannot refrain from commenting that the last sentence is one of the greatest distortions of fact ever uttered about audiovisual materials. Scientifically, it is about of a level with the idea that too much vinegar in the salad dressing will cause softening of the brain. The ideological implications of the other statements are obvious.

[11]Friends of mine at the University of Illinois inform me that Bestor likes to consider himself a liberal. Perhaps this consistent approval of him by the right wing foreshadows some sort of rapprochement between right and left which will confound political scientists.

[12]Hildebrand, in one of his recent speeches, was loud in his praises of the school board of the city of Santa Ana, California. The board issued a statement concerning basic education, to Professor Hildebrand's great delight. Included in the statement was the fact that Americanism was to be taught in Santa Ana according to Clarence Manion's book, *The Key to Peace*. This is another example of the strange connections between the CBE and the extreme political right.

committee—each man will work alone. That, of course, puts quite a burden on each specialist, but, apparently, men like Hildebrand have absolutely no qualms about their omnipotence. The anti-committee bias of CBE also leads one to speculate whether the "Founding Group," as it is deferentially referred to (Bestor, Smith, Whitman, Fuller, et al.) was frightened by a committee somewhere between conception and gestation of the CBE. Or it may be that Mortimer Smith accepted William Whyte's book, *The Organization Man*, as an apocrypha to *The Restoration of Learning*.

One of the more interesting aspects of the CBE is its organizational structure. The policy-making body is called The Senate and consists of four groups. The officers and directors are one group; representatives of affiliated organizations are a second; a third group is elected by the membership. The fourth (actually the first in the list) is "The Founding Group," whose members hold *permanent* office in The Senate. Thus we see how a new priesthood can spring into being full blown. Whether or not Professor Bestor will later be elevated to a position of infallibility with a title something like *Cantankerous I* remains to be seen.

In general, the *Bulletin* reviews books, presents "programs in practice," comments on current articles in magazines, rounds up "trivia of the month" (example; a Minnesota high school building a barbecue), reports on CBE activities, includes quotable quotes attacking professional educators, and comments generally on the sad state of affairs in education or the great courage of the San Francisco Teachers Association in fighting Harold Spears.

Examination of some 18 copies of the *Bulletin* reveals a pretty consistent line. Throughout the issues can be found this position: "We are for more money for education, *BUT* . . ." By the time the BUT is taken care of, there seems little need for more money. The *Bulletin* constantly distorts the teacher certification picture and gave much space, for example, to the recent Lydia Stout article in the *Atlantic* on that subject. It appears to be opposed to teacher tenure. Anything even mildly praising current efforts in education is pounced upon. Thus James Bryant Conant and the Rockefeller Brothers Fund have both been scolded recently for their findings of things good in American education. The NEA is constantly under such vicious attack that the only conclusion that can be drawn is that Bestor intends to destroy it. Other red flags include testing, counseling, anything smacking of methodology, the whole field of psychology,[13] and federal aid.

A final word should be said about style. Credit must be given both to Clapp and Smith. They have kept the style of the *Bulletin* consistent through two regimes. At best, its style could be described as "cute." In general, it is a talk-down, let's-laugh-at-the-peasant-educationists, throw-barbs-in-'em approach. Consistently, the editors have undertaken deliberately to use emotion-loaded words while castigating the educationists

---

[13]The CBE is anxious, however, to locate "pure" psychologists, who don't follow what they consider the party line, in order to *prove* certain contentions regarding intelligence and ability already agreed upon in Sacred Council. Unemployed "pure" psychologists, please note.

for doing the same thing.

As an example of the use of emotive language, the February 1958 *Bulletin* is revealing. The lead article is entitled "The Seven Deadly Dogmas of Elementary Education" (now available in reprint form at 10 cents apiece, of course). The "dogmas" treated are readiness, interest, the whole child, freedom, integrated subjects, scientific knowledge (about children and learning), and proféssionalism (anyone who has studied "education" has no right to have anything at all to do with education). You have to hand it to Smith; "deadly dogmas" is a great phrase.

Another example is from a review of the American Council's 1955 edition of *Educational Measurement* (appearing in the June 1958 issue). Samples: ". . . the teacher . . . becomes a mere mouthpiece of the professional educational measurer . . ."; ". . . realistic academic standards [have] been lost in a mass of pseudo-scientific educationist mumbo-jumbo . . ."[14] These statements reflect careful attention to the examples prepared by Mortimer Smith, as when he said, "That prophet of distorted Deweyism came and displayed his usual ragbag of vague educational clichés . . ." (he was speaking of Kilpatrick at Pasadena) (Smith, 1954), or to a line from the Second Gospel of *Cantankerous I,* "The state-enforced requirement in pedagogy is the taproot of the great educationist upas tree" (Bestor, 1955, p. 167). Such careful, stylistic objectivity, of course, produces no emotive wording. This is in contrast to Mr. Lynd's discovery that in professional educational literature "I found a great many juicy words"—which he then quotes (Lynd, 1950, p. 31). After the quote, he refers us to review by Roelofs of a typical educationist book in which the good Professor is quoted with approval as saying, "From the introduction to the final page the writing is replete with emotive, question-begging words and phrases" (Lynd, 1950, p. 103).

We are dealing here with stylistic immorality, or, if the reader is more charitably inclined that the writer, a double standard in style in which it is perfectly proper for the CBE, its editors, writers, and scared Founders to spray literary insults, epithets and adrenalin-arousing words and phrases at every professional educator who sticks his head up; while, at the same time, the educator is blasted for referring to some past practice in education as "stultifying" or "sterile" on the grounds that he uses forbidden emotive words or gets himself entangled in pedaguese.[15] The CBE has, in effect, invented a new ecclesiastical language reserved for the priesthood. There must be a better name for it, but, at the moment, I find myself stuck with Bestorese—measured, on the average, by six dirty names to a sentence

---

[14]The reviewer, one Edmund A. Gibson, listed as an "Educational Specialist" with the U.S. Navy Publications Center in Washington, is to be commended for his courage. It's a wonder that Admiral Rickover didn't give him the Navy Cross. He managed to review a highly technical volume of 20 chapters written by 51 people in about 1,000 words with no reference whatever to the great volume of statistics with which the book is concerned. This is almost as great an achievement as sailing under the North Pole.

[15]One is inclined to wonder how men of the caliber of Richard Hofstadter and Howard Mumford Jones (both listed as CBE charter members) can associate themselves with such practices.

(dirty, but by all means cute, too, remember, fellows).

The cores we have examined from the world of pamphlets and reprints are, it must be emphasized, merely test cores from a mountain. The serious student can soon fill several filing cabinets with additional samples. What happens to them? Who knows for sure? I caught my own lawyer mailing some of them to the members of the school board in the district in which my children go to school.

And, having looked at a portion of the world of critical books and having tested the world of pamphlets and reprints, where are we? We have a long way to go. There remain more books and pamphlets, the whole gamut of magazines, essay collections, editorials, newspaper stories, speeches, letters to the editors, radio, television, motion pictures, and more. We have not looked, for example, at people like Chet Huntley (television), Dore Schary (motion pictures), the Luce organization (*Life* stories and editorials), Admiral Rickover (speeches), David Riesman (lectures and essays), and Edward P. Morgan (radio). It should be emphasized, too, that not all of these people and other representative critics in these various media are of the same critical mind or of the same orientation as those treated in this article (thank goodness!). The critical products of these other media need careful study in and of themselves.

In this article, dealing a little with books and even less with pamphlets and reprints, some attempt was made to point out the "bad guys" and distinguish them from the "good guys." If the results seem to show a preponderance of "bad guys," that's the way it is. Books that have good things to say about our society do not sell very well; very few orders for pamphlets and reprints are received when the pamphlets and reprints have something worthwhile to say in favor of a person or an institution. It's small consolation, but the last best seller that sang praises was probably the Book of Psalms.

At the moment, there's money, prestige and almost sure publication without much chance of getting slapped back awaiting anyone willing to kick an educationist in the stomach. Aggressions are also relieved. By creating a new minority group (the educationists) to push around, most of the critics treated in this paper (but not all critics) have discovered a form of therapy that brings wonderful release . . . and money. Lecture on education, anyone?

References

Alesen, Lewis Albert. *Mental Robots.* Caldwell, Ida.: The Caxton Printers, Ltd., 1957.

Allen, Mary L. *Education or Indoctrination.* Caldwell, Ida.: The Caxton Printers, Ltd., 1955.

Amory, Cleveland. "Trade Winds," *Saturday Review,* September 5, 1954, pp. 5-8.

Bell, Bernard Iddings. *Crisis in Education.* New York: McGraw-Hill Book Company, Inc., 1949.

Bell, Bernard Iddings. *Crowd Culture: An Examination of the American Way of Life.* New York: Harper & Brothers, Publishers, 1952.

Bestor, Arthur. *Educational Wastelands.* Urbana: The University of Illinois Press, 1953.

Bestor, Arthur. *The Restoration of Learning*. New York: Alfred A. Knopf, 1955.

Brogan, D. W. "The Eggheads and Ike," *The New Republic*, Vol. 135, No. 11 (September 10, 1956), pp. 9-11.

Caley, Percy B. *A Teacher's Answer*. New York: Vantage Press, 1955.

CBE Bulletin, Vol. 2, No. 11 (June 1958).

"The Council's Basic Curriculum Study and Related Plans for the Future," *CBE Bulletin*, Vol. 2, No. 12 (July 1958), pp. 1-5.

Dreiman, David B. *How to Get Better Schools*. New York: Harper & Brothers, 1956.

The Elementary Schools (chart). New York: The Long House, Inc., 1954.

Eliot, T. S. *Notes towards the Definition of Culture*. London: Faber and Faber Limited, 1948.

Facts in Education, Vol. 5, No. 6 (November-December 1957).

Facts in Education, Vol. 6, No. 1 (January-February 1958).

Flesch, Rudolf. *How to Make Sense*. New York: Harper & Brothers Publishers, 1954.

Flesch, Rudolf. *Why Johnny Can't Read—And What You Can Do About It*. New York: Harper & Brothers Publishers, 1955.

Hand, Harold C. "Black Horses Eat More Than White Horses," *Bulletin*, AAUP, Vol. 43, No. 2 (Summer 1957), pp. 266-279.

Hildebrand, Joel H. *Science in the Making*. New York: Columbia University Press, 1957.

Hunter, Evan. *The Blackboard Jungle*. New York: Simon and Schuster, 1954.

Hutchins, Robert M. "T. S. Eliot on Education," *Measure*, Vol. 1, No. 1 (Winter 1950), pp. 1-8.

Hutchins, Robert M. *The Conflict in Education*. New York: Harper & Brothers Publishers, 1953.

Hutchins, Robert M. *Freedom, Education and the Fund: Essays & Addresses 1946-1956*. New York: Meridian Books, 1956.

Keats, John. *Schools Without Scholars*. Cambridge: The Riverside Press, 1958.

Kirk, Russell. *The Conservative Mind*. Chicago: Henry Regnery Company, 1953.

Kirk, Russell. *A Program for Conservatives*. Chicago: Henry Regnery Company, 1954.

Lieberman, Myron. *Education as a Profession*. Englewood Cliffs, N.J.: Prentice-Hall, Inc., 1956.

Lynd, Albert. *Quackery in the Public Schools*. Boston: Little, Brown and Company, 1950.

Manion, Clarence. *The Key to Peace*. Chicago: The Heritage Foundation, Inc., 1951.

Melby, Ernest O. and Morton Puner (Eds.). *Freedom and Public Education*. New York: Praeger, 1953.

Nathan, Beatrice Stephens. *Tales of a Teacher*. Chicago: Henry Regnery Company, 1956.

Neatby, Hilda. *So Little for the Mind*. Toronto: Clarke, Irwin & Company

Limited, 1953.

Nock, Albert Jay. *The Theory of Education in the United States.* New York: Harcourt, Brace and Company, Inc., 1932.

Nock, Albert Jay. *Memoirs of a Superfluous Man.* New York: Harper & Brothers Publishers, 1943.

Research Division. *Ten Criticisms of Public Education.* Washington, D.C.: NEA, Vol. 35, No. 4, 1957.

Riesman, David. *Constraint and Variety in American Education.* Lincoln: University of Nebraska Press, 1956.

Scott, C. Winfield and Clyde M. Hill. *Public Education Under Criticism.* New York: Prentice-Hall, Inc., 1954.

Scott, Virgil. *The Hickory Stick.* New York: The Swallow Press and William Morrow & Company, 1948.

"The Seven Deadly Dogmas of Elementary Education," *CBE Bulletin,* Vol. 2, No. 7 (February 1958), pp. 1-9.

Shafer, Paul W. and John Howland Snow. *The Turning of the Tides.* New York: The Long House, Inc., 1953.

Skaife, Robert A. "Neo-Conservatives Are on the March With 'Sound Education' as Battle Cry," *The Nation's Schools,* Vol. 59, No. 5, May 1957, pp. 54-56.

Smith, Mortimer. *And Madly Teach.* Chicago: Henry Regnery Company, 1949.

Smith, Mortimer. *The Diminished Mind.* Chicago: Henry Regnery Company, 1954.

Smith, Mortimer (Ed.). *The Public Schools in Crisis.* Chicago: Henry Regnery Company, 1956.

Stout, Lydia. "What Strangles American Teaching," *Atlantic Monthly,* Vol. 201, No. 4, April 1957, pp. 59-63.

Wilson, Charles H. *A Teacher Is A Person.* New York: Henry Holt and Co., 1956.

Whyte, William H., Jr. *The Organization Man.* New York: Simon and Schuster, 1956.

Woodring, Paul. *Let's Talk Sense About Our Schools.* New York: McGraw-Hill Book Company, Inc., 1953.

Woodring, Paul. *A Fourth of a Nation.* New York: McGraw-Hill Book Company, Inc., 1957.

Wylie, Philip. "America—The World's First Pediarchy," *The Pocket Book Magazine,* No. 1, September 1954, pp. 1-20.

Some Notes for an Essay on Griswold and Reading

Peabody Nightingale, late Professor of Education
Thomas Edison Teacher Training Institute
Bull Moose, Montana

William H. Allen, Editor
Audiovisual Communication Review

Dear Bill:

As you know, there has been some discussion in audiovisual circles over the last few years concerning the anti-audiovisual bias of certain members of the "literary set." I refer here to writers like Joseph Wood Krutch, an associate editor of the *Saturday Review* and one of the most famous of American critics. This dislike of audiovisual communication devices is not new with Krutch (Paul Reed did an editorial on his attitude years ago in the *Educational Screen*), and is by no means his exclusive domain, as his dislike of the audiovisual is apparently shared by a group of classicists, writers and critics.

The newest development in this direction is that, in addition to the usual sniping remarks that appear from time to time in various literary publications (publications that may, in the same issue, carry other departments reviewing films and records), the literary anti audiovisual attitude has been picked up by some of the professional critics of education that are making so much hay these days.

Thus, we find that Mortimer Smith, spokesman for the Bestor-inspired Council for Basic Education and the most vicious educational critic writing today, is now waving the anti-audiovisual flag. Smith's influence cannot be ignored because, in addition to his monthly diatribe in the *CBE Bulletin* which reaches many important opinion-formers in the United States, he also has access to the columns of quality magazines like the *Atlantic Monthly*—magazines which, for the most part, are closed to us.

As an example of the position of Smith and the CBE, the June 1958 issue of

Reprinted from *Audio-Visual Communication Review*, Vol. 7, No. 2, Spring 1959, pp. 111-121.

Notes for an Essay on Griswold

the *Bulletin* is instructive. With approval, Smith quotes Joseph Wood Krutch as follows:

> Give people a picture if they can't read and you merely encourage them not to be able to read. If I were the dictator of the educational system..., I would abolish almost the entire system of visual education. I think it is not true that a picture is better than a thousand words. I think in many areas, a thousand words is [sic] better than 10,000 pictures.

Now what does all this have to do with Peabody Nightingale? As you know, I am a native of Montana and got to know the old gentleman during his last few years at Bull Moose. For some reason, he was taken with me and, after his recent death, I was pleasantly surprised to find that he had left me his important papers. They are voluminous, and I have had time to do little but riffle through them. However, the enclosed, almost completed essay caught my eye and I took time out to read it. The manuscript bore the title as I have listed it, "Some Notes for an Essay on Griswold and Reading." Professor Nightingale deals, in these notes, with this problem of the literary anti-audio-visual bias, and I thought they ought to be presented to the profession as soon as possible. I have full permission from Professor Nightingale's heirs for the reprinting of these notes and, hence, am submitting them to you.

The reference to Griswold is, of course, to his widely publicized speech at the National Book Awards dinner of 1952 in which he restated the literary versus the audiovisual position. However, as will be seen from the notes, Professor Nightingale did not stop there, but had in mind dealing with the topic in general. I think we can find in his reasoning some important answers to this recrudescence of anti-audiovisual criticism which, paradoxically, is rising during a great expansion of the audiovisual movement.

The manuscript is substantially as I found it. I have added explanatory footnotes here and there where I thought they would help. I hope you can find room to print it.

Cordially,
—James D. Finn, Professor of Education
University of Southern California

I was right in thinking President Griswold's speech on the occasion of the National Book Awards was important when I caught that phrase, "We have traded in the mind's eye for the eye's mind," in the *Saturday Review*. I'm very glad that *Harper's* decided to print it in full.[1] But that "eye's mind" phrase still bothers me. Griswold's essay could stand some examination.

---

[1] The speech may now be found in Griswold, 1954, pp. 65-72. Griswold, of course, is the president of Yale.

Notes for an Essay on Griswold

The personal essay is supposed to be out of fashion. Yet Griswold was personal enough. He even brought his bedroom into the speech. And I have always liked the personal essay. If I decided to use it, how would I begin to examine his propositions? Somewhere I have read that many personal essays begin, "Some of my best friends have been . . ." (in this case, "college presidents"). That's partly true. One of my best friends is a college president emeritus; I have known several fairly well, and I know of two others that keep such Olympian distances between themselves and their faculties that they could sleep on a bed of nails and the professors would never know it, let alone the contents of the night stand alongside the spikes.

Griswold certainly doesn't belong in this last category. Judged by his essay, he is very human, intelligent and urbane. And he is obviously a man of conviction that would not be swayed too easily by the wind blowing from the nearest alumnus or Women's Breakfast Club. Much of what he has to say is sound and thought-provoking, and I would certainly like to add his name to my small list of "some of my best friends." But there are still one or two things in what he says . . .

The essence of President Griswold's argument is this: (a) The people of the United States are suffering from an all-consuming disease which he calls technological illiteracy. By this he means that the average citizen no longer reads widely in informative, challenging or beautiful literature. (b) The Founding Fathers of our country were great thinkers, statesmen and creators because they did read widely everything that was in print in the Western world, proving the usefulness of such a pursuit. Furthermore, these same Founding Fathers forged the materials they read into new literature by writing vigorously and prolifically all of their lives, thus developing discipline in thinking. (c) Today we have technological illiteracy created by television, ghost writers, opinion-samplers, etc. "We have traded in the mind's eye for the eye's mind." The implication is that we have substituted the picture for the word. "[Reading] . . . hangs on in competition with more efficient methods and processes, such as the extrasensory and the audiovisual." (d) The great amount of leisure time now available to all men, if used for picture-watching instead of for reading and reflection, will create a nation of empty men, of muscle and mob-men.

President Griswold's thesis is alarming. But, as a political scientist, I suppose he would be the first to ask whether or not such alarms have been sounded before. Is this another case of the boy, the sheep, and the wolf? At least, we can establish quickly that other people have been worried about the same problem over the last 150 years. Clifton Fadiman, in a very penetrating essay in the *Saturday Review* called "The Decline of Attention,"[2] developed this idea completely. He covered more ground and, as my friend Professor Edgar Dale says, "uncovered more ground" on the same subject than did Dr. Griswold.

Fadiman noted that Henry James was worried about this problem in

---

[2] Now to be found in *The Saturday Review Reader.*

Notes for an Essay on Griswold

1902 when he said, among other things, that "the *faculty of attention* has utterly vanished from the . . . anglo-saxon mind, extinguished . . . by the . . . newspaper and the *picture* (above all) magazine" [italics are those of James]. And he [James] went on to use words like "Illustrations, loud simplifications, and *grossissements*." Fadiman also cited Wordsworth 100 years earlier in 1802 to the effect that " . . . a multitude of causes . . . are now acting with a combined force to blunt the discriminating powers of the mind . . . to reduce it to a stage of almost savage torpor." Words- worth blamed " . . . a craving for extraordinary incident which the *rapid communication of intelligence* hourly gratifies" [italics mine].

However, Fadiman, for all his literacy, missed when he didn't quote Wordsworth's poem, "Illustrated Books and Newspapers" written in 1846:

> DISCOURSE was deemed Man's noblest attribute,
> And written words the glory of his hand;
> Then followed Printing with enlarged command
> For thought—dominion vast and absolute
> For spreading truth, and making love expand.
> Now prose and verse sunk into disrepute
> Must lacquey a dumb Art that best can suit
> The taste of this once-intellectual Land.
> A backward movement surely have we here,
> From manhood—back to childhood; for the age—
> Back towards caverned life's first rude career.
> Avaunt this vile abuse of pictured page!
> Must eyes be all in all, the tongue and ear
> Nothing? Heaven keep us from a lower stage!

It would certainly seem that Griswold and Fadiman have not laid hold of anything new. It would take a finer literary sense than mine to judge whether Griswold's "mind's eye" phrase was better or worse than Wordsworth's "Must eyes be all in all" written 106 years before. Apparently, literary men in general do not like the "rapid communication of intelligence," especially if pictorial means are used. "In the beginning was the Word." Apparently at the end, there must the Word be also. It does seem a little bit like the boy, the sheep, and the wolf.

However, I suppose the pragmatic question would still be, is it really the wolf after all these years? Are we now so illiterate? Have we reached the point described by Fadiman where "the goal of the word is to approach the condition of the picture . . . [the picture] which attracts at once; it induces an immediate stimulus, and it is for- gotten directly. It is the ideal medium of communication without real connection, so ideal as to make it inevitable that the two great com- munications inventions of our time—the radio and the movie—should somehow copulate and engender television."

Why do literary men object to the rapid communication of intelli- gence? President Griswold even brings in the extrasensory as objection- able, and I was not aware that Dr. Rhine and his associates had so far been able to achieve a degree of communication that would compete with books. There is a deep-seated fear at work here. Could it be that these literary men are mental flagellants? Are they searching for means to make the mind suffer for what it gets? Do they hold that "prose and poetry of meaning and substance" constitute a series of mental weights to be lifted until one's brain is muscular enough to show well in a competition for Mr. Cloister of 1952?

One of the first difficulties with this position is that from the point of view of the literati only "reasonably complex works of literature and speculation" (Fadiman's phrase) are useful for mental weight-lifting. As I understand this phrase, the social sciences, the natural sciences, and technology would have to be classed as mental featherweights (except as in their upper regions the sciences cross over into philosophy, thus permitting the literary gentlemen to comment wisely on Einstein, and so on).[3] Disregarding the fact that the theory of mental discipline has long since been exploded in psychology, our scribbling friends would be hard put to establish a scale of hardness—and that's what they are talking about in part—which would place factor analysis, foot-lamberts, the measurement of social class, and circuit feedbacks down with McGuffey's Readers, and Publius, Shaw, Plutarch, and the "new criticism" in the graduate school.

But our typewriter-pounders do not really mean to establish a scale of hardness. That is ridiculous. What they mean to set up is a sort of pecking order in which reasonably complex works of literature and speculation are *better* than works in the sciences, technology and social sciences. And they defend this order by implying that reading the former requires more "mental stretch" and by maintaining that it is the only way to the "human" viewpoint. All other ways, they imply, lead to Orwell, *1984* and technological illiteracy. So I conclude that the rapid communication of intelligence is objectionable to the literati because it makes things easier than they have any right to be, and that, in making things easy, the pecking order of literature on top and everything else below is somehow threatened.

President Griswold implies that the cause of technological illiteracy is the technology of communication; Mr. Fadiman, probing deeper, absolves advertising, education and picture magazines of the blame and finds the cause in the industrialization and urbanization of society which has changed values and our way of life. In a sense, both men are wrong. If I may use a comparison which will probably strike horror in the minds of these literary gentlemen, they are both somewhat like a photographer setting up his lights to take a picture of a cube. The location of his light or lights will determine exactly what the cube will look like in the finished picture. Both these essays are brilliant and illuminating, but they only highlight one side of the cube and give

---

[3]The post-Sputnik science binge outdates this concept to the extent that the scientists and engineers have now been accepted by the literary set. The degree of acceptance is still, however, very questionable, public statements to the contrary.

a flat picture. Solid geometry, on the other hand, informs me that a cube has six sides and is three-dimensional. What are some of the other dimensions of this problem?

President Griswold and Mr. Fadiman have, in their strictures upon the taste and reading behavior of the public, dodged completely one of the most important issues of our time. This issue is that *the rapid communication of intelligence to all the people is an absolute requirement for survival.* Ogden and Richards (a source not too far outside the mainstream of literary thought) have said in their great work, *The Meaning of Meaning* (p. xxix):

New millions of participants in the control of general affairs must now attempt to form personal opinions from matters which were once left to a few. At the same time the complexity of these matters has immensely increased. *The old view that the only access to a subject is through prolonged study of it, if true, has consequence for the immediate future which have not yet been faced. The alternative is to raise the level of communication.* . . . [Italics mine.]

Does President Griswold honestly believe that the level of communication can be raised on atomic energy, flood control in the Missouri Valley, and the problems of a turbulent Orient by reference to a reading list of Thomas Jefferson's circa 1800? Or would he add all the books in print in the Western world since that time—a collection which could not be housed in all the buildings of his own university, let alone in its magnificent library? Who would do this reading? And when?

And then there's the matter of understanding what is read. Reading with understanding involves the manipulation of meanings—meanings of concepts and generalizations. Without meanings, reading is mere verbalizing, a skill that can be taught to a crow. Do Mr. Fadiman and President Griswold naively assume that meanings spring full blown from words? Only a reader with a relatively high degree of sophistication *in the field in which he is reading* can derive meanings from context and words representing concepts for which he has a poor stock of meanings. Even President Griswold, for all his erudition, might not achieve too much communication with the author of a monograph on physical chemistry or on an investigation into the autokinetic phenomenon.

How, then, do concepts and generalizations become meaningful for an individual? The answer lies in the experience an individual has had. Thus, in America; elementary terms like *dog* or *automobile* are completely meaningful because almost everyone has seen, touched and otherwise experienced these things. But what of *Chile*, and *Vietnam*, and *radioactive isotopes*? Before meaningful reading there must be experience, and can an aircraft worker in Burbank, or an advertising executive in New York, or a professor of English in Ohio experience all these things directly? Is President Griswold prepared to say that a documentary film on Chile, a television program featuring travelers from Vietnam, and a motorized exhibit on radioactive isotopes would not provide meanings which would help all Americans what they could then read on these subjects?

Or would he prefer that we make decisions on these matters based on a series of words which we have half picked up from newspapers and books about nitrates (what is a nitrate?), temples (what kind of a temple and for what?), and thyroid glands and iodine (what do they do and why?). The only alternative is a near lifetime of study on each subject, developing the meanings slowly, laboriously and, if carried on without any reference to experience or representations of experience, even wrongly. Contrary to our literary fretters, I conclude that our forests of television masts, our thousands of projectors, our techniques of quick display and communication are an absolute necessity, first to raise the level of communication, and second to actually achieve President Griswold's own objective of having people read with intelligence.

We come now to the heart of the problem—*time*. Dr. Griswold speaks of the Colossus of leisure time that will create muscle and mob-men if they do not use this time for reading and reflective thinking. I concur with his basic idea that people must deal with serious and beautiful matters instead of trivial and trashy ones if we are to achieve true humanity. But I submit that the leisure time of which he speaks is more of a delusion than a reality. The gentlemen who had the leisure to read were Hamilton, Jefferson and the others—there was real leisure time with servants, cooks and clerks to take care of the basic business of living while the master read, wrote and thought. There were exceptions, of course, but these could not have been too numerous.

In an urban society today, "leisure" time is used up in getting to and from work, in gardening and maintaining a house, in shopping, and in a hundred other ways. Most women in households with moderate income do their own work which, while lightened by labor-saving devices, still takes time. Time demands are greater than money demands today. A parking ticket received while trying to snatch a little time will take several hours out of what a former newspaper publisher used to call the *Time Bank*. The machines have brought us "leisure" which we must expend in keeping alive in a machine age. And in addition to keeping alive, the average person makes some sort of a try at a family life, a small amount of social communication, and, perhaps, some community service in an organization or two.

Don't forget, too, that the machine age is exhausting. An hour in heavy traffic and Aristotle is not too appealing, even, I suspect, to a graduate of St. John's. The dullness of a factory job has to be relieved by some escape after hours, escape from the exhaustion of machine boredom. No, President Griswold, the wonder is that we do so much intellectualizing in the time at our disposal.

Literary Cassandras would do well to examine the hundreds of study groups maintained by the PTA, for example—study groups that are not painting on some sort of an artificial veneer of culture, but that are making a sincere attempt to understand one of the most difficult phases of human relations—the rearing of the young. Our culture is sprinkled throughout with groups of individuals studying world government, conservation plans, or numismatics. Writing off America on the basis of the total misuse of an illusory leisure time is not justified by the facts. President Griswold himself admits he has no time to read and he doesn't blame that on pictures.

If the problem is time, it is also amount—the amount of material that must be known today. We have to know about India, about world trade, about heart disease. Time is short and the amount of knowledge continually increases in geometric progression. It follows that we *must* use all the means possible to distribute this knowledge. We must, in fact, use "more efficient methods and processes" where they will do the job. And here Mr. Fadiman and President Griswold, in their fear of audiovisual communication, react like any fearful person—they show signs that their fear is derived from lack of knowledge. Mr. Fadiman, for example, speaks of a picture as "directly forgotten." Yet everything we know from research tells us that exactly the opposite of this is true. What is forgotten most readily are concepts *that are held only at the verbal level without meanings in support.* Pictures always come out best in tests of remembering.

Our literary friends are also apparently not familiar with research in communications behavior. The fact that most adults do not read books is tempered by the fact that there is a marked tendency for those who don't read also not to attend movies or listen to the radio. Charles F. Hoban pointed out in an address before the Fourth Annual Michigan Audio-Visual Conference in 1949 that " . . . there is a relationship between participation in any of the modern media and participation in all modern media. Rather than fight each other—as it has been supposed—radio, movies, books, and magazines tend to reinforce each other, so that those who abstain from one tend to abstain from the others, and those who patronize one tend to patronize the others."

When we examine what adults voluntarily like in the form of reading, combinations of text and pictures, such as *Life* magazine presents each week, rank very high. Furthermore, research done at Harvard in 1933 by Rulon established the fact that *combinations* of film and text rated superior on every count (recall, etc.) for general science subject matter *three months after exposure.* Here are implications for President Griswold, the book publishers, and Mr. Fadiman! More efficient methods and processes are needed, and here are some clues as to what might be done.

It may have escaped the literati, but the Fund for Adult Education of the Ford Foundation has been aware of these implications and is doing something about them. Right now the Fund is very successfully using a combination of specially written essays and films on all types of serious topics with adult groups. Reports on the use of these combined materials at weekly meetings of Rotary Clubs, labor unions, and so on indicate that the method is producing astounding results in the form of intelligent discussion and broad thinking.

After all this, it would seem that I am in violent disagreement with President Griswold. And yet I am not. I abhor what he abhors—trashy television, mammary novels, the half billion or so comic books sold in the United States each year, the general absence of interest in ideas, the venality of much of the press (Dr. Griswold should live on the West Coast!), and X-rated double features. But Griswold (and Fadiman—and John Mason Brown and many others) offers no solution.

Notes for an Essay on Griswold

We must first have faith that the people are not as bad as they are painted. They are a long way from being beasts and "thing-men." There are a lot of beasts and "thing-men" in advertising agencies, in publishers' offices, and in radio, television and motion picture circles. But, on the other hand, for example, examine the quality of the articles in the *Ladies Home Journal* as compared with 20 years ago and ask if this does not represent progress; so the *Reader's Digest* has maneuvered into the ridiculous position of becoming a "medical" journal; what about the excellent syndicated articles recently appearing in many newspapers on medical topics. Given help, the people will seek quality.

Second, modern life is putting an intolerable burden on everyone in terms of what is to be known and thought about, the difficulty of the concepts involved, and the time necessary to do the job. Intellectuals of good will must bend every effort to solve this problem. All means, including the audiovisual (and the extrasensory, if possible), must be used to help people develop meanings so that they may think clearly and have the desire and ability to read widely. Then and only then will be achieve Dr. Griswold's objective.

Third, we must adapt television, films, radio, even comic books and displays to the rapid communication of intelligence where reading is not possible or profitable in terms of time, energy or ability. We must wage constant war on the "thing-men" in the advertising agencies and other places they control the media of communication so that real ideas and materials can be transmitted instead of trash. This war cannot be waged directly in most cases, but by example and creative effort. The first question I would ask President Griswold is, "What is Yale University doing about television, radio and audiovisual processes?" What Yale does in its community and in the nation to use the mass media to better ends, and what P.S. 29 does, and what the Plumbing Society of South Paducah does will make the difference. Castigation is only helpful when followed by creative effort.

Finally, all of the ringers of the belles-lettres had better take a good look at themselves. As I said before, basic to their whole viewpoint is a scale of values which insists on putting the reading of certain works on a higher value plane than a thousand other rigorous intellectual activities now pursued by man. Literary gentlemen should recognize that many people other than themselves accept the same basic values they do, but realize that these values may be achieved in many ways. Intellectual activities may be diverse without being ranked; humaneness may be achieved through science or cookery or even educational method; and communication can take place in efficient ways without ruining the ideas communicated in the process.

References

CBE Bulletin. June 1958.

Fadiman, Clifton. "The Decline of Attention," *The Saturday Review Reader.* New York: Bantam Books, 1951, pp. 23-36.

Griswold, Alfred Whitney. *Essays on Education.* New Haven, Conn.: Yale University Press, 1954.

Notes for an Essay on Griswold

Hoban, Charles F. Address at Fourth Annual Michigan Audio-Visual Conference, 1949.

Ogden, C. K. and Richards, I. A. *The Meaning of Meaning. A Study of the Influence of Language upon Thought and of the Science of Symbolism.* London: Kegan Pond, Trench, Trubner & Co., Ltd., 1923.

The Sound and the Fury of Rudolf Flesch

By the time this editorial appears in print, the new book by Dr. Rudolf Flesch, *Why Johnny Can't Read*, will have been discussed, distorted, deified, and damned. On the one hand, those who are convinced that the schools are teaching nothing—certainly not the three Rs—will set the eminent Viennese pedagogue on a pedestal and use him to fire superintendents, harass teachers, and demand that school boards abolish "fads and frills." On the other hand, the anointed, the white-robed, among the curriculum specialists, reading supervisors, et al. will pontificate from their high mountains that the good doctor is someone who accidentally got left up in the trees when the rest of us came down and began to walk upright.

Why Johnny Can't Read, in case some our readers have missed it and the excitement it is bound to cause, is a book which says that American children are two years behind their European counterparts because all reading instruction in the United States is by the "sight" method and that the only way to teach children to read is by phonetic analysis, or the "phonics" method. There are, says Dr. Flesch, no remedial reading cases in Europe because all reading instruction is by phonics. Dr. Flesch details his particular version of phonics instruction after a long attack on the sight method. He carefully ignores the "experience" method, devoting approximately two pages to dismissing it.

It is very hard to be objective toward a book like *Why Johnny Can't Read*. Dr. Flesch is an expert communicator with words, and in this book he is an advocate, a messiah of the first order. His extremely argumentative approach gives rise to what the psychologist Thouless called "argument by extension." This meant that, when someone takes the extreme of one side, it forces the person arguing with him to take the extreme of the other side, when the middle ground may be the only intelligent position. Now, it may be that Dr. Flesch was forced into his extreme position regarding phonics by the hostility of the advocates of the sight method toward any suggestion of phonics. At least that is indicated in the first chapter of *Why Johnny Can't Read*.

We will try, in the next few paragraphs, to reduce this conflict between extremes to something that approaches objectivity, at least in a few spots. It would take another book as long as that describing the difficul

Reprinted from *Teaching Tools*, Vol. 2, No. 3 (Spring 1955), p. 91.

ties of Johnny to analyze the Flesch thesis point by point. Let us look for a moment at the good doctor and what might be expected as to his reliability.

Rudolf Flesch has long been known as a student of language. He is most famous for having published several formulas for the measurement of readability of written material. He has a doctor's degree from Teachers College, Columbia University, and is not without sophistication in the fields of education and psychology. He has been very successful in the publications field, both as an author of books on language and as a consultant in the problems of readability. He has been unfairly attacked, we think, by the literati of the country for his attempts to reduce writing to a formula. Of his books, in our opinion, *How to Make Sense*, a discussion of communication, is by far the best.

Rudolf Flesch, however, is not without his faults. He is not the most careful scholar in the world, as scientific as his work sounds to the uninitiated, and he is a simplifier. With these two danger flags let us take a look at one or two items in *Why Johnny Can't Read*. Dr. Flesch says, on page 72, "Phonics—*any* kind of phonics—before second grade is too much for a child, the educators say: they consider it an established fact that six-year-olds cannot learn phonics. I have seen this statement repeated— and explained at length—in every single book on teaching reading that I have studied."

Now Dr. Flesch, according to a direct quotation on page 97, read Paul McKee's well-known book, *The Teaching of Reading*. And so it will be easy enough to find out if what he says is true. Either McKee said six-year-olds couldn't learn phonics or he did not. On page 243, McKee says, "The elements to be taught for phonetic analysis at the first grade level are those most needed in attacking the strange words included in the vocabulary of the reading matter that the child is most likely to meet in that grade." In the next few pages, he lays out five principles of teaching phonics to six-year-olds. Following this discussion, he develops the techniques of structural analysis of words for six-year-olds. On pages 248 through 252, as an example he gives a detailed method for the direct teaching of the single consonant "t" in the initial position. On pages 254 and 255 he identifies all the consonants, consonant blends, etc. that should be taught directly by phonetic analysis to six-year-olds. Dr. Flesch's statement is obviously not true in the case of Paul McKee, one of the leading authorities in this country in the field of teaching reading. Either Dr. Flesch has deliberately falsified for the purpose of building his case, or he is guilty of questionable scholarship.

Anyone who makes a statement about *all* educators, even to sell books, has lost all claim to objectivity. Such statements abound in the book. "*All* European children can read better and earlier than American children." "There are *no* remedial reading cases in Europe." "*All* instruction in reading in America is by the sight method except for a Chicago suburb and some parochial schools." "There is *no* such thing as reading readiness." "There are *no* emotional problems in reading." Etc., etc., etc. Such an all-or-none technique is fine for a political pamphlet, but we seriously doubt if it belongs in any kind of printed material purporting to be even

The Sound and the Fury of Rudolf Flesch

halfway scientific.

On the other hand, we don't want to be guilty of a Fleschian approach to this problem. As we said above, there are many people in education who have reacted the other way with the same all-or-none formula. "Phonetic analysis is *all* wrong." "*Only* remedial cases should use it." "Our system is *the* answer." "Dr. Flesch and questioning parents are benighted heathens."

The plain truth of the matter is that reading is one of the most complicated of all human activities. It cannot be explained in the over-simplified terms of Dr. Flesch's phonics any more than it can be explained in terms of some county supervisor's independent activities materials in visual discrimination. What Dr. Flesch has ignored or forgotten is that reading is for the purpose of getting *meaning*; what many soft-headed reading consultants have ignored or forgotten is that the English language has a structure with some logic to it. Flesch is right, for example, in his attack on some of the inane reading materials supplied to children; his opponents in the reading field are right in that phonics alone is a good thing for parrots.

We suspect that both extremes are neglecting the key issue: meaning. Dr. Flesch proudly cites the fact (page 23) that he can read the Czech language aloud and not understand a word. What use this particular skill would be except to an actor like Sid Caesar is certainly questionable. On the other hand, the drivel like "Come Jane, come Dick, come, come, come" for constant sight practice is practically as meaningless and precisely as inspiring. And here is our middle ground in this matter.

Once many years ago, Horace Mann tried to find out "how far the reading in our schools is an exercise of the mind in thinking and feeling, and how far it is a barren action of the organ of speech upon the atmosphere." And he came to the conclusion that "the ideas and feelings intended by the author to be conveyed to, and excited in the reader's mind still rest in the author's intention, never having reached the place of their destination." Good reading is the reverse of this: it conveys meaning; it enables the reader to think critically about what is read; it involves concept formation and concept juggling.

Phonics alone won't convey meaning. It can make possible the barren actions of the organs of speech upon the atmosphere. Flesch makes a good case that the repetition of insipid materials viewed for sight identification won't convey meaning either. A good reading program must be based on experience that helps make meanings clear. This means that the teacher must use the tools at her disposal to build meaning so that the words can be understood in context and thought about in relationships.

Dr. Flesch would have been truer to the ideal of scholarship had he investigated the work with films and storybooks that Clyde Arnspiger did for Encyclopaedia Britanica Films a few years ago. A film and a book on the Navajo had an interesting story and it made meaningful a word like "hogan." You could sound "hogan" for 10 years and get nothing out of it if you didn't understand it. The materials made the meaning clear and developed an interesting story. Nothing then would prevent the teacher from introducing phonetic analysis.

The Sound and the Fury of Rudolf Flesch

As a matter of fact, the scientific nature of language as analyzed by Bloomfield and others is a part of experience when handled correctly. With films, filmstrips, recordings, and other tools of teaching to help develop concepts, with other reading skills like structural analysis and visual discrimination not neglected, phonics can find a rightful place in the reading program. The pragmatic test of any method of teaching reading ought to be: *Does reading help the student achieve constantly greater meaning and maintain his interest so that he becomes increasingly independent?* Teaching tools will help do this job; phonetic analysis will help do it; sight methods will help do it. Everybody ought to get off their respective soap boxes and get to work.

References

Flesch, Rudolf. *How to Make Sense.* New York: Harper & Brothers, 1954.

Flesch, Rudolf. *Why Johnny Can't Read—And What You Can Do About It.* New York: Harper & Brothers, 1955.

McKee, Paul. *The Teaching of Reading in the Elementary School.* Cambridge, Mass.: The Riverside Press of Houghton Mifflin Company, 1948.

Conventional Wisdom and Vocational Reality

There is, I believe, little reason for this meeting today. We are supposed to discuss, within the context of California higher education, the role of vocational-technical-professional education. However, thanks to Thomas Braden (a newspaper publisher), William Norris (an attorney), Bishop Gerald Kennedy (the Methodist Church), and several other influential citizens, including Mrs. Talcot Bates (a housewife), it has been promulgated with the force of law by the State Board of Education that vocational-technical-professional instruction is not academic; by virtue of this definition these people on the Board have condemned all such learning by claiming it is nonlearning, and that anyone seeking to instruct the young in the mysteries of such crafts, arts and applied sciences relating to human work and welfare are hereafter condemned to be second-class citizens of the educational community within the state of California. Certainly, this stance on the part of Mr. Braden and the rest of his cohorts on the Board requires that we dare not consider vocational-technical-professional education as part of a program in *higher* education within this state.

Nothing I have just said should be construed to mean that I am therefore sympathetic with Mr. Rafferty in his occasional jousts with Mrs. Bates and her friends. To give our great superintendent his due, he has objected in print to second-class citizenship for people in business administration, homemaking, etc. However, since he has no alternatives to offer except a thorough textual criticism of *Silas Marner* for everyone (providing the proper patriotic slogans are recited fore and aft of the exegesis), Mr. Rafferty merely contributes to the reasons why we should not have this meeting.

One is tempted to go further in suggesting a nonmeeting for a non-academic topic. The dangers of vocational specialization are apparent in the situation at Berkeley: We can all see what happens when political science and sociology are applied to a social problem, when demonstrations have been removed from the lecture to the street. One might add that there is some favorable evidence that the state colleges are shaping up to the Braden *zeitgeist*; at least they seem no longer to be teaching accounting. I have not dealt here with the private system of higher education in the state, principally because I represent this sector and we have

Paper presented at the Fourth Annual Conference on Higher Education, San Francisco, California, May 7-8, 1965.

enough trouble as it is. However, I do feel that Cal Tech's decision to move into the humanities is trying to tell us something; it could be that Cal Tech has decided to become academically respectable at last.

Perhaps you do not care for my attitude; if you do not, you do not understand the Irish. We laugh that we might not cry, and the question of status for vocational-technical-professional education is enough to make a laughing hyena break out in tears. We have today reached a height of pseudo-snobbism vis á vis practical education which is ruinous. Any sober consideration of the problems of such instruction in the innovative society must take that into account.

The great irony in this status-ridden situation is that, while the disciplinarians fiddle, a great social protest is burning through America demanding more, better and infinitely more functional vocational-technical-professional education (Swanson, 1965). Many years ago, it was no accident that the Charles River was used to separate the Harvard Business School and M.I.T. from the Yard; Harvard is Harvard and there the painter has no status—only the art critic. It is now, however, a horrible political joke that in California, perhaps *the* leading industrial state in the Union, with the greatest system of higher education of any, legislators, boards and elected and appointed administrators who may disagree among themselves unite to downgrade all vocational-technical-professional education into the nonacademic ghetto. Until this situation is resolved, all discussion of this problem is, if you will excuse the expression, ladies and gentlemen, academic.

In the search for ideas for this symposium, I took a look at the conventional wisdom in vocational field and found that there were, in fact, two categories of conventional wisdom. One might be called the old-line or Smith-Hughes syndrome which equates all vocational-technical-professional education with vocational agriculture circa 1918. While this point of view apparently is still very effective and influential in the fields of secondary education and apprenticeship, I doubt if it is making much impact in higher education these days except, perhaps, in some technical institutes in the east.

The second category of conventional wisdom is the current intellectual "in" view. It might be called the IBM syndrome, since it is based on concepts of structural unemployment, need for more advanced technicians and professionals, and the changing structure of the work force, work space, and work time, as America, with startling rapidity, becomes an advanced technological society, automating as it goes along.

Let us examine some of the propositions of this newer conventional wisdom and comment upon them as they appear. One of the first has been well developed by Edward T. Chase and it simply says that the old ideas of vocational education have to go (Chase, 1963). Taking some care to recover any infants that might have sailed out in the bathwater, we can, I believe, say amen to this principle.

There is a cluster of "in" conventional wisdom principles surrounding the general notion that an individual's job may disappear three or four times in his lifetime due to technological change; that he, therefore, must have a very flexible, innovative-oriented, vocational-technical-professional background; that on-the-job training is the answer to all problems of skill performance. Mr. Hutchins, with his usual economy recently summarized this viewpoint in his syndicated newspaper column (Hutchins, 1965).

The trouble with this general viewpoint is that it lacks scales—both time and degree scales. The logic is beautiful but the present relation of this logic to reality is open to serious question. Let me try to illustrate what I mean. There is no question that people will be automated out of jobs— more so in time to come; there is no question that the nature of work will change, etc. etc. *BUT*, this general phenomena is not going to occur everywhere at the same time and with the same force. It will occur on some kind of a time scale.

At this point in our history, such an idea suggests two approaches to the vocational-industrial educational problem. First, any viable vocational curriculum must do the best it can for students *presently in school* and not prepare them to function in a theoretical world that is 10 years away. Second, the world, the student and the technological change process must be monitored by the preparing institutions with a rigor never before attempted. This monitoring (work force analysis, estimates of rates of change, economic growth, etc.) should do two things: (a) it should feed back necessary changes in procedures, direction, etc. which need to be acted upon immediately in the form of curricular change, and (b) it should provide long-term directional soundings which can infuse the same curriculum with some sort of insurance for the people going through it that they can meet, preferably without anxieties, an innovative future, as well as doing a little innovating themselves. This requirement is not easy. It demands a design for constant and perhaps accelerating reconstruction of the vocational-technical curriculum.

When I referred to degree scales as well as time scales a moment ago, I was referring to the fact that not all sectors of the economy change at the same rate, and therefore, their *degree* of change must be measured by the suggested feedback loop and this degree must be reflected in the curriculum.

This general notion of a short-term, measured approach to the realities of vocational-technical-professional education also has merit for other areas of education. Some years ago, Howard Mumford Jones, in suggesting a program for general education somewhat opposite in point of view to the famous Harvard Report, pleaded for a program that would get us through the next 20 years, not forever (Jones, 1946). Needless to say, his plea went unanswered as all general education programs are designed for eternity.

So far I have been using a sort of Korzybski term to cover this whole field—the term "vocational-technical-professional." This usage is meant to imply that such preparation for the life of work exists on a continuum; that, in a certain sense, preparations for becoming a plumber, interior decorator, or brain surgeon are the same order of events; that the trade school, terminal junior college, and university professional school have, or ought to have, certain common principles in their orientation. This view may bother some purists, but I will further confuse the issue by pointing out that there is no more pristine example of the vocational school in

America than the graduate school of arts and sciences in the university. Dilettantes who think otherwise probably also believe in the Easter, as contrasted with the Playboy, Bunny. The difference between graduate school vocational training and the other types is that the professors in the graduate school, engaged in reproducing their own kind, know exactly what they are doing.

Returning to the concept of the continuum of vocational-technical-professional events—and the new conventional wisdom treats it this way—the next proposition is that such training ought to be compounded of two and only two parts. These two parts are (a) a study of the disciplines backing up the vocational-professional activity, and (b) some kind of on-the-job training or internship. The third element of such preparation—which lies in between (a) and (b) and which has something to do with application, method, technique, experienced-passed-on-down—is dismissed as both useless and dangerous.

This two-pronged curriculum design of discipline and internship really exists as a theory and it might be called the amateur theory of vocational-technicial-professional education. The idea is to encourage amateurism and to save us from vocationalism or professionalism or method or bag-of-tricks or "simple" devices.

Within higher education there are three areas of vocational training which are exceptions to this general rule. In the truly crucial portions of life—keeping out of jail or writing a contract, trying to stay healthy, and trying to keep from going to hell, or at least staging a proper marriage ceremony for one's daughter—men are not willing to trust their fate to amateurs trained only in the disciplines with a little on-the-job training on the side. Hence Law, Medicine and Theology, three faculties of the universities of the Middle Ages, continue to have a healthy slug of professional content and are exceptions to the rule.

Elsewhere—in the training of plumbers, nurses, architects, and oboe players—the theory may be found full blown in the literature, if not actually operating. It is an interesting theory—and more interesting because it isn't true. Any junior engineer will tell you, for example, that he cannot deduce his craft on some sort of magic one-to-one basis from the physical sciences. I am not trying to deprecate the advantages of knowing a range of disciplines; I do say that disciplines plus internship are not enough.

The trick, of course, is to decide, not only on a constantly updating vocational-professional content, but a system of teaching this content and of relating it to the disciplines and the job. It is here that I wish I had the time to indicate what I consider the role of an intelligently applied instructional technology to such a mission, for, after all, that is my field. Perhaps, however, we may bring these ideas out in the discussion and use the remaining time to explore this confused area further.

The third precept of the conventional "in" wisdom concerning vocational education that I would like to examine is one I heard enunciated by Mortimer Adler many years ago at a lecture at USC. It is now very popular. It might be called the Greek City State or Athenian theory of vocational education and is a variation on the theme of education for leisure. Adler's argument was that the Greeks built their great culture on slavery; today, particularly with automation, machines are men's slaves and, hence, everyone should have a liberal arts education similar to the Greeks as the only vocational-technical-professional education. More recent expressions of this point of view might disagree with Mr. Adler on the details—i.e., the liberal arts content—but the basic concept is much the same. A new element has been introduced in the idea that, since work as we know it is disappearing, the puritan work ethic must be replaced by a leisure work ethic involving hobbies, crafts, arts, life-long learning, etc. This is, in another way, the Grecian ideal.

There is much to recommend this viewpoint and I believe much of it to be valuable and useful. There is no doubt we must move in this direction. The direction, however, must again be placed on a time scale. While the move toward nonwork is speeding up, it is by no means here and, certainly, will not hit us on an across-the-board basis all at once.

However, my objections to this Athenian approach are really deeper. For some time I have had the feeling that this argument does not quite square with reality, even though there is a great deal of truth in it. This incongruence, if it exists, must be dealt with in any program for preparing people for a world of work.

There are, in fact, three fallacies within this concept:

1. The reduction of the hours worked per week since 1850 is not nearly so drastic as the statistics indicate.
2. The amount of leisure time contained in the reduction that has occurred is illusory.
3. For certain segments of the population—professional and managerial—the work week has increased.

It is not possible to develop here the case for the relatively small decline of the hours worked in a week over the past century. Sebastian de Grazia, in his very perceptive work, *Of Time, Work, and Leisure*, proves the point conclusively by showing that the statistics themselves are questionable (moonlighting, for example, does not show up) and that other work-like activities such as the long journey to and from work have been added. Any advocate of immediate education for leisure will have a sobering experience reading de Grazia.

Secondly, the time made available by the shortened work week is by no means all leisure time. I have had a suspicion for some years that this is true, and, upon analysis, I came to a personal conclusion that the entire world of automation was a great conspiracy to make *me* do the work formerly done by other people. In this sense, much of automation is a fraud. It merely spreads the work so that I have to serve myself in a hundred places such as the supermarket; I have to assist the insurance company when it bills me; I must never fold, staple or mutilate; and I must help my wife service labor-saving gadgets which made us do the work that other people did for her mother. Finally, a lot of unexamined praise has been heaped upon the "do-it-yourself" movement. It does not, as claimed, for the most part represent a return to the values of the guilds; rather it is part of the great conspiracy of automation to spread the work and, in fact, decrease leisure time. Nonwork is not leisure time; nonwork for the most part now consists of doing the work for General Motors,

Safeway, and the Bank of America. I recently found scholarly confirmation for this hypothesis. Heckscher, writing in the *American Scholar*, said:

. . .it is plain that the organization of the modern world is systematically directed toward putting upon individuals who are presumably being served an ever heavier burden and a more complex responsibility. Today's telephone system, like today's supermarket is designed to make a man do the work he is under the illusion of paying others to do for him. Some remnants of an obsolete civility remain, as in the telephone operator who comes to one's aid after a dozen digits have proved too much for the mind to hold or the fingers to manipulate. (Heckscher, 1964, pp. 575-576)

The third fallacy in the Athenian theory is that the blessings of leisure apply equally to all segments of the population. Heckscher also destroyed this point when he said (and all of you here know this instinctively):

The part of the population that once enjoyed leisure, the scholars, the wise men, the managers of wealth and the rulers of the social order, are today the ones who are working long hours. The bifurcation of society into a hardworking minority and a mass condemned to the pursuit of unrelieved recreation is not wholly a nightmare of the future. (Heckscher, 1964, p. 572)

From all this it should follow that facile generalizations about education for the new leisure and the provisions for the innovative, creative life will take some doing. We are probably going through a passing phase in this time of nonwork that generates toil. However, unless we wish to fail several generations of students, we must take it into account in our planning and not surrender entirely to the blandishments of wishful-thinking poets who wish to hear America singing beginning tomorrow.

Finally, I would like to examine the tomorrow beyond tomorrow. As I indicated earlier, at all times, our soundings should take into account both the short-range and the long-range situation. Some of the short-range problems will change drastically and others will disappear as we move into the future. Let us hope a good feedback system can handle this. Now let us take complete leave of the new conventional wisdom and try our own crystal ball.

As many observers have pointed out, a technological-scientific society is an increasingly abstract society.[1] It increasingly removes itself from flesh and blood, from sunsets and perfume; it replaces love with the study of love; it substitutes the obtuse, the obscure, and the generally incomprehensible for the real, the vital, and the present.

Since men, however, live both in a personally real and an objectively real world at one and the same time, this academic construction of reality handed us by science, technology and even the humanities and the arts sets him into conflict within and without himself. The young person today faces an abstract world of large-scale organization vibrating with the rhythms of industry, and governed by the atomic clock, the punched card, and equations describing the frequency of four-letter words in James Joyce.

---

[1]The French sociologist, Jacques Ellul, for example.

Conventional Wisdom and Vocational Reality

This situation of an abstract, bloodless environment has set up a number of great dialogs. There are dialogs between violence and nonviolence; between person and corporation; between book and nonbook; between commitment and noncommitment, etc. Involved, too, among college students these days, is a thorough questioning of the nature of existence itself. I am sure many of you have experienced these things directly and know the nature of the personal, upsetting experiences now being undergone by many students. A good summary of the current status of these matters may be found in a recent article by Gray (1965) and in Goodman's book, *Compulsory Mis-Education* (1964).

For more evidence, we may turn to a recent European observer who quotes Walter Kerr to the effect that:

> . . . the conclusion [is] that all value resides in the abstract ratios, the theoretical equations, of our cerebral manipulation of things, and that no value is to be discovered by merely playful contact with the things themselves, whether those things be stone or salt water or flesh. (Von Borch, 1962, p. 193)

Since this is the case, it is surprising, I think, to find little about it in the literature of vocational-technical-professional education. Perhaps most people dealing with these matters believe that such concerns may be safely left to the academic disciplines, particularly the humanities and the social sciences. This is a naive view if I have ever heard one. The academic disciplines are the fountainhead of abstraction. Their goal is to leave life as far behind as possible. Analysis is replaced by micro-analysis which in turn is replaced by micro-micro-analysis using obscure symbol systems.

In the meantime, humanity—man—cries out for something more. I do not believe in turning the disciplines aside from abstraction; we must, however, add a new dimension to educational experience or the whole thing may fall apart long before we are inundated by either knowledge or population.

All the projections into the future suggest that the work force, in the long run, will be less and less concerned with work per se. The solutions to this problem are, for the most part, unimaginative. We must encourage creativity, we must know what to do with leisure, etc., etc., etc. The evidence clearly suggests that, on the other hand, people living in an advanced technological society of large-scale organization are alienated, and that the proposition that the world is absurd is the only one that makes sense to them. Their problems appear to be personal, human and concerned principally with relating.

If the projected technological trends continue, it may be that we will, in fact, be split into two types—a relatively small class of professional thinkers, managers, advanced technicians who manipulate abstracting, etc., and the rest of the population. The problem of the training of the elite, while difficult, should not be unsolvable.

What, however, of the rest? Vocational education in this sense may turn out to be primarily an education in living. Since most people have to live with human beings, it ought to follow that an education for living would be an education in *relating*, in becoming a person. Acceptance of oneself as a person then permits hospitality toward innovation and ingenuity in adversity. This might be a strange direction for vocational education to go. However, even if the bifurcation I suggest above does not come about, the problem of the person remains and the vocational-technical-professional area at least has some relationship with human reality. It seems to me that the world will turn out absurd or not as we make it so.

It would have been comforting to me as well as to you, I am sure, if I could suggest more definite answers to these problems. I have, however, learned long ago one great lesson about education. There is, as an industrialist friend of mine once told me, no such thing in education as a silver bullet. What we have to look out for is witchcraft.

References

Chase, Edward T. "Learning to be Unemployable," *Harper's Magazine*, Vol. 226, No. 1355, April 1963, pp. 33-40.

de Grazia, Sebastian. *Of Time, Work, and Leisure.* New York: The Twentieth Century Fund, 1962.

Goodman, Paul. *Compulsory Mis-Education.* New York: Horizon Press, 1964.

Gray, J. Glenn. "Salvation on the Campus: Why Existentialism is Capturing the Students," *Harper's Magazine*, Vol. 230, No. 1380, May 1965, pp. 53-59.

Heckscher, August. "Reflections on the Manpower Revolution," *The American Scholar*, Vol. 33, No. 4, Autumn 1964, pp. 568-578.

Hutchins, Robert M. "Vocational Training—'Dumping Ground' of the School System," *Los Angeles Times*, April 19, 1965, sec. 2, p. 6.

Jones, Howard Mumford. *Education and World Tragedy.* The Rushton Lectures. New York: Greenwood Press, Publishers, 1946. Reprinted by permission of Harvard University Press.

Swanson, Gordon. "Action in Vocational Education Considered as Social Protest," *Phi Delta Kappan*, Vol. 46, No. 8, April 1965, pp. 353-354.

Von Borch, Herbert. *The Unfinished Society.* New York: Hawthorn Books, Inc., 1962.

An Uncomfortable Rejoinder*

*Some Comments Directed at W. H. Ferry's Proposition Concerning Vocational Education in Junior Colleges

My instructions are to prepare a short memorandum criticizing W. H. Ferry's proposition to reorganize the junior college system of this state as presented in his paper of May 1 and in his additional comments presented in a memorandum to the participants of this conference. I find this assignment difficult, and all sorts of phrases of the variety of "mixed emotions," "ambivalent attitudes," etc. continually crop up.

First I would like to enter a disclaimer. I am not an expert on vocational-technical education. I happened to be asked to address a meeting and Mr. Ferry got hold of the speech. Since professional education is my general field, I, like all such optimists, am prepared to make a speech on any phase of it at the drop of a hat, and Mr. Ferry caught the hat. I probably should have stuck to my own field of instructional technology—and probably should do so today—and show you a film or something. However, since I love confrontations, let us proceed, ready or not.

Mr. Ferry proposes that we eliminate the terminal (vocational) programs from the junior college curricula and remake these institutions exclusively into academic two-year colleges. He is not clear, or at least did not say, whether he makes a distinction between a terminal general education program and the two-year transfer curriculum. This omission may or may not be crucial. Further, Mr. Ferry proposes to create a new institution—a regional technical institute with intimate relations established with the various sectors of interest in vocational-technical training (industries, unions, etc.) within the region. Since this proposition has just been advanced, it lacks flesh for its bones and, understandably, questions of who should attend, govern, pay for, etc. this new regional institution are left unanswered.

The arguments advanced for the proposition, as I understand them, are about as follows:

1. **The liberal education argument**—The two years of academic junior college experience are necessary for future generations to (a) start on a lifetime program of general education, and (b) adjust to and control for the better a new and complex world.

2. **The work/leisure argument paradox**—Vocational education tends

Prepared for the participants in the Conference on Technical Training in Junior Colleges, Center for the Study of Democratic Institutions, Santa Barbara, California, June 25, 1965.

to become obsolete before it is undergone; men must learn to live with the new leisure without work; in addition, paradoxically, in a technological society there will obviously be a need for vocational-technical training that should be efficient and not obsolete.

3. **The educational reform argument**—The university has abdicated its responsibilities in the field of teaching; the logical institution to pick up this ball is the junior college, providing it is totally oriented to an academic program.

In an attempt at a rejoinder, or, as is more likely, a somewhat diffuse and confused discussion of this proposition, I shall deal with it under three headings:

1. The problem of organization
2. The problem of teaching and general education
3. The problem of philosophical position.

1. The Problem of Organization

Certain trends in the organization patterns of junior colleges in California lend support to the Ferry position. Apparently, the tendency is toward organizing the community colleges into separate districts and to remove them from districts in places where they now exist as part of a unified school district. The opinion in Los Angeles, for example, is that such a reorganization will occur for the seven colleges now part of the Los Angeles Unified School District. Obviously, larger separate districts and separate boards (with, perhaps, some overall state control later) would make it easy to convert the system into the total academic system for students and adults as envisioned by Mr. Ferry.

Further, the idea of regional technical institutes is appealing to administrative theorists on the grounds of efficiency alone. It is thought that costly duplication of equipment and services could be avoided; that once vocational students were out of the way in the academic colleges, management problems would be eased and savings instituted.

On the other hand, we have to face up to the old rule that a hypothesis (in this case, a proposition) ought to account for all the facts. In one of his papers Mr. Ferry makes some reference to his difficulty in adjusting to California freeways. From this remark, I take it that he is not too familiar with California. His proposition certainly suggests a lack of familiarity with the California system of higher education, as he has not accounted for the gigantic state college system which lies in between the university of which he despairs and the junior college system which he intends to reorganize. The state college system—a series of stubborn facts if there ever was one—must be accounted for in any proposed reorganization of higher education within the state.

Does Mr. Ferry assume that all 15 or 18 (or however many there are) state colleges become additional branches of the University of California? Or does he assume that the state college system will become another university system, also concentrating primarily on graduate work and research? What is wrong with the idea of the state colleges in the first place? Some were re-created from the old teachers' colleges and others were founded to carry out this teaching-education function within a reasonable distance of a student's home; in addition, of course, they have a vocational-technical-professional function.

It is tempting, of course, to push Mr. Ferry's argument vis á vis vocational-technical education to the brink of absurdity. For, if we knock it out of the junior colleges, why not the state colleges? We could even threaten the medical and law schools of the University of California, and we would be right back with Robert Hutchins in 1936 when he was going to reduce all professional schools to sort of loose attachments to the university providing they behaved themselves.

On the other hand, if we extend a little more charity to Mr. Ferry, it can be argued with some force that the state college system is in fact becoming a junior university system that is growing up fast, if not absurd. This being the case, why not make the University of California solely into an international research institution, convert the state college system to the state university system, and expand the existing junior colleges into four-year colleges (that will, within five years after that, want to award the masters degree)? One of my best friends and colleagues—Dr. Robert Hall of California State College at Hayward—thinks this is the answer.

On balance, I find I must take the conservative view—a position a little uncomfortable for me. I do not believe that the organization-efficiency arguments are forceful enough, and I find the neglect of the possibilities of the state college system to be decisive. From the point of view of organization, I feel that more attention should be paid to the state college system and that this system should be developed into the best possible teaching-learning entity available to the young people of California. The junior college system can then remain to perform its classic three- or four-prong function and can be thought of, as many people do think of it today, as occupying the position that the high school did 75 years ago.

2. The Problem of Teaching and General Education

The second portion of Mr. Ferry's argument has two parts, both of which could stand further specification. His new junior college is to provide something called *education*. It is not clear exactly what this education is, except that its products are to be able to think, to face to the uncertain cybernated future with flexibility, begin a program of lifelong learning, and, if Mr. Sheinbaum can ever forgive me, adjust.

It seems to me absolutely crucial to state what it is intended this program of education be before we can decide to adopt or reject an organization pattern designed to convey it. To say that there should be two years of liberal arts is to say nothing. To follow the transfer pattern so that students may enter the University of California without penalty is, of course, the easiest. This means that the program for the two years will be one of distributed requirements—so much of this and that—as developed by logrolling between departments in Berkeley and Los Angeles. I will state categorically that such a program will not help very many students achieve Mr. Ferry's admirable objectives.

What such a program will do, of course, will be to eliminate many students who, for one reason or another—status ambitions of parents, general peer-group pressures, inadequate guidance, etc.—were pressured into a situation they should not have been in the first place. There is, however, one important difference. With the Ferry reorganization plan, *these students will have no place to go.* Transfer between curricula is easy enough in the junior college, although the academics make it as hard as possible on the vocational student moving into the academic curriculum. Transfer to a regional institute in another location and obviously with a lower status would be difficult for some and impossible for others. (I will have more to say on this question of status below.)

To return to the question of general education—the proposed academic curriculum—what is it that is proposed that will accomplish the stated objectives? Until a reasonably satisfactory answer is developed to this question, the Ferry proposal must be rejected out of hand.[1]

The question of teaching which Mr. Ferry raises is certainly an important one. The students at Berkeley—and at many other places in this land—are saying that university teaching is pretty bad. You all know the story. What has not been remarked upon is the great irony this story, this movement, contains. It is, to my mind, the greatest irony present in American education in this century.

A man sitting in this very conference, one of the representatives of the institution whose hospitality we are enjoying, made very great contributions to the passage of a bill through the California legislature now known as the Fisher Bill. The Fisher Bill was based on the proposition that there is nothing to the teaching-learning process that couldn't be picked up in a few weeks by someone who knew his subject practicing the craft of teaching under some kind of unspecified supervision.

The California State Board of Education has since implemented Dr. Sheinbaum's theories for the education of all teachers in California. In the meantime, Berkeley and a hundred other institutions have provided testing grounds for the Sheinbaum position.[2] The results, I think, speak for themselves. The irony is there for all to see, because we are now being asked to reorganize the California junior college system, partly to put students back into contact with *teaching,* by the same institution that, but a short time ago, helped develop a legal position that there was nothing to teaching in the first place.

The way to improve teaching at the college level is to improve

---

[1]And, please, in suggesting a curriculum, don't say "liberal arts," or "academic." The liberal arts albatross has hung too long around the educational neck. It would have been much better, I think if old Martianus Capella's little allegory had remained undiscovered in the Middle Ages. As it is, most spokesmen for the liberal arts don't know that geometry in his trivium-quadrivium included botany and the medical properties of herbs. This ignorance is due, of course, to the fact that the history of education is not kosher from the academic point of view.

[2]Not all academicians agree with Dr. Sheinbaum, by the way. I can think of two pretty fair ones off-hand—Gerald Holton and Jerrold Zacharias (and they don't agree on too much else).

teaching, not necessarily to redesign the gross system of higher educa-tion in California and redefine functions. If the University of Cali-fornia is supposed to teach, then it should teach. I am not claiming that it would be a simple thing to force the faculties of higher institutions to the level of at least a workmanlike job at what is supposed to be part of their profession. I am not suggesting that conveying to these people "a few tricks of the trade" (a phrase much loved by the critics of pro-fessional education) will solve the problem of good teaching. I am say-ing it can be done and it seems to me a much better attack on the prob-lem than Mr. Ferry's.[3]

We have been skirting questions of educational philosophy for the last few pages, and it might be better if we made a frontal assault on these questions in the last section.

3. The Problem of Philosophical Position

The question of dividing the junior college into an academic system and regional technical institutes is, of course, the same question as to whether or not a community ought to have comprehensive high schools or academic and vocational-technical high schools. Up to this point in time, the American people have generally opted for the com-prehensive high school. The recent Gross report for New York City rec-ommended the conversion of vocational-technical schools into com-prehensive high schools. (England, as is quite well known, is also moving in this direction.) Since the rise of public high schools after the Civil War, the comprehensive high school has been the mainstay of American secondary education.

While there are many arguments on both sides, the ultimate argu-ment has to be philosophical, and it is related to the concept of democracy. Proponents of division of any educational system into academic and nonacademic sectors argue that there is no democracy of the mind, that the gifted are held back by being placed with slower students and that this, too, is undemocratic. Other arguments in this category are similar; for example, it is stated that giving a slower learner a "watered-down course" in an academic subject or a course in fly casting is, in itself, a perversion of the democractic process.

This argument may be dealt with on three levels. The first is social. Any such proposal for division into academic and vocational-techni-cal immediately creates a class situation, a status-elite situation. This is the beginning of discrimination. As an example, due to the pro-visions of the Fisher Bill and subsequent interpretation by the State Board of Education, no man who has had a physical education or

---

[3]I have covered only a part of the irony. The explosion of knowledge has not shunned the field of professional education, much as many crusading academics wish that it would. A great deal has been learned about the teach-ing-learning process in the last 20 years and more will be learned in the future. Academicians of the type under fire, however, insist that such in-formation does not exist and by no means should be communicated to the prospective teachers.

business administration major as an undergraduate in college may now become a school administrator in this state. Merely by virtue of, so to speak, his educational religion, he is barred forever; this, to my mind, is unconscionable discrimination. It is a direct result of the status-elite pattern of thinking.

Further, in addition to providing a ready-made second-class citizen, the proposed division prevents students from social contact with each other at a time when they can still learn to appreciate human variability. The corporate structure of our society already puts too much emphasis upon status and has made alienation the order of the day. To add to this artificially is to compound the problem. We need more democracy, not less.

Second, there are some psychological aspects to this philosophical problem. The typical academic instructor in the junior college English department already discriminates against the vocational student because he is teaching first-year Berkeley English, and the vocational student doesn't exactly set the world on fire with his verbal ability. The assumption on the part of the instructor is that verbal learning and the ability to manipulate verbal symbols are the only tests of thinking and expression. Vocational instructors, on the other hand, will tell you that such is not the case—that, in many instances, these young people are intelligent and simply think in other patterns.[4]

Such a bifurcation would be further emphasized by the separation suggested. Any chance that the vocational-technical student could improve his verbal skills would probably go down the drain. There is, of course, the other—and perhaps more important—side. The academic student would be cut off further from the world of reality and pushed into the world of abstraction. He would be driven to abstraction, so to speak. Abstraction is one of the characteristics of our age, and it certainly has its bad side.[5] Thomas Huxley summed up this problem in 1877 when he said, "...if encouragement is given to the mischievous delusion that brainwork is, in itself, and apart from its quality, a nobler or more respectable thing than handiwork—such education may...lead to the rapid ruin of the industries it is intended to serve" (Huxley, 1896, pp. 415-416).

Finally, we may return to the social-political level with the simple observation that the feelings of the American people on education have generally been pretty sound over the years. There is no doubt a philosophical division possible on that bold assertion, but I will stand on it. And I don't think the American people or the people of California will buy Mr. Ferry's idea, strangely enough because it is against our best traditions.

---

[4]There is not time nor space to go into this matter here. There is a tremendous literature relating to the subject in art and psychology. For a quick overview, the reader is referred to Rudolf Arnheim, "Visual Thinking," in Kepes, 1965.

[5]It is from this point that I could develop my *ad hominem* argument about general education.

References

Arnheim, Rudolf. "Visual Thinking," in Gyorgy Kepes (Ed.) *Education of Vision.* New York: George Braziller, 1965.

Huxley, Thomas. "Technical Education," *Science and Education.* New York: D. Appleton and Company, 1896.

Kepes, Gyorgy (Ed). *Education of Vision.* New York: George Braziller, 1965.

New Uses for Old Clichés

Mr. President, members of the graduating class of 1965, members of the faculty, honored guests, ladies and gentlemen:

I recall, with a great deal of fondness, a commencement speech given right after the war by one of America's great women medical scientists, Dr. Florence Sabin. Dr. Sabin had just retired from her life of research and had accepted, as sort of a fifth career, a position of responsibility in public health in Colorado. As she had had a lifelong interest in Colorado State College at Greeley, she agreed to do a commencement speech. In her opening remarks she made one unforgettable point. She said, in effect, no one listens to a commencement speech and no graduate can ever remember what is said at his own graduation.

I think her observation very true; we occupy here a few minutes devoted to a spoken ritual between the parade in, the diploma presentations, and whatever parties you intend to have later. If it is true, it ought to be a great comfort to a commencement speaker; except for the vested interest of the college news service, a speaker could recite all 47 verses of "Abdul-a-Bul-Bul-a-Mer," and we could get on to the business of the evening.

Such an idea is, of course, wishful thinking. For you have the speaker of the evening to reckon with. And he at least will want to listen to the speech. My hope, as I suppose all such speakers hope, would be to produce a few remarks that are fresh, memorable and completely clear of clichés. Yet, when you think about the assignment, this is practically impossible. What, really, can be said at a time like this that will not, in one form or another, present the warmed-over wisdom of a thousand commencement speakers who have addressed a thousand graduating classes.

"You, the class of 1965, stand at a threshold..." Certainly you do. You worked hard enough to get this far and, in these times particularly, it *is* only a threshold. "You, the class of 1965, are now at the beginning..." Right again. As you should know, if you've thought about it, that's why this ritual is called commencement. "You, the class of 1965, must go forth and do something about the world..." Again, what else can be said? The world does need a little working on, you know. Finally, "You, the class of 1965, must perfect yourselves and develop all virtue..." This simply represents the age-old need of your elders to give you one more admonish-

---

Address delivered at Los Angeles Valley College, June 18, 1965.

ment so that you will, we hope, end up somewhat better than we have—when, in times of self-examination we admit quietly that we have been less than virtuous.

Such a collection of clichés, I suspect, would get us exactly nowhere. You would accuse me of not being with it and, I'm afraid, I would return the compliment. It might be instructive for us to examine why this is the case.

Of all the wordage that came out of the affair up at Berkeley, the most insightful to me was the report by William Trombley of the *Los Angeles Times* that a significant minority of the students—and usually the brighter ones—indicated that they simply didn't trust anyone over the age of 30. And commencement clichés have obviously been generated by men who have long since passed this outer age limit.

While this generalization probably doesn't apply to all of you, it follows that many of you simply have little inclination to buy what I'm selling in the way of graduation advice. At least in terms of age, I represent the generation that happens to be running the world right now—and many of you suspect this generation. One has to admit that there is some merit to the proposition that the world might be run somewhat better than it is. Whether 29-year-olds would do it better, or whether or not things have always been more or less in this state and you are merely reflecting what all generations have thought—reflecting a little louder, perhaps—I will not introduce into the discussion.

Rather, I will accept, for now, the proposition that we have put the world into its present mess and that, therefore, many of you do not trust us. Advice, under these circumstances from my generation—and with some exceptions, as from the late President Kennedy—is pretty superfluous.

Secondly, many of you are concerned with the moral issues of our time—with the rights of men as men, with the problems of death and capital punishment, with war and its destruction, with the effects of poverty and the needs of vast numbers of men submerged in the underdeveloped nations of the world. No thinking man, of course, can argue with these concerns, although there are some of my generation that would wish you would return to swallowing goldfish.

I flatter myself by thinking I know a little something about your generation; and in that connection I am not so naive as to believe that all of you share completely these humane concerns we have been hearing so much about lately. There are those among you who intend to concentrate on making the almighty buck, who share no great worry for the welfare of mankind. I do think, however, that the flavor of the times is in the direction of caring about mankind. Somehow, many of your age group feel that you care more than we do.

There is, however, another generalization that I think applies across the board to the members of this graduating class. All men are concerned and, to some degree, anxious about themselves. The members of this class, however, if you are representative of your age and education levels, are very deeply concerned with personal problems. It is no accident that existential philosophy is so popular these days; you are seeking to find your role as you enter a large-scale, impersonal corporate society, to find yourselves as persons. You fear and fight alienation.

The advances of modern psychology and psychiatry help you see the dimensions of your personal problems even if they have not provided insight. The generation to which I belong, surprisingly a part of the same human race, has, no doubt, similar problems. These problems are not so near the surface, however, and are handled by different adjustment mechanisms. Since personal problems do not show quite as much in the culture of my generation, this probably creates another argument for lack of trust. For who can relate, to use your language, if you can't feel the other person's problems?

This, then, is the situation as I see it. You, the members of the graduating class of 1965, in looking over the world into which you are about to take a next step, view the controlling generation with suspicion; a substantial number of you care deeply about the moral issues of our time that surround the question of man as man and you may feel that we do not care; and all of you, wrestling with the problems of alienation in a large-scale society, have anxieties at the personal level which are difficult to share with my generation.

All this adds up to great problems of communication between the generations represented here tonight. If it is difficult for me to dispense commencement advice, how much more difficult is it for your parents and your teachers to communicate with you—especially those of you who can't trust anyone over 30? If you have the sensitivity you claim to have, you must realize that this is very tough on parents and teachers—who are finding it impossible to understand anyone between the ages of 10 and 40.

Consider some of the obvious things. Take music. And I'm not going to worry about rock and roll, although as an old Dixieland buff I resent what they do with the beat. Let us consider Bob Dylan. Again as an old musician, the less said about his musicianship the better; but that is not why Dylan is popular. It is what he says that counts. And somehow, he strikes a great responsive chord in your generation. Your parents, and I suspect many of your teachers, have great difficulty in understanding what this metaphysical poet of the street is saying (if they have ever listened to him). Perhaps you do not understand him, either, but you feel him and that makes all the difference.

You can apply the same test to art and poetry. The other evening, a friend of mine who has one of the galleries on La Cienega needled a group of somewhat older university graduate students for not being with it; for holding up the contemporary art of 1910 and 1920—modern art?—as being contemporary today. You don't need to be told about pop and op art, while the older generation is still struggling with the early Picasso. To some degree this difference in knowledge and taste between generations has been true since Adam, but today the gap is much larger.

Finally, many in my generation simply do not understand this moral commitment to the generality of mankind that many of you have. In specific terms, for example, they do not understand the horror most of you have toward capital punishment and many of them still believe in it. The message of Camus has not come through. Morality to many older people does not mean Mississippi; instead it has something to do with sex and drinking—issues which you consider as belonging to the past.

What we have here is a great gap between generations—a gap that seems to be making communication increasingly difficult. Viewed from either side, yours or mine, this gap is no good. To communicate you must trust and be trusted. This is the problem facing all of us.

What can we do about it? All advice that can be derived from experience is not bad; all clichés and old ideas have some merit, too. The trick, it seems to me, is to adapt, where possible, the voice of experience to contemporary problems—to find new uses, if you will, to some, not all, of the old clichés. And then to mix them in with new ideas. This, I suppose, if I were to give you a charge at all, is the charge I would give you.

If we can fix responsibility this way, we have not yet outlined a strategy of closing the gap. First, all of us, including all of you, must become truly sensitive to the causes of this gap. These causes hinge primarily upon the speed of change; change has always been with us, but its rate accelerates daily, and what was fixed yesterday is gone today and will be replaced again tomorrow. We cannot reach for the moon and expect the world to remain the same. Further, change is communicated almost instantaneously; no longer are Dubuque or Accra or even Mars islands of untouched stability; overnight pop art becomes advertising for Father's Day neckties. Those who fail to stay with these lines of communication lose ground and the gap widens.

A second movement contributing to this situation is that humanity is rising and on the march throughout the world, from your own neighborhoods to the jungles of Africa and the tundras of the north. This really makes a difference as, since the history of man began, it has been assumed by those in power that not all of humanity will, at the same time, attempt to sit in the sun as men.

Third, knowledge grows at an enormous pace; this growth leads to specialization which tends to do several things: Each man, individually, grows more ignorant every day; people tend to be separated according to ability and knowledge; specialization increases the problems of communication.

Knowledge is part of the production of a technological society which in turn must be a large-scale corporate society regardless, incidentally, of the form of government adopted. Large-scale organization, specialization, etc. create alienation. In the meantime, however, along with other knowledge, our knowledge of ourselves—personal knowledge— also increases and we have had, in recent years, an emphasis upon irrationality in a rational, abstract society as we learn more about it.

These movements are complicated and not easy to understand, and yet we must understand them a little if we are to close this gap. If I were to sum up the final argument before I make it, I would say that we need, in the future, more marginal men—marginal between the generations. Some of the graduating class of 1965 must become marginal men; some of the parents and teachers sitting here tonight must also become marginal men with an ability to operate both in the future and in the past.

There are many things that I could urge you to do as you leave these exercises tonight. It would make your parents and your teachers pleased if I would urge you to be both virtuous and successful; it would make some of you pleased if I were to suggest, as they used to say, that the world is your oyster; it would please others to hear a battle cry for freedom and personal commitment.

I wish you all of these things and many more as they fit these times. I feel, however, that the message I would like to leave is contained in the old cliché about the relay race. From time immemorial, commencement speakers must have referred to the relay race—to the passing of baton or torch. In a way, as is painfully obvious, we are passing the baton tonight.

In an effort to revive an old cliché and infuse it with a message for the present, let us examine the metaphor of the relay race. While each runner must do his best, the crucial moment occurs in that short space and those fleeting seconds of time when the baton passes from one runner to another. If either runner blows it, the race is lost. We have about reached this point on the track—your generation and ours. What happens now may determine the future of the race—and you may take that phrase with all of its meanings.

My generation has run better than you think; but it is not enough to finish at our present pace. We—your parents, your teachers, your governors—must make more of an effort. Because, again as should be obvious, this small space for baton-passing, this short time for baton-passing is the only time and space we have for communication, for gap-closing between generations.

And your generation, on the other hand, dare not start too soon or run too fast at the outset. Without the baton, the race is also lost. This requires an intelligent effort and sensitive pacing on your part; you, too, must help close this gap or all is lost. While many, should I say advanced, adults would like to go another round of the track, they can't. Only you can.

This means, in less metaphorical terms, I suppose, that we all must listen more. Some of the old clichés still have merit; these days history is not too accurate a guide for the future, but experience, properly evaluated, is, in many things, about all you have. Then listen—not slavishly, but critically. If you become doctrinaire in refusing experience, you ask for defeat.

To the generation of parents, governors and teachers, I also say "Listen." Listen and try to understand. These young people are the owners of the future; they are sensitive in many cases—and here comes another cliché—to a different drummer. You cannot be sure that you are right. Unless you give them a chance to communicate, not only will you lose them, but you may cause them to lose later by having set a bad example. For if you think our gap is large, wait a few years. With the acceleration of change, when this fine group of young people here tonight occupies your position a few years hence, the communication problem will be infinitely greater. This is a lesson they must learn now, from you.

Perhaps in dealing with this message of personal communication between generations, I have been less than inspirational and I certainly have not avoided clichés. I would, however, suggest in closing that we can turn to still another cliché for inspiration. Many of you, in the class of 1965 have thought of the statement I'm about to read as connected only with Mississippi or Vietnam or Hungary. I suggest that we give it a new twist and consider the possibility that it could be applied on both sides to the generations present here tonight. The author, of course, is John Donne. Think of each other as we read once again:

No man is an *Iland*, intire of it selfe; every man is a peece of the *Continent*, a part of the *maine*; if a *Clod* bee washed away by the *Sea*, *Europe* is the lesse, as well as if a *Promontorie* were, as well as if a *Mannor* of thy *friends* or of *thine owne* were; any mans *death* diminshes *me*, because I am involved in *Mankinde*; and therefore never send to know for whom the *bell* tolls; It tolls for *thee*.

(From John Donne's "The Tolling Bell—A Devotion")

A Walk on the Altered Side

One thing that is new is the prevalence of newness, the changing scale and scope of change itself, so that the world alters as we walk in it, so that the years of man's life measure not some small growth or rearrangement or moderation of what he learned in childhood, but a great upheaval.
                                                            —J. Robert Oppenheimer

As I write these lines, I am traveling 35,000 feet above the Grand Canyon of the Colorado River at a speed in excess of 600 miles per hour. A voice comes on the intercom—a mild technical miracle in itself. It is the voice of the pilot relaying, in a matter-of-fact way, one of the greatest stories of this generation. For above me—150 miles or so—a Marine colonel by the name of John Glenn is traveling in a space capsule at a speed of 17,000 miles per hour. While I have been reaching Arizona from Los Angeles, he has come halfway around the globe. Glenn has been twice around the world since his flight began; the decision has just been made to try for the third orbit.

Below me the Arizona desert—dimly seen through a covering of white clouds—sits ancient and quiet. Memories are buried here—of the conquistadores who explored it, of the Indians, and, before them, of geologic time. A contrast, heightened by the middle ground, the limbo, the partial ascent to the stars symbolized by the magnificient aircraft in which I ride—a contrast greater than man has ever known—exists in this relationship cf ancient desert and the capsule called Friendship 7. Between them rides the jet, symbol of our generation. For we, truly, must be the midwives of the new era.

The conquistadores were inevitable after Columbus and Magellan; John Glenn was inevitable only after Newton, Einstein, Planck, Helmholtz, and generations of other scientists and unknown but dedicated engineers, technicians and inventors. The distance between the caveman and Magellan is as nothing compared to the distance between Magellan and Glenn. Most of us do not live in the world of John Glenn; many of us cannot or will not; many of us do not take kindly to the role of midwifery in this birth of newness. Returning to the Oppenheimer metaphor, the world does, indeed, alter as we walk in it; the world of John Glenn, not our transitory world, is

Reprinted from *Phi Delta Kappan*, Vol. 44, No. 1, October 1962, pp. 29-34. This article was delivered as a paper before a meeting of the John Dewey Society at Las Vegas, Nevada, March 3, 1962.

the world of our children. Their side is the altered side, where, as gap bridgers, as educators, we must learn to walk.

It is in this context that I should like to remark upon a relatively new relationship in the world of education. A new world, symbolized at least to a modest degree by the flight into space, seems to be forming within the educational society. This world is technological in nature. Men are seeking to solve some of the problems of education by technological means. Technology is not, as many of the technically illiterate seem to think, a collection of gadgets, of hardware, of instrumentation. It is, instead, best described as a way of thinking about certain classes of problems and their solutions.

This view of technology when applied to education becomes a legitimate object of concern for the educational philosopher. We are met here in the name of John Dewey, in the name of educational philosophy. I would like, in the short time at my disposal, to outline some of these philosophic concerns as I see them from the point of view of a student of instructional technology.

The Revolution's Potential

Perhaps it would be useful to indicate briefly some of the dimensions of this *potential* technological revolution in education. I emphasize potential because it has not yet happened; it may never happen; education may remain the only natural (primitive) sector of our culture, but I strongly doubt it.

Since about 1930 we have had a slow development of a group of tools for communication and teaching, and a program of research into their use. These instruments and materials include what today we call conventional audiovisual devices—the sound motion picture, various forms of projected still pictures, recordings, etc. Since 1950, this arsenal has expanded to include television, electronic learning laboratories, teaching machines of various kinds, and, recently, computers. Accompanying these devices and materials, again, has been a vigorous program of research into the nature of learning and communication, supported by a rapidly growing body of theory derived from experimental and social psychology and related disciplines such as linguistics, criticism and engineering.

In other sectors of the educational enterprise, other technological innovations are being tested. The work of Lloyd Trump and his associates in school organization, the various attempts at team teaching, and experimentation with new school environments are examples. I shall confine my remarks principally to the main line of instructional technology—audiovisual materials and the so-called newer media and their intellectual bases.

I mentioned a slow development of these approaches to instruction. More important, this development was almost discontinuous with any main lines of growth of American education during the past 30 years. It was so little connected with progressive education, for example, that I found the word *films* only in a footnote relating to Alice Keliher in Lawrence Cremin's book (Cremin, 1961, p. 257). It has not notably influenced the theory of school administration or the education of superintendents. In preparing this paper, I had occasion to examine a number of recent books on educational philosophy. With one or two exceptions, they are so little concerned with these developments that, using the philosophers as a source, one must conclude that a technology of instruction does not exist. The Association for Supervision and Curriculum Development, until very recently, also turned a blind (and horrified) eye in the direction of instructional technology.

Within the last five years this discontinuity has ended and the possibilities of a technology of instruction have suddenly thrust themselves into the educational mainstream. The philosophers have begun to cluck, if not in books, then in speeches, articles and conversations; the curriculum specialists have been seen running about throwing up barricades to protect the child from the machine monster; and educational statesmen have managed to raise the adrenaline of their constituents with speeches that sound as if they were ghostwritten by Ned Ludd or Jean Jacques Rousseau. We are urged to destroy the weaving machines and return to nature—all in the same breath.

I am not concerned here either with educational statesmen or curriculum specialists. I would like to concentrate on the philosophers. For I come not to defend instructional technology, as I am sure our chairman would like me to do, but instead, I come to indict. I feel that many educational philosophers have lost the way and that they have committed an even worse crime—they have failed to understand.

From Apathy through Antagonism

First, there is a generalized, nonspecific attitude that holds that instructional technology is both trivial and, at the same time, dangerous. This position is well stated in Van Til's excellent paper (Van Til, 1962) in which, on the one hand, he dismisses technology as mere tinkering when compared to the real concerns of education and, on the other, sees it as a threat for mind control of Orwellian proportions. In fact, I get the impression that in some philosophical and curriculum circles the attitude toward instructional technology runs all the way from apathy through antipathy to antagonism.

This negative approach is not surprising. The intellectual has, for the most part, always hated the city which makes his intellectuality possible. And a technological civilization is an urban civilization. Thoreau, of course, was the great prototype of the intellectual who hates the city. He refused to stay in the city and once said, "The only room in Boston which I visit with alacrity is the gentlemen's room at the Fitchburg depot, where I wait for cars, sometimes for two hours, in order to get out of town." There is something comically ironic—perhaps Freudian—about this. For it has always seemed to me that plumbing—and I assume they had plumbing in the gentlemen's room of the Fitchburg depot even in those days—is a rather appropriate symbol of a technological society.

Opposition to Scientism

Morton and Lucia White remind us that John Dewey himself, between 1899 and 1927, developed the same attitude toward the city and its industrialization. They said, "Instead of taking the city as the model *for* the progressive school, he almost speaks as though the urban community should be modeled *on* the progressive school. . . . At the end of his life Dewey seemed to conclude every speech with the words, 'Divide the cities into settlement houses' " (White and White, 1961, p. 176). And, a bit later, White and White summarize: "For functionalism, like pragmatism, is one of a complex of American ideas that could not exist in a nonurban society, and yet its greatest spokesmen seem to hate the American city" (White and White, 1961, p. 176). I suggest that educational philosophy has reached the stage when, in order to remedy what is a special case of the same general syndrome, it should cease hating the city.

Second, to this day there is in educational philosophy a distrust and a strong antagonism to what in the 30s was called *scientism* in education. Charters, Judd and Bobbitt all felt the hot-tipped shafts of Dewey, Kilpatrick, Bode, and Childs. Kilpatrick was even spanked for too much attachment to Thorndike's connectionism, although, for the life of me, I could never see it in his project method.

Scientism in those days was Charters and educational engineering and activity analysis; scientism today is B. F. Skinner and pigeons and programed learning. Charters was demolished for inventing a system of curriculum-making designed—so it was charged—to preserve the social status quo, and the measurement movement was subject to blast after blast. Today philosophers and curriculum specialists make jokes about pigeons not being people—neglecting, by the way, many other forms of learning research and programing theory.

For those of you who follow Dewey, Bode and Kilpatrick, whose god was the method of science, this is, indeed, a strange attitude. It was strange when they had it; it is stranger now.

Take only one facet of this scientism—Charters' theories of analysis. The philosophy group at Ohio State University ridiculed them. Yet, today, those theories are being used for identical problems by psychologists who never heard of Charters or of his contemporary in the industrial field, Allen. For analysis is needed in all sorts of programming, in the statement of objectives, and throughout the developing technology of instruction.

Analysis, in the sense that Charters used it and as it is being used today in a hundred ways, is, in part at least, the discrimination of details. As Gerard Piel points out, J. Bronowski, the British mathematician, in commenting on the contributions of Leonardo da Vinci, said, "[He] gave science what it most needed, the artist's sense that the detail of nature is significant. Until science had this sense, no one could care—or could think that it mattered—how fast two unequal masses fell or whether the orbits of planets are accurately circles or ellipses" (Piel, 1961, p. 208). I suggest that, because of a social bias characteristic of the 30s, the great exponents of the scientific method in education successfully struck down one of the great educational scientists of that generation and prevented a generalized scientific technique from becoming more effective in education. I suggest that the educational philosophers of this generation ought to avoid such a mistake. I can tell them this: Even if they do not, their strictures will not have the same effect.

A Walk on the Altered Side

The Question of Means and Ends

A third point on which I should like to offer advice is in the other direction. I think current educational philosophy should pay more instead of less attention to Dewey on the questions of means and ends in education. Mr. Van Til continually restates the point that until the question of aim is settled by philosophic reflection, it does absolutely no good to consider the means necessary to achieve those ends. The means in this case, of course, are the devices, materials and approaches of a technology of instruction. Bode made the same point somewhat more succinctly when he said many years ago, "Unless we know where we are going there is not much comfort in being assured that we are on the way and traveling fast" (Bode, 1921, p. 241).

There has probably been more confusion on ends and means, method and subject matter, than on any other point in educational theory these last 30 years. We need, first of all, clarification once again of this question; and clarification, I take it, is one of the jobs of the philosopher.

The confusion began with Dewey, who has to be read very critically in order to determine when he is talking about the method of science as such and educational method as such. Kilpatrick tossed in the notion of concomitant learnings, which introduced so many variables into the learning process that method, subject matter, student, and the school flagpole got mixed into a great ball of fuzz. When the curriculum specialists got through with it, curriculum was defined as no less than life, in which there was really no method except, perhaps, the pursuit of happiness. While all of these views had much to recommend them, and all had laid hold of a piece of the truth, the usefulness of these generalizations for intelligent action had been reduced, not only to zero, but into the negative.

On the other hand, a much more naive view about method also still prevails, fostered in such intellectual circles as the Council for Basic Education and the California State Legislature. This view holds that method is a mere manipulation of a few variables such as good enunciation on the part of the teacher—skills that can be learned in less time than it takes to learn to drive a car.

We Still Need Direction

The ultimate result of the development of the Dewey position through Kilpatrick to the ASCD was that method, on the one hand, reduced itself to a worship of group dynamics while pacifying the god of child-individuality; on the other hand, it was completely subordinated to something called aim, which was the result of a process of navel contemplation, either on the part of educational philosophers or of curriculum committees. In either case, we have received little help in doing our job from these statements of aim, most of which degenerate into generalized propositions from which no action may be deduced. The ultimate result of the other concept—that method is nothing but the manipulation of a few tricks—is, first of all, ignorance; and, secondly, a perilous bypassing of both individuals and values.

The suggested direction for clarification of this long-suffered problem is to return to Dewey and work from there. Because he believed in a unified, nondualistic universe, Dewey maintained that method could not, *when in use*, be separated from subject matter; that eating and food were inseparable, for example. However, for purposes of study and control, he was equally firm on the point that method had to be teased out of this universe, examined, analyzed and put to work in the most intelligent way possible.

Now, add to this idea the fact that, as early as the time of writing *School and Society*, Dewey suggested that we live in a technological, industrial culture and that technology was, in fact, the main determinant of its direction. The school, he felt, should reflect this. Such a view could be considered a special case of the general law of pragmatism—that ends and means are inseparable, that ends become means to further ends.

If you now consider technology from two perspectives—the entirety of technology that has transformed our society in about 200 years and the special application of technology to the instructional process—it is possible to indicate the direction the educational philosopher must go to clarify the problem of method in relation to aim.

First, as the perceptive students of general technology continually insist, technology in society is an organic process. This concept is central to Hannah Arendt's *The Human Condition*. In it she said, "As matters stand today, it has become as senseless to describe this world of machines in terms of means and ends as it has always been senseless to ask nature if she produced the seed to produce a tree or the tree to produce the seed" (Arendt, 1959, p. 133). Slightly later, she quotes Werner Heisenberg to the effect that general technology is no longer "'the product of a conscious human effort to enlarge material power, but rather like a biological development of mankind in which the innate structures of the human organism are transplanted in an ever-increasing measure into the environment of man'" (Arendt, 1959, p. 133).

The first obligation of the philosopher is to understand these concepts—a task that is not easy because the views they represent, as Kurt Marek has pointed out, are qualitatively and psychosomatically different from any ever held before. They do not fit into the tight, abstract, three-dimensional world of Euclid and the present educational philosopher. They would, strangely enough, fit into Dewey's world—a world of organic unity, although, as I indicated before, if his later work is a clue, he would probably not have been happy with the consequences of his own thought.

Taking technological development as the central organic process of our society, the implications, as this process invades education, are interesting indeed. The process does not destroy aim and its role, but it binds aim inevitably to technology. For technology is an aim-generator as much as purpose or philosophy is a technical direction-giver. Each conditions the other and is not, as Mr. Van Til maintains, arranged in a hierarchy with aim on top and method at the bottom.

An example might serve to throw some light on this relationship. In any number of technological approaches to instruction—programed learning, the use of massed films, or in the developing instructional systems, for example—there is one unvarying requirement. That requirement is an absolutely clear statement of objectives. The general statements of the philosopher and the curriculum specialist are not good enough. Objectives must be developed from general aim statements as experiments and hypotheses are developed from general scientific laws. There is no guarantee in either case that the specific will correspond completely with the general. The specifics are conditioned by the instructional reality—a condition that philosophers abhor but that scientists, in the case of scientific laws, do not worry about.

At any rate, objectives can only be developed in this sense by a thorough analysis heretofore rarely applied in education. This is where a technology gets its direction. It is hard work to create such objectives and, if the philosophers resent everything else, they should see that such a procedure, in fact, brings philosophic thinking into practice—more than a hundred generations of philosophers have been able to do.

Further, I should like to remind you of my introductory point that technology is, fundamentally, a way of thinking. As such, it inevitably will play some role in the development of educational aims. Once a technology exists, certain aims dreamed of in philosophy may disappear. As Marek has said of technology in general:

> In our technological age, man can conceive of nothing that he might not invent. A magic carpet is no longer a scientific problem, but only a problem in construction. All the pipe dreams of the old high cultures can today be made to come true, but some of them are so primitive (like the magic carpet, for example) that it is no longer worth the trouble. The pipe dreams of the men of the old high cultures appear to have been consummated in the same historical period as the old high cultures themselves. (Marek, 1961, p. 43).

Take care lest the educational philosophy you are preaching does not meet the same fate.

In discussing the hatred of the city, the attack on scientism, and the problem of ends and means, we have only scratched the surface of the great job of readjustment needed in educational philosophy as technology, reflecting the increasing technical complexity of our culture, invades the instructional process. In the space left, all that can be done is to list some other tasks which, I suggest, should occupy the attention of educational philosophers.

These include a thorough study of the process of technology in our culture as a whole. Outside of some indication in Phenix's new book (1961) and in Thelen (1960), I find little evidence that this is going on. What has gone on, incidentally, is inadequate; philosophers have presumed too much. They have presumed that they can study technology in a vacuum without, for example, the cooperation of engineers and without the mastery of certain languages and concepts. In order to look properly at the altered side, such multi-disciplined study is necessary. To give the philosophers something to think about, why is it that recently the greatest visions (in, it is true, a somewhat restricted sense) of what might be possible in education have come from Simon Ramo, a technologist (1960, p. 367-381)?

What are the Myths about the Machine?

Educational philosophers should spend some time in examining current myths and destroying them. Many educators have demonized the machine in the manner of witch doctors. Concerning this practice generally, Marek said somewhat bitterly, "The machine is demonized only by those who feel helpless in its presence. Where such demonization occurs today, its authors are neither scientists, nor engineers, nor managers, nor workers, but only outdistanced philosophers and writers sulking in their historical corner." Shades of C. P. Snow! And there are other myths of our age of midwifery that must disappear under the hand of the philosopher.

Another challenging area opened up by the organic processes of technology includes both technical and practical problems. For example, who among you will follow Bode's great example and, continuing his work which stopped in 1940, relate concepts of mind in their newer sense to educational theory? To do this today you would have to consort with cyberneticians, electrical engineers, and neurologists. Related to this is the general problem of knowledge—its nature, its size, its structure. At the level of so-called practical problems, I *urge* you to face the economic and productivity problems inherent in an attempt to educate all Americans. These things have never concerned you centrally. They must now.

Finally, you must face the consequences of the generation in which you have been born and the world in which you live. You cannot deny technology on arbitrary, literary, uninformed grounds. If you deny the teaching machine, the computer, television, and the motion picture, if you deny new ways of teaching and learning, you cannot stop until you deny yourselves fire, the wheel, and even the very language which you speak. For, as Karl Jaspers so well put it, "A denial of technology's last step is equivalent to a denial of the first" (Jaspers, 1961, p. 192).

And, as Max Lerner reminds us, we Americans have not sold our souls to the devil of technology in a Faustian bargain. It is as true of education as of society as a whole that "truer than the Faustian bargain . . . is the image of Prometheus stealing fire from the gods in order to light a path of progress for men. The path is not yet clear, nor the meaning of progress, nor where it is leading: but the bold intent, the irreverence, and the secular daring have all become part of the American experience" (Lerner, 1957, p. 263). Does this not imply that, as midwives to the new era, as conductors to the altered side, the vista of educational philosophy is more exciting than ever? I think John Dewey would have liked that.

References

Arendt, Hannah. *The Human Condition.* Garden City, N.Y.: Doubleday Anchor Books, 1959.

Bode, Boyd H. *Fundamentals of Education.* New York: The Macmillan Company, 1921.

Cremin, Lawrence A. *The Transformation of the School: Progressivism in American Education, 1876-1957.* New York: Alfred ·A. Knopf, 1961.

Jaspers, Karl. *The Future of Mankind.* Trans. by E. B. Ashton. Chicago: The University of Chicago Press, 1961.

Lerner, Max. *America as a Civilization. Life and Thought in the United States Today.* New York: Simon and Schuster, 1957.

Marek, Kurt W. (C. W. Ceram). *Yestermorrow: Notes on Man's Progress.*

Trans. by Ralph Manheim. New York: Alfred A. Knopf, 1961.

Phenix, Philip H. *Education and the Common Good: A Moral Philosophy of the Curriculum.* New York: Harper & Brothers Publishers, 1961.

Piel, Gerard. *Science in the Cause of Man.* New York: Alfred A. Knopf, 1961.

Ramo, Simon: "A New Technique of Education," in *Teaching Machines and Programmed Learning: A Source Book.* A. A. Lumsdaine and Robert Glaser (Eds.). Washington, D.C.: Department of Audiovisual Instruction, NEA, 1960.

Thelen, Herbert A. *Education and the Human Quest.* New York: Harper & Brothers, 1960.

Van Til, William. Presentation at March convention of ASCD in Las Vegas, N.M.

Van Til, William. "Is Progressive Education Obsolete?" *Saturday Review,* February 17, 1962, pp. 56-57+.

White, Lucia and Morton. "The American Intellectual versus the American City," *Daedalus,* Vol. 90, Winter 1961, pp. 166-179.

A Revolutionary Season

> To every thing there is a season, and a time to every purpose un-
> der the heaven: . . . a time to keep and a time to cast away . . . a time
> to keep silence; and a time to speak.
>
> —Ecclesiastes, III, 1-7

The city of Detroit was hot on July 4, 1963. Even so, 10,000 or so American schoolteachers and schoolmasters were celebrating Independence
Day by attending educational meetings, for the occupation of educator is,
perhaps, the last stronghold of the puritan ethic in America, and the NEA
Convention was in full swing. Duty called, and most of the delegates and
other educators present answered the call and went to the meetings.

Some, but not all of the meetings. Around a thousand delegates resolute-
ly gathered themselves in groups of 10 or more and discussed NEA organizational problems; others distributed themselves generously throughout
several different groupings—except one in the ballroom of the Sheraton-
Cadillac Hotel, carrying the title, "The Technological Revolution in Education." This revolutionary gathering—an audiovisual presentation of some
of the findings of the NEA's own Technological Development Project—
made an attendance record in reverse. Forty-two individuals from that
population of thousands of America's leading educational politicians attending the convention managed to find a seat in the ballroom. (Actually,
it was less than 42, as several friends of mine showed up, as well as the
families of the graduate students from Wayne and Michigan State who
were assisting us. The total delegates contacted would run no more than
25.)

At this point, some cynic might suggest that the attendance was a good
measure of the drawing power of the performers. This argument is un-
answerable. I would merely counter with the fact that the performers were
probably not a consideration as, I suspect, we[1] were unknown to over 90
percent of the conventioneers. Further, where else at 2:00 in the afternoon
of a hot July Fourth could a delegate have seen a mammoth three-screen
show using slides, film, tapes, records, a tote board, and dramatic anti-
phonal lighting in an *air conditioned* room? And have the puritan con-

---

[1] Dr. Robert O. Hall of California State College at Alameda and the writer.

Reprinted from *Phi Delta Kappan*, Vol. 45, No. 7, April 1964, pp. 348-354.

science salved at the same time—because it *was* an educational meeting?

I have told of this incident, not out of pique, but because I believe it to be a fair measure of the value placed, by organized education, on instructional technology and the potential or actual technological revolution about to occur or now occurring in American education.

No one who knows anything about the history of American education should be surprised at this inherent conservatism. The popular press, Admiral Rickover, and academics who can't read have created a myth since World War II which labels organized education as *progressive, radical,* etc. Nothing could be further from the truth. Other than espousing general federal aid to education (hardly a radical idea, except in the hands of a Kirk or a Goldwater) and giving some useful support to efforts at international peace and understanding, organized education has never been even mildly progressive. It is one of the most conservative forces in America.[2,3]

This conservatism, I believe, is an honest reflection of the conservatism of the average American teacher and administrator. And I further believe that current attitudes toward instructional technology expressed by organized education also are mirror-images of the general attitudes to be found in the field. Instructional technology is thought of at one and the same time as (a) completely peripheral to the problems of education, (b) a passing fad which will go the way of sloyd, (c) something invented to get money out of the federal government and to prevent the passage of a general aid bill, (d) a threat to teachers' jobs, (e) a movement to dehumanize education and make robots out of little children, (f) gadgets, (g) an antonym (bad) for something called creativity (good), (h) something that will never replace the horse, etc.

Conservatism in education is nothing new. No educational reformer in the history of Western society ever got beyond first base; and I include the range of educational thinkers from Comenius to John Dewey. The advances in education that have been made have occurred by extremely slow accretion, forced on the system primarily by social events—and to which the educational system responded with a reaction time of about 50 years from the onset of the events to change in the system.

There is absolutely no reason to suppose that, given things as they are, there would be a change in this inherent conservatism merely because the possibility of instructional technology has appeared on the scene. There is, however, one catch to all of this. *Things are not as they are (or, rather, were) and have not been for some time. Hence, the basis for educational conservatism no longer exists.* Before examining this proposition in detail, it might be well to look briefly at the case for educational conservatism.

Years ago, classes in the philosophy of education often began with the proposition that formal education had two functions:

---

[2]Test this thesis on any school superintendent by suggesting to him that George Counts had something when he wrote *Dare the Schools Build a New Social Order?*
[3]Organizations such as the NEA are exceeded somewhat in their conservatism by splinter educational groups, such as the American Federation of Teachers and the Council for Basic Education. The AF of T is so conservative its leadership appears to lack any connection with either modern trade union or educational thinking, and yet, in many cases, maintains an appearance of liberalism.

1. To conserve and pass on the cultural heritage.
2. To act as an agency of progress and reform.

Except in the sense that progress and reform is accomplished in our society by the process of creating "better" men, the second function has never been fulfilled directly. This means that American schools and colleges and the local communities that support them carry out (and are required, almost exclusively, to carry out) the conservatism function only. Over 200 years or more, this process inevitably conditions the entire profession into a conservative attitude.

Liberal—as the opposite of conservative—can also mean the generous spending of money. American education has never, in its long history, been in that position. It has never had quite enough funds to do the job. The motto has been "make do." There is nothing like the requirement to pinch a penny to generate conservatism in a system.

Further, a powerful argument can be made to the effect that a conservative educational system is necessary for stability in a democracy, and that a conservative profession—school teachers and administrators—is necessary for the functioning of the system. Another telling argument revolves around the proposition that serious moral questions as well as social questions can be raised if experimentation with children is involved. The conservative position is therefore the morally correct position.

Sociological and political reasons can also be adduced which bolster the conservative orientation of the educational system and the concomitant professional organization. In spite of "Great Cities" and "Higher Horizons" projects, in spite of slums and suburbs, in spite of miniature educational pentagons occupying decaying central cities; in short, in spite of the fact that most of America lives in the city, our educational philosophy, much of our theory of educational administration, and, in fact, our whole educational orientation, as Kimball and McClellan (1962) have shown so well,[4] is based on the myth we all live in a small town or farm in central Iowa and are educated in the adjacent schools. The whole business is probably unconscious, but school boards think this way; superintendents, in particular, think this way; principals think this way; teachers, although they are slowly changing, think this way; custodians think this way. And the way of the farm and the small town is pristine conservatism.

Further (and if any teacher readers are still with me, they will not like this[5]) the American teacher, somewhat like his counterpart in the rest of the world, works in and is part of a bureaucracy. Sociologist Sloan Wayland has pointed out that:

> . . . the teacher is a functionary in an essentially bureaucratic system. As such, he is a replaceable unit in a rationally organized system, and most of the significant aspects of work are determined for him. (Wayland, 1962, p. 43)

Bureaucracies are notoriously resistant to change. This is true of govern-

---

[4] In my opinion, the best book on education published in the last two decades. Title: *Education and the New America*.

[5] My enemies, by now, should number into the thousands. However, if they are angry enough to stick with the argument, I think I can show them a way out.

ment bureaucracy, industrial bureaucracy (in spite of propaganda to the contrary), union bureaucracy, and school system bureaucracy. The fact that school systems individually and considered as larger wholes are bureaucratically organized explains Henry Brickell's findings that educational innovation outside of the individual classroom is most often successfully introduced by an administrator. In a smoothly operating bureaucracy, that's about the only way it can happen. Even there, one is tempted to add, the system may defeat the innovator in the long run.

In summary, a conservative occupational group (teachers and administrators) exists to serve an institution (schools and colleges) which is conservative in self-image, which has existed on make-do funds, which is not designed for extensive experimentation, which is oriented on the small town-farm cultural pattern of the Midwest of 1880-1900, and which is built into a bureaucracy.[6] It is not, therefore, surprising to find that organizations of educational workers, except for certain specialists, do not reveal any bounding enthusiasm for the developing instructional technology and the changes it implies.

All of this would be very well, perhaps, if James Watt had decided to distill Scotch whiskey instead of making a steam engine, if Einstein had confined his talents to the fiddle, if Henry Ford had never put together an assembly line, if Roentgen had never had a Crookes tube to play with, and if that inventor of long ago had confined himself to a throwing stick instead of stringing sinew to a bow—in short, all would be well if things stayed the way they were supposed to. They didn't. The whirlwind of change has come in the middle of *our* century; it has affected many other aspects of our lives. Only the presumptuous, only god-players, can claim education, and the educational workers are immune.

This is indeed the revolutionary season, a time of change, a time when the educational worker as an individual and the organized groups to which he belongs must decide what to keep and what to cast away. And what to add and what to change. "To every thing there is a season, and a time. . . ." The measured prose of *Ecclesiastes*, however, does not communicate how incredibly swift is the passage of time which has brought us to the brink of great educational change—to the revolutionary season. Organized education must now, and in the foreseeable future, face the greatest internal conflict of its distinguished career—the conflict between its inherent and built-in conservatism and the pressure for changes resulting from the revolutionary season of science, technology, cultural upheaval, and democratic aspiration.

Change surrounds the teacher, the administrator, the supervisor in many

---

[6]Nothing has been said here about teacher education institutions and schools of education. Although there are now some signs of change, radical educational thought is about as foreign to those groups as the heresies of Calvin are to Jesuit doctrine. The current "radical" statements of the professors were part of accepted thinking in the early 30s and were not exactly new then. That is why an exception, such as Myron Lieberman, stands out when he sees the professor of education as a critic who refuses to criticize the system and merely adds to the forces maintaining the status quo (See Lieberman, 1963).

of the aspects of life. Change which is spectacular is sometimes brutal in its effects. It should never be forgotten that now change is all-pervading and there is no vaccine which education can take and become immune.

What, then, of change in the educational process itself? What are the sources of the conflict between conservatism and change? And what is the specific nature of the revolutionary season in education?

Space does not permit the necessary detailed exposition of all the pressure points for educational change, but brief consideration may be given three—the population explosion, the knowledge explosion, and the even more basic "long revolution." These are examples of the sources of conflict between educational conservatism and educational change.

The population explosion, of course, has economic aspects as well. Rather than develop this concept once again, reference can be made to a succinct statement of the problem by John O'Toole. He said:

National commitment to more rapid economic growth and long-range programs of space exploration and foreign assistance will impose pressures on schools and colleges for significant improvements in educational content and productivity, and will require increased public expenditures over an extensive period of time. If projections of school enrollments by 1970 are correct, college attendance will nearly double, and secondary schools will increase enrollment by 50 per cent. The current rate of U.S. population growth shows an annual average increase of approximately 1.6 per cent, which would amount to a total U.S. population of about 245 million by 1980, with a resulting tidal wave of students for the schools. How will these students be educated when the nation's schools are already overcrowded, understaffed, and inadequately housed? . . .

In 1962, the U.S. Office of Education estimated total expenditures for all public and private education, including capital outlay, as 25.2 billion dollars. School costs are expected to increase, on the average, at least one billion a year during the next decade to at least 45 billions in 1970, and perhaps to exceed 50 billions. Can the national economy withstand such expenditures, and if so, with what sources of support? How should financial resources for education be allocated so as to derive maximum return from the investment? (From an unpublished paper by John O'Toole, Systems Development Corporation.)

The second pressure point, the explosion of knowledge, has existed almost unnoticed for about three centuries. Recently, a portion of the problem (science knowledge) has been under careful study by one group of knowledge producers—the scientists—and, to a much smaller degree, by a segment of the knowledge warehousemen—the librarians. Knowledge distributors—teachers, administrators, etc.—on the other hand, have had to be content with infrequent and vague second-hand references to this other work.

Professor Derek Price of Yale has done most of the analysis of the growth of knowledge in science, and his original work should be a "must" for educators interested in curriculum and teaching problems. Price calls our attention to the fact that the increase in knowledge is not a new problem by citing Barnaby Rich, a scholar writing in 1613:

A Revolutionary Season

> One of the diseases of this age is the multiplicity of books; they doth so overcharge the world that it is not able to digest the abundance of idle matter that is every day hatched and brought forth into the world. (Price, 1963, p. 63)

Price's studies show that scientific papers have been published since the beginning of science. Since the growth of scientific knowledge is exponential, the number of papers doubles every 10 years. This means that, at the present time, about 600,000 new papers are being added to the stock every year. There are about 30,000 current scientific journals. The 10 million science papers produced since the middle of the 17th century have been written by about three million authors—of whom *somewhere between 2.4 and 2.7 million are alive today.*

The sheer total of knowledge is not too meaningful in the abstract. A sense of the speed and variety of knowledge development in one field has been described by the biologist, Bentley Glass (1962). A biology textbook of 1900 would have had no reference to genetics, to biochemistry, blood groups, viruses, or specific vitamins. A similar book of 1930 would, of course, have covered all of these fields and more. In 1960, however, it would have to add the whole DNA-RNA story, the cracking of the genetic code, the mapping of the protein molecules, the discoveries relating to metabolism and ATP, antibiotics, insecticides, etc. Such an analysis could be repeated in field after field.

It is sheer cowardice for the standpat educator to say, "So what? . . . We aren't teaching facts anyway." Does anyone seriously propose that these materials should not be covered, to some degree, not only in biology courses but in health units and in other pertinent places? If they are covered, they represent an addition to the burden of content to be taught, even if judicious pruning has been done and all references to phlogiston, the humors, and the aether are erased—which is highly unlikely.

Many teachers today know only too well that the knowledge explosion exists. They have sweated through National Science Foundation workshops, language institutes, summer courses in the "new mathematics." Even that sacred of sacreds—English—is being assaulted by structural linguistics to the acute discomfort of many a "Received Standard" grammarian. And curriculum revisions in physics, biology and mathematics are pouring out of Cambridge, Boulder and Stanford—revisions which, incidentally, are probably already behind the times.

The educational system, then, is ill with an inflaming manpower problem consisting of a rapidly growing student population, a shortage of qualified teachers, increasing demands for space, rising costs, potential ceilings on funds, and accelerating demands from teachers and from the culture, and, in addition, suffers increasingly from an indigestion problem brought on by the exponential growth of knowledge and the variety of demands from the society for educated manpower and a decrease in the uneducated.

These twin explosions must be seen, however, in the light of the third pressure point mentioned—the truly fundamental pressure point of which the other two are results. I refer to what Raymond Williams called, with great insight, "The Long Revolution." He said:

A Revolutionary Season

It seems to me that we are living through a long revolution, which our best descriptions only in part interpret. It is a genuine revolution, transforming men and institutions; continually extended and deepened by the actions of millions, continually and variously opposed by explicit reaction and by the pressure of habitual forms and ideas. Yet it is a difficult revolution to define, and its uneven action is taking place over so long a period that it is almost impossible not to get lost in its exceptionally complicated process. (Williams, 1961, p. x)

Williams goes on to identify the long revolution as having three parts: the democratic revolution, the industrial-scientific revolution, and the cultural revolution. And he concludes his discussion of this point by saying:

. . . we must keep trying to grasp the process as a whole, to see it in new ways as a long revolution, if we are to understand either the theoretical crisis, or our actual history, or the reality of our immediate condition and the terms of change. (Williams, 1961, p. xiii)

The population and knowledge explosions, then, although they directly bear upon what organized educators must do *now*, are part of this larger, extremely complicated longitudinal and revolutionary development of our culture. They interact with one another and with the larger revolution and reflect back constantly into the educational enterprise down to the single classroom and the first-year teacher.

I will state flatly that unless educators, as individuals and within their organizations, begin to sense this whole pattern of revolutionary development and devise ways of living with it, they will either go the way of the dancing master and elocution teacher or become second-level technicians in an enterprise operated by other elements of our society.

There is no reason to suppose, for example, that the growth of knowledge stops at the borders of what we call "Education." In fact, there is reason to believe that the rate of acquisition of knowledge in the behavioral sciences as a whole (which has, until recently, been much slower than in the physical and biological sciences) is now speeding up. Within this development, increasing attention is being given the problems of education by psychologists, sociologists, engineers, and educators. This has resulted, in the last decade, in an accelerating increase of what might be called "hard knowledge" in the field of education and the development of manpower and techniques which will speed this process further in the years immediately ahead.

In contrast, and no doubt as an absolutely necessary social preliminary to the current scientific aspects of education, the progressive education movement of the first four decades of this century generated the opposite of hard knowledge. Kimball and McClelland suggest that the great contribution of educational progressivism was moral exhortation. "It [progressive education] asked a commitment which, in its extreme form, assigned to the schools the moral responsibility for the entire society" (Kimball & McClelland, 1962, pp. 109-110). Such preaching created a conscience among educators for which the country must be thankful, but preaching is not hard knowledge, nor technique, and yet the residue of those moral imperatives has passed for such the last 30 years. (For evidence, take a look at most books on curriculum today.) This, incidentally, is the source of much criticism of education as the moral imperatives of the giants such as Dewey, Kilpatrick and Counts were vulgarized and ground into a sterile jargon by their followers and then used in methodological discourse.

Problems—including educational problems—are only solved in a technical context. And for years the majority of our most capable people in education avoided technique and technical theory. *By some strange transformation, technique was thought to be high moral purpose and high moral purpose was thought to be technique.* Specifically, among a large number of educators the party line in curriculum and methods became confused, consisting of snippets of perception, gestalt and personality psychology, highly selective items taken from social psychology and sociology relating to group dynamics, various statements of educational purpose and moral imperatives, a large dose of Rousseau seen through the work of the followers of Kilpatrick, a good-sized slice of successful practice handed-on-down, and 30- to 50-year-old concepts of the several intellectual disciplines.

There was an exception to this trend which complicated and made matters worse. This was what Callahan has called the "Cult of Efficiency," adapted to education by school administrators and administrative theorists beginning about 1912 and reaching its apex in the middle or late 20s, but continuing long after that. This adaptation was made from the work of Frederick Taylor, the famous (or infamous) time and motion study man who introduced the "efficiency expert" into American industry. Although it has long been forgotten, the schools were under vicious attack by chambers of commerce and industrial associations around 1912 for waste, inefficiency, etc. The response to this attack was to move into something called "scientific management" in education. Educational scientific management produced such gems as:

> . . . 5.9 pupil-recitations in Greek are of the same value as 23.8 pupil-recitations in French; that 12 pupil-recitations in science are equivalent in value to 19.2 pupil-recitations in English; and that it takes 41.7 pupil-recitations in vocal music to equal the value of 13.9 recitations in art . . . *we ought to purchase no more Greek instruction at the rate of 5.9 pupil-recitations for a dollar. The price must go down, or we shall invest in something else.* (From a 1913 speech by Frank Spaulding, superintendent of schools at Newton, Massachusetts, as quoted in Callahan, 1962, p. 73. Italics are Callahan's.)

Such thinking, of course, was the source of the trend that ended up with masters degree projects in school administration counting toilet seats, the installation of unreliable teacher-rating scales throughout the country, and the complete separation of school administration from the conduct of learning. It is forgotten by many critics, however, that these crude efficiency concepts became educational reality because of the pressure from the American business community and the mass media of the time, such as the *Ladies' Home Journal.*

Callahan vigorously criticizes the cult of efficiency in educational administration, essentially from the high moral purpose position of the educators described above, and points out that we live with a dangerous residue of this viewpoint today. However, he failed to analyze and discern clearly two important things: (a) That the cult of efficiency was a sort of cancerous by-product of the legitimate measurement or scientific movement in education. (b) That *circa* 1910-1930 education had neither the science nor the technology available to make the concept of efficiency in education both viable and humanistic.

We are now ready to return to the long revolution as it relates to American education and the historial factors we have discussed. Almost overnight, in its uneven course through modern history, the scientific-technological phase of the long revolution appears to be turning up within American education, bringing us to the revolutionary season with which this piece began.

A scientific revolution is a compound of several events, including the adoption, by the majority of workers in the field, of a model of what the world is like, the conjunction, at a given point in time, of several different and sometimes opposite trends of thought and investigation, and the rise of a high technology which makes certain scientific activities possible (for example, the development of the microscope).[7]

The nature of technological revolutions is not so well understood. Extremely complex social and economic factors are at work here and the history of technology is difficult to locate and generalize. White said:

> As our understanding of the history of technology increases, it becomes clear that a new device merely opens a door; it does not compel one to enter. The acceptance or rejection of an invention, or the extent to which its implications are realized if it is accepted, depends quite as much upon the condition of a society, and upon the imagination of its leaders, as upon the nature of the technological item itself. (White, 1962, p. 28)

Technology, because it involves men, machines, patterns of organization, and economic feasibility, and because it interacts at any given time with the science of that time, is imbedded in the general social condition. In addition, like its Siamese twin science, the exact effects of technology can never be predicted—for example, it is now believed that such things as the expansion of population, the rise of cities and industrial production, and the development of commerce in northern Europe in the Middle Ages was due, for the most part, to the system of three-crop rotation which produced vegetable protein and increased the energy factor in the diet. As White says, " . . . the Middle Ages, from the tenth century onward, were full of beans" (White, 1962, p. 76).

It seems that all of the factors necessary for a scientific revolution and all of the factors necessary for a technological revolution have come together, under the umbrella of Williams' long revolution, to forever and irretrievably change the face of the American educational enterprise. It might be useful to list some of these under the headings of science and technology.

Science

1. A large segment of psychological scientists interested in the problems of learning have an agreed-upon model of what their scientific world

---

[7]Good discussions of these and other aspects of scientific revolutions may be found in Kuhn, 1962 and Price, 1961.

is like—centered upon the operant conditioning concepts of B. F. Skinner and, more generally, upon the postulates of psychological behaviorism.[8]

2. Increasingly, this model, or at least many of its postulates, has spilled over into research relating to the use of the so-called "new media" of instruction—films, television, language laboratories, and, of course, teaching machines and programed learning. This develops a close relationship between what Lumsdaine and others call the "science of learning" and the "technology of instruction."

3. The methods, attitudes and paradigms of all of the rapidly growing behavioral sciences are increasingly feeding into educational thinking and research. For example, concepts from research in mass communication, the management sciences, anthropology, etc. are all making themselves felt at the conceptual level in professional education. To date, these concepts have not widely affected practice, but such effect may be expected.

4. Science itself, in its traditional assault upon the unknown, has developed salients within fields once confined to philosophy, crude observation, and traditional common sense, with the result that can always be expected from such invasions—change in concepts. These salients relate more and more to education. The increased understanding of the brain, the beginning solutions of the problems of creating artificial intelligence, the suspected relationships of metabolic and other chemical processes within the body to behavior, the application of statistical techniques to whole new classes of social and psychological phenomena, all are examples of the trend toward the creation of an entirely new scientific basis for the educational process.

5. Quietly, the scientific movement in education is in the process of rebirth. This time it comes upon the scene with a much more adequate general scientific base and a functioning technological base. As evidence, consider the recent *Handbook on Research in Training*, published by the American Educational Research Association. This time the portents for survival of the scientific movement in education are much better.

6. As our whole culture moves in the science-technology direction, as noted above, the importance of abstract knowledge increases—almost geometrically. One of the main characteristics of such knowledge is that it can be communicated and used in a variety of ways. This has two effects upon professional education. What is to be taught is changed; abstract, replicable knowledge about the educational process itself can be communicated and replaces, in many cases, existing folklore.

Technology

1. Instructional technology must be viewed within a setting of a general growth of technology (in both extent and sophistication) within our entire culture. Thousands of examples can be adduced from such various fields as transportation, production of drugs, communication, etc. which have surrounded the lives of citizens with entirely new environments and patterns of thinking and have given them entirely new capabilities. Per-

---

[8]This does not discount other psychological developments, particularly those new notions relating learning, brain functioning, and information theory. In fact, I suspect a new psychological paradigm is about to descend upon us.

haps the best single example of all this is the sudden growth of computer technology.

2. Within this general technological growth there has occurred a pale (but improving) reflection in education, as instructional technology begins to speed up its growth. Essentially, this modern growth has had three periods. It began about 1920 and its first period extended to approximately 1955. During this period the whole conventional arsenal of audiovisual devices and materials was developed—films, filmstrips, recording devices, etc. The second period, 1955-1965, can be considered transitional, embracing near its beginnings television, the language laboratory, and teaching machines and programed learning. As the decade approaches its end, other additions include 8mm sound film, classroom communicators, multi-media presentation techniques, test-scoring equipment, more complicated teaching machines, and computer-controlled instructional systems.

The third period, extending into the future, will see further developments in computers and educational data processing, information storage and retrieval systems, satellite communication, and adaptations to educational communications of such devices as the optical laser. Incidentally, none of this should be taken to mean that the educational system, as presently constituted, is approaching saturation, even with the conventional audiovisual technology of the first period.[9]

3. Economic and social factors are at work within and without education to force change in the direction of increased applications of instructional technology. First is the general technological environment of the whole culture, from which education cannot forever hide. Second is that, as man-machine systems continue to be used more frequently in education (more motion picture projectors and films, for example) a technological milieu is created which makes it easier to adopt more technology.[10] Other factors include the National Defense Education Act (response to Sputnik), the efforts of the Ford Foundation and others, increased pressures of economic-manpower-knowledge problems, and the existence of great training and education problems and efforts in industry and the military.

4. An advanced technological society is characterized by a large, expensive and complex research and development effort. Billions of dollars are now being spent in the U.S. on research and development. All modern achievement—industrial and military—is inextricably linked to research and development. This accounts, of course, for many of the fantastic accomplishments of the last two decades. Until recently, the research and development effort in education amounted to so little (particularly, when thought of in present terms) that it produced very little. That picture is now changing. Several of the foundations and the U.S. Office of Education, particularly through its Cooperative Research and Title VII Programs, have put fairly substantial funds into research. Further, there is a fall-out of research results from both the military and industry which is benefiting educational science and technology.[11]

---

[9]See for example, Finn, Perrin and Campion, 1962.

[10]For an elaboration of this theory, see Finn, Perrin and Campion, 1962.

[11]An excellent discussion of this point may be found in Glaser, 1962. The past role of educational "research" is discussed in this book with great insight by Robert M. W. Travers.

A Revolutionary Season

J. Robert Oppenheimer once said that great advances in science occur when several strains come together and form a "lacework of coherence." Derek J. de Solla Price says of scientific revolutions: "Whole new sciences have arisen as the result of the confluence and interlocking of previously separate departments of knowledge" (Price, 1961, p. 22).

While a professional field such as law, public administration, or education cannot have the same *raison d'être* as a science because of its intimate relation with society and social need, nonetheless I believe the evidence is conclusive that revolutions in professional fields occur in the same manner. This is to say that "confluence and interlocking of previously separate departments" must occur; that a "lacework of coherence" must suddenly be created. The difference between a revolution in science and a professional field may merely be in the number and kind of factors required for conjunction. American education as a public enterprise and as a professional field has reached such a point of multiple conjunction.

These are the revolutionary forces of change arrayed against the conservatism of organized education. A great confluence has occurred in American education—the probable confluence of a revolution. The inherent conservatism of the system—a great stabilizing influence for decades—has been based on the assumption that things would remain pretty much as they had been. This assumption is no longer true, inside or outside of education. What, then, should be the response of organized education?

Space does not permit a discussion of that response now. It should be remembered, however, that the traditional defensive responses—"There is nothing in all this about creativity—and that's our important task." or "Everything you have been talking about has to do with the teaching of facts and what we are now teaching is 'structure' of knowledge." or "You will never get love and understanding out of your machines and your science." or "It's too expensive and, anyway, Strayer and Englehardt got along fine without any of this stuff because *they* knew how to deal with school boards."—such traditional defensive responses *simply will not do. None of them fit any more.*

It follows that teachers and administrators, as individuals and as represented in organized education, *must* develop appropriate responses to this challenge of educational revolution. Since the time of Darwin, it has pretty well been established that the road to oblivion for any species is littered with the debris of responses that did not fit.

We can then return to the main theme. Organized education faces the time of revolution. It must seize on and keep (and change in form, if necessary) those values most necessary for ultimate survival; it must cast away much of what has been useful or comforting or status-giving in the past. It must take on new responsibilities, patterns of action, knowledge. We are presented, I believe, with the great opportunity of converting all the way from a subprofessional occupation into a true profession. This is what the revolutionary season has brought to us—opportunity. It is of this opportunity that it is time to speak.

References
Callahan, Raymond E. *Education and the Cult of Efficiency.* Chicago: The

University of Chicago Press, Phoenix Press, 1962.

Counts, George. *Dare the Schools Build a New Social Order?* New York, John Day Co., Inc., 1932. John Day Pamphlet No. 11.

Finn, James D., Donald G. Perrin, and Lee E. Campion. *Studies in the Growth of Instructional Technology, I: Audiovisual Instrumentation for Instruction in the Public Schools, 1930-1960, A Basis for Take-Off.* Occasional Paper No. 6. Washington, D.C.: NEA, Technological Development Project, 1962.

Glaser, Robert (Ed.). *Training Research and Education.* New York: John Wiley & Sons, Inc., Science Editions, 1962.

Glass, Bentley. "Information Crisis in Biology," *Bulletin of the Atomic Scientists,* Vol. XVIII, No. 8, October 1962, pp. 6-12.

Kimball, Solon T. and James E. McClellan, Jr. *Education and the New America.* New York: Random House, 1962.

Kuhn, Thomas S. *The Structure of Scientific Revolutions.* Chicago: The University of Chicago Press, 1962.

Lieberman, Myron, "Professors of Education as Critics of Education," *Phi Delta Kappan,* Vol. 44, No. 4, January 1963, pp. 164-167.

O'Toole, John. Unpublished paper. Systems Development Corporation.

Price, Derek J. de Solla. *Little Science, Big Science.* New York: Columbia University Press, 1963.

Price, Derek J. de Solla. *Science Since Babylon.* Clinton, Mass.: The Colonial Press Inc., 1961.

Travers, Robert M. W. "A Study of The Relationship of Psychological Research To Educational Practice," in Robert Glaser (Ed.), *Training Research and Education.* New York: John Wiley & Sons, Inc., Science Editions, 1962.

Wayland, Sloan R. "The Teacher as Decision-Maker," in A. Harry Passow (Ed.), *Curriculum Crossroads; a report of a curriculum conference.* New York: Teacher's College, Columbia University, 1962, pp. 41-52.

White, Jr., Lynn. *Medieval Technology and Social Change.* London: Oxford at the Clarendon Press, 1962.

Williams, Raymond. *The Long Revolution.* New York: Columbia University Press, 1961.

The Franks Had the
Right Idea

The acceptance or rejection of an invention, or the extent to which its implications are realized if it is accepted, depends quite as much upon the condition of a society, and upon the imagination of its leaders, as upon the nature of the technological item itself. ... the Anglo-Saxons used the stirrup, but did not comprehend it; and for this they paid a fearful price.... it was the Franks alone—presumably led by Charles Martel's genius—who fully grasped the possibilities inherent in the stirrup and created in terms of it a new type of warfare supported by a novel structure of a society which we call feudalism.... for a thousand years [feudal institutions] bore the marks of their birth from the new military technology of the eighth century. (White, 1962, pp. 28-29)

If you are an American teacher and your world of work echoes with statements concerning language laboratories, flexible scheduling, teaching machines and programed learning, closed-circuit television, and (perish the thought!) computers used in connection with teaching, how do you feel about it all? If the latest institute or workshop speaker's rhetoric ends with a ringing statement that we are in the midst of a technological revolution in education or if your audiovisual director waxes sarcastic because his national professional group tells him that what he is really doing is working with instructional technology, are you concerned, indifferent, happy, curious, or belligerent?

Assuming the institute speaker was right, what is it like to be in the middle of a technological revolution? After all, if we're talking about a technological revolution in *education*, the direct effect, in a large part at least, will be upon teachers, teaching and the professional organizations of teachers. Perhaps no one knows what it is like to be in the middle of any revolution. It may be difficult to sense precisely what is going on.

The Frankish knights of Charles Martel's time, the coal miners of James Watt's time, the stockbrokers of Morse's time, the physicians of Lister's time probably sensed stirrings and changes, alarums and excursions, without being able to put their fingers on causes and transformations—certainly, for the most part, without being able to understand, control and make the most of the process of change.

Reprinted from *NEA Journal*, Vol. 53, No. 4, April 1964, pp. 24-27.

The Franks Had the Right Idea

Each time of change has had its prophets and its individuals who have understood the implications of the process through which they were living and have operated accordingly. In a broad sense, it has always been true that, in time of change, the world belongs to those who can grasp the nature of that change and fashion their life and culture to make the most of it. As White shows, because of what Charles Martel and his people did in the eighth century (which the Anglo-Saxon did *not* do), the victory of William I at Hastings in the 11th century was assured. In simple terms, the Franks took the stirrup, changed society, and controlled the world. The Franks understood that the stirrup not only enabled the rider to keep his seat but also to deliver a blow with his lance that had the combined weight of the rider and his charging horse.

The only danger in applying this generalization to what is happening to education today, particularly in the field of instructional technology, is that we may underestimate the speed and extent of the change which is now taking place. For this revolution is merely a minute reflection of the great force of the scientific and technological revolution that sweeps through Western culture at an ever-accelerating pace and moves now to engage the world.

Man proposes to change human genes and control heredity; to cross the United States in an hour; to redirect human behavior; to land on the moon by 1970. Can the American school teacher honestly believe that the scientific-technological change which permeates the whole culture will stop at the borders of education?

Perhaps at this point, it would be well to bring the micro-revolution that represents the application of science and technology to education into better focus. It might be helpful, for clarity, to list some of the technological developments that are the signs of change. For two-and-a-half years, a project known as the Technological Development Project of the NEA, funded by the U.S. Office of Education, undertook to describe these developments, to predict trends, and to make some kind of assessment of the effect of this developing technology upon education.

A brief summary of trends as analyzed by the Project might prove helpful to this discussion. Central to technological change in education have been developments in both hardware (devices, machines) and materials (films, programed learning materials, etc.). These developments fall into a three-part chronological pattern.

1920-1955

These years marked the development of conventional audiovisual devices and materials; i.e., films, filmstrips, slides, recordings, tape, and the projectors, players, etc. necessary to use them. This phase reached its climax shortly after the end of World War II and leveled off by 1955, at a point much too low to have made a really effective impact upon the educational process. Educational administrators, school boards, and other decision makers did not place a high enough value on these developments.

The Franks Had the Right Idea

1955-1965

The years between 1955 and 1961 saw the introduction into the schools of television, language laboratories, and teaching machines and programed learning. As the decade nears its end, 8mm sound film for individual viewing, large-group multi-media presentation devices, classroom communicators, computers, and other advanced devices and materials are appearing on the experimental scene.

This is the decade that has produced the expressions, *new media, educational automation, instructional technology,* and *technological revolution in education.*

With the exception of the language laboratory and, to some extent, television, these devices and the materials and techniques they imply have not yet been applied on any scale. However, since 1958, the amount of audiovisual equipment and materials has been increasing, as a result of Title III of the National Defense Education Act.

1965-

The next period promises technological developments even more spectacular than those of the past. The items beginning to show now—computers for educational data processing and as controls of teaching systems, instructional systems themselves, etc.—will be followed by sophisticated information storage and retrieval systems, the communication of visual and other forms of data from computers to people, new forms of long distance communication and control, and machines that can learn.

The foregoing summary presents but one facet of the technological revolution in education. Technology is not just hardware—or even hardware and materials. Technology is a way of organizing, a way of thinking, involving at the center, to be sure, man-machine systems, but including systems of organization, patterns of use, tests of economic feasibility. And in the times in which we live, there is a high-order interacting system between technology and science.

Let us consider some of these other elements in the pattern of educational technological change.

For instance, a whole system of thinking has developed in the field of programed learning which relates a science of learning (behavioristic psychology) to a technology of instruction involving specification of objectives in behavioral terms, frame writing in programed instruction, task analysis as a basis of lesson planning, and the like.

Various staff utilization plans, such as team teaching and other organizational change proposals, lie on the periphery of this technological revolution along with new concepts of school buildings, all of the new curriculum developments (the "new" math, physics, etc.), and even the great interest in studying the problems of educational innovation itself.

Other pieces of this complicated picture include the efforts of the great foundations, the money power of the National Defense Education Act directed toward the new media and the provision of teaching equipment for the schools, the general interest of researchers in learning and instrumentation, and the press of industry to diversify into the educational market.

The Franks Had the Right Idea

So much for the specifics of technological development in education. Before returning to the basic proposition that, in a time of change, success comes to him who grasps the significance of that change, let us take one step further back and examine the socio-economic-political pressures surrounding the educational system to which this micro-system of educational technological development is a partially realized response. Bear in mind the one brute fact that all of this is occurring in the context of the swiftly accelerating general scientific-technological revolution in our culture, embedded in a setting of world tension and upheaval.

The educational world is faced specifically with two thorny, continuing and growing problems. One is related to knowledge; the other to population. A technological culture is a knowledge-generating culture; for reasons of survival (both economic and national), research and development—the information-producing mechanism—is now built into everything we do. This presents the schools with the fantastic problem of deciding what to teach from the exponential growing body of knowledge and how to provide more learning in less time.

The second problem—the problem of extraordinary population growth—continues to fill the schools with more students to be taught, housed and provided with learning materials. It would appear that the only answer to the pressures on the schools caused by the population and knowledge explosions is to turn to technology, at least in part.

In recent years, the American teacher has been facing the chaos of knowledge growth, the threat of population inundation, and the increasing speed of change. And what, up until now, has been the response—your response?

Any representative of organized education (such as an association executive), many school administrators, and even some perceptive citizens will tell you that teachers today are restless indeed. Social and economic problems—salaries, unionism, sanctions—seem to occupy the center of attention along with questions of school integration and other sticky problems.

Viewed objectively, and excluding a few problems such as integration, it is fairly safe to say that the major concerns are with such personal and welfare items as salary schedules, working conditions, and the degree to which teachers and teacher groups are involved in educational decision making. These areas of restlessness, these concerns, are responses to deep-seated events in our culture, but they are not adequate responses to a needed technological revolution. A profession cannot be built on welfare principles and political power alone; this is the Anglo-Saxon approach to the stirrup.

The educational future will belong to those who can grasp the significance of instructional technology. Many years ago William Wrinkle pointed out that the history of the secondary school showed conclusively that when one institution, e.g., the Latin Grammar School, could no longer meet the needs of a society, it was replaced by another, e.g., the Academy. I believe this generalization holds good throughout education today.

Unless present educational workers and their organizations can find the means to deal with problems of knowledge and population growth, and related problems of urbanization, skill demands, etc., within the present school and college context, society will somehow invent other institutions and other means to deal with these problems.

This challenge to the teaching profession requires a creative response for survival and presents unprecedented opportunities for growth and advancement of the profession. In many cases, the response can be related to the equally legitimate welfare demands, for American education has, since its inception, been predicated upon a coolie labor theory. Either the teachers or the children did the work, or it wasn't done.

Certainly no other profession in the United States has put such a low value on its human power. Whether this is the fault of school boards, local control, school administrators, teacher timidity, inadequate pay-as-you-go financing, or combinations of these and other factors is now beside the point. The coolie labor theory of operating the educational system in this country has got to go.

Since 1850 and before, this country has elected to seek technical solutions to its problems of creation and production in order to increase productivity, save manpower, and (to a greater extent than is realized) dignify man and improve the quality of his life. As a result, in many areas of life, the citizen is surrounded by a technological milieu that he takes for granted. This milieu functions as an educator, so that when a new kind of washing machine comes along, the housewife welcomes it.

American education has not yet developed such a technological milieu, and this is as much the fault of teachers and their organizations as of any other group. Schools do not have enough conventional audiovisual equipment or materials, enough books, enough laboratory equipment, let alone the newer devices and materials.

Teachers and administrators are not used to thinking of solving their problems by technological means because so few of their problems are now solved with anything but sweat and, as Anna Hyer has said, "new uses for old coat hangers."

The best possible example of this is the relatively inexpensive test-grading machine. Such machines, certainly a boon to teachers, have been possible since the middle 1920s. The profession has consistently failed to demand them and has opted for the red pencil, elbow grease, and hours of time. Not until 1962 did such machines become available and not because of teacher demand, even then, and very few schools have them even yet.

The absence of a technological context within the educational system and the general hand-labor orientation of teacher and administrator are, of course, in the process of change because of the pressure mentioned above. However, there is no sign yet of the development of a creative response to this technological revolution on the part of American school teachers, considered either as a group of individuals or as individuals whose will is expressed in the actions of local, state and national associations.

If the collection of associations, together with the supporting teacher education institutions, can be considered "The Establishment," can the Establishment adjust and make the creative response to the technological revolution in education? Since the Establishment in the long view really expresses the will of its members, will the American teacher creatively respond?

We can return to the stirrup. The stirrup created the knight and the whole feudal system designed to support him, and the Franks had the vision to seize the idea and use it. I would argue that the same situation holds today with instructional technology and the teaching profession.

Programed instruction, film, television, and computers can be looked upon as noncreative gadgets; and that might be called, in White's context, the Anglo-Saxon view. Instructional technology can also be viewed as the best means of making teaching truly a profession—the Martel view. The second view is, it seems to me, the only source of solutions to some of the difficult problems of education and, hence, the means of survival.

White concludes his discussion of the stirrup with a quotation from Denholm-Young, which sums up his whole argument: "It is impossible to be chivalrous without a horse."

I would conclude in the same way—it will be impossible to be a professional teacher without instructional technology.

References

White Jr., Lynn. *Medieval Technology and Social Change.* London: Oxford at the Clarendon Press, 1962.

AV 864 vs A.D. 1970

In 1958, the Congress of the United States passed Public Law 864, referred to as the Hill-Elliott bill. The official name of the law is the National Defense Education Act of 1958. P.L. 864 has eight "Titles" or sections dealing with:

1. General provisions.
2. Loans to students in college.
3. Assistance to the several states in strengthening instruction in science, mathematics and modern foreign languages.
4. Fellowships for graduate students.
5. Grants and contracts to improve counseling and guidance programs.
6. Modern foreign language development.
7. Research in more effective use of teaching tools.
8. Grants to improve state vocational programs in the training of technicians.

The provisions of the Act cover four years, with authorization by Congress (but not appropriation) for the expenditure of about $90 million; when state matching funds are added, the figure will exceed $1 billion. The 1958 congressional appropriation was about $40 million, approximately half of which was earmarked for the third section—strengthening instruction in science, mathematics and foreign languages. The amount appropriated to date is, of course, far under the amount authorized. It is expected that the new Congress will add to these appropriations.

P.L. 864 has caused a great stir in audiovisual circles as well as among those interested in science, foreign languages, and guidance. Four sections of the law apply directly to the use of the tools of teaching—the section on strengthening science, mathematics and language instruction where great emphasis is put on the provision of adequate audiovisual materials and equipment, the section on language development stressing the concept of the language laboratory, the special section on research as to the effectiveness of audiovisual materials, and the section on improving instruction for technicians which also encourages the use of modern teaching equipment and materials.

In this connection, the staff of the National Audio-Visual Associa-

Reprinted from *Teaching Tools*, Vol. 6, No. 1, Winter/Spring issue, 1959, pp. 4-5.

tion (the organization of producers and distributors of audiovisual equipment and materials) has done an excellent job in preparing a booklet on the audiovisual provisions of P.L. 864. Someone with a touch of genius came up with a great title for the booklet, *AV 864*. Essentially, the National Defense Education Act of 1958 is AV 864. It extends massive recognition to the idea that our readers and supporters have been pushing for many years, namely that instruction can be measurably improved and hastened by the wise application of a large variety of the modern tools of teaching.

AV 864, then, represents a great victory for the audiovisual idea. Congress itself, through its Committee on Education and Labor, was very clear on this point. The Committee report on the law stated, among other things, that "there is need for modern laboratory equipment, including audiovisual materials and equipment such as motion pictures, slides, filmstrips, transparencies, disc and tape recordings, still pictures, models, globes, charts, and maps in elementary and secondary schools, if instruction and learning is to be improved."

To a dedicated group of people committed to the audiovisual way, to thousands of hard-working teachers desperately in need of better teaching tools, and to concerned administrators attempting to improve instruction, AV 864 is manna from heaven, the oasis in the desert, the top of Everest. The National Defense Education Act not only provides badly needed federal aid, but it supplies it in areas where it can do the most good with the least harm.

But when we have said all this about AV 864, we would like to state to our elected representatives, to the teaching profession, and to workers in the audiovisual vineyard that we have some serious reservations about the National Defense Education Act of 1958. We hasten to add that these reservations do not apply to any provisions for the improvement of instruction with the tools of teaching.

Our reservations about P.L. 864 go much deeper—they go to the intent of the law, to the intent of Congress, and to the intent of many of the people who testified in favor of the bill in committee hearings (published in a 1,602-page book). In the first place, Congress and all the scientists and educators involved are dealing with the future when they provide for education. Any such act cannot be compared to a decision to buy so many Atlas, Jupiter, Thor, or Nike missiles. Such hardware can be assembled in a given period of time with predictable (within reason) results. This is not dealing with the future, it is dealing with the present. In contrast, education is man's way of attempting to control his future, and, even in the provisions of AV 864, we are dealing with 1970 and beyond. What kind of a country are we providing for? What kind of decisions must be made? What kind of a world are we constructing? What kind of people are we creating?

At this point, we are in jeopardy of being called soft-headed. Isn't the National Defense Education Act a hard-headed, practical solution to the problems of defense here and now? Isn't it what we need to counteract Russian leadership in the missile or any other technical field? We won't have any future if we don't have a defense which includes an adequate supply of scientists, technicians and linguists. So the argument runs. The Russian Sputniks and moon rockets, the scare-type headlines created by the Admiral Rickovers as to the state of our education, and the general education panic whooped up by our mass media of communication make this argument almost invulnerable. P.L. 864 is a step in the direction of creating a scientific defense for the United States in an age of technology and science. Any suggestions that the humanities, the arts, or the social sciences are being neglected is impractical and soft-headed.

This is the point at which we have to take issue with the intent behind P.L. 864, Admiral Rickover, and the bulk of "educational" opinion expressed in the press. We contend that the current general and exclusive emphasis on science and mathematics and the limitations of the National Defense Education Act to science, mathematics and modern foreign languages is unwise. While no doubt contributing something to our future defense, this exclusive technical orientation also leaves gaping holes through which the Russians, the Martians, or the Mau Maus will be able to penetrate with ease.

Further, in the future (and here we are really getting soft-headed) we need to build a culture *that becomes increasingly defensible.* Such a culture would have more art and less juvenile delinquency; such a culture would allow no second-class citizenship to exist within our boundaries; such a culture would reward a creative achievement in political science or sociology equally with one dealing with the efficiency of rocket motors. It should be emphasized that such a culture would also be increasingly technological and scientific in nature and that science and mathematics would play a greater and greater role; but that role would not be the *exclusive* concern of governmental or private agencies.

Finally—and these are but examples—such a culture would help man relate creatively with man all over the world from the rice paddies of the Orient to the ice pack of Antarctica. The future being prepared by the National Defense Education Act can only contribute to one side of this culture. The direction for 1970 is being warped out of joint.

Let us return to the other point, the areas the National Defense Education Act leaves undefended. Regardless of pronouncements from on high, our failure in the missile race was not a failure of education or of American science and technology. It was a failure, if anything, in the sciences and technology of man—in management, in administration, in political science, in organization and communication. The problem of inter-service rivalry, for example, will not be solved by creating countless additional Wernher von Brauns for the Army, Navy, Air Force, and civilian agencies to hassle about as if they were college football coaches looking for players. And we hold that inter-service rivalry with 57 varieties of missiles had more to do with our time-lag relation to the Russian program than all the alleged shortage of high school physics classes and students in the last 10 years.

Our readers can no doubt supply countless other examples dealing with failure in administration, social relationships, etc. We do not care to go into them here, partly because many of them have political implications. Can any admiral, ex-senator, history professor, or news magazine editor demonstrate how the present P.L. 864 can help this problem? Is it, in this sense, making a future contribution to our defense in a world in which the problems of organization and social relationships will become ever more complicated?

Even this situation is minor when cast upon the world scene. Our basic world-wide defense problems are in the social-cultural arena, not in the science arena. Will the problems surrounding China, Southeast Asia, Africa, and the Middle East succumb only to rocket technology and the tensor calculus? Or will they also have to be approached with better ideas—better ideas about man's relationship to man, about man's aspirations, about the way he organizes to better his lot and the lot of his children? Will physical chemistry or strength of materials help here? Has P.L. 864 plugged these gaps? A serious question is whether or not the Russians are better prepared to operate in the social-cultural fields than we are.

Again, we would like to make clear that we are enthusiastically behind everything that AV 864 is attempting to do. The point is that it does not do enough; it is a one-sided statement of the educational objectives of our country. We need more and better teaching tools for the arts, the humanities, and the social sciences as well as for the sciences. We need a more inclusive AV 864. In fact, we have always had more and better science materials than any other kind. These science materials were never good enough, extensive enough, useful enough, it is true; but they were better, more plentiful, and more easily available than teaching tools in many other important areas. Now we continue to widen the gap.

AV 864 was a reaction. It was a reaction to the artifically induced fear that Russian education was superior to American education and that this educational superiority caused the Russian Sputniks. Even so, it will result, we think, in many spectacular improvements in American education—and that is all to the good. But it is high time we stopped reacting to Russian Communism and began presenting the Russians and the rest of the world with ideas and actions which will require *their* reaction. We can do this only with an educational system that develops a superior culture for the future. And a superior culture does not rest solely on the sciences and mathematics. Let the Congress, the Administration, and assorted educational Messiahs think on these things.

Section II

Automation and Education

Automation and Education

"We are on the edge of a breakthrough. It will, in fact, be forced upon us by circumstances if we do not make it ourselves. A new content for education as a discipline needs to be made."

When he made this probing statement, James D. Finn was most concerned regarding the apathetic plateau from which professional educators regarded technology in this age of automation. He saw the need to close the gap between the rocket-like progress of technology in society and the snail's pace of technology in serving man's needs through education. He also saw the need for students of audiovisual communication and technology to provide leadership. In a series of three articles he stated how he proposed to examine the general relationships of automation and education.

Completed over a decade ago, these statements are as timely today for a large segment of our profession as when they were written. After discussing the general aspects of automation and its relationship to education in the first article, Finn next surveys the background of the efforts to automate the classroom in terms of the organizations leading the movement and the theoretical justification for the proposed changes in educational procedure. In the third paper, Finn's scholarly discourse on technology and his creative descriptions of the roles, developments, trends, and implications of technology in the instructional process provide the reader with a statement which pushes outward on the frontiers of educational knowledge.

In the opening article, "Automation and Education: 1. General Aspects" (1957), Finn describes automation as an industrial process and as an intellectual system. He outlines five characteristics of automation and its associated processes and systems, not as mechanical or technical in a machine sense, but rather as conceptual and having educational significance.

The five characteristics isolated for discussion were: the concept of systems, the flow and control of information, scientific analysis and long-range planning, an increase in the need for wise decision making, and a high-level technology. In analyzing and relating these characteristics, Finn concurred with leading writers that automation brings about conditions which cannot be accounted for in the experience of mankind. He regarded automation not only as a product of technology but also as a source of fresh creative ideas.

His discussion on the effect of automation on education ranges through such areas as general education, group dynamics, vocational education, sciences, humanities, and the application of useful concepts and processes from the new technology. He states that the complexity of the technological society, its need for specialists, and the unbelievably rapid growth of knowledge are problems we must recognize and come to grips with. In addition to a thorough revision of content there must be development of a means to teach more to more people in less time.

In the second article, "Automation and Education: 2. Automatizing the Classroom—Background of the Effort" (1957), Finn attempts to make the audiovisual field aware that the ground rules for development are rapidly changing. Sensing the fallacy of a blind faith that the audiovisual tradition will continue to grow and influence education as it has since the 1930s, he contends that the audiovisual movement has arrived at a point of serious crisis.

His sober deliberations reveal that the crisis is accentuated by the increasing influence of the Ford Foundation. This influential institution had a specific program for changing educational procedures, methods and organization in the direction of classroom automation with audiovisual communication devices; but its program and policy were both outside the mainstream of the audiovisual movement.

From a general historical perspective, Finn compares and contrasts the developments of the 1930s and up to 1950 with the developments in the 1950s. The difference he found to be both quantitative and qualitative. He shows how the foundation-supported research programs of the 1930s led to the emergence of such leaders in the audiovisual movement as Charles F. Hoban Jr., Francis Noel, and Floyd Brooker, who each provided leadership in the audiovisual programs of the Army, Navy and U.S. Office of Education in World War II.

After differentiating the styles of foundation support, Finn discussed the Ford Foundation's theoretical position for automatizing the classroom. The thinking of Ford Foundation leaders is summarized in seven statements which are followed by a discussion of certain factors implicit in the concept of automation.

"The 'good old days' are gone; approached with intelligence and zest, the days of the future will be better." With this sentence, Finn summed up "Technology and the Instructional Process" (1960), one of his major contributions to education. A new thrust in the form of the National Defense Education Act had radically reorganized the educational setting. The temporary position Finn had taken in the previous paper because of the apparent pervasiveness of the Ford Foundation has been transformed into one of positive leadership and a philosophy of action. Finn does not wring his hands because further development of science and technology is the number one national objective. Instead he challenges us to see that technology is more than invention, more than machines; that it is also a process and a way of thinking.

Turning to education he alerts his readers to the pre-technological position occupied by educationists and the consequent problems of unpreparedness which leave the door open for the neo-technocrats to move into the field of instruction.

In the rationale he proposes, Finn recognizes the accompanying stresses technology introduces, anticipates the need for new theories, and states the need for vigorous analysis of roles and consequences of technology in education.

Two trends are identified, one toward mass instruction and the other toward individualized instruction, with a combination of these trends providing the possibility of total educational automation.

Finn foresaw the current experimentation toward eliminating teachers and school systems through automation. He raised serious questions on these matters and emphasized the potential danger of man as the slave to the machine:

Instructional technology is, no doubt, here to stay. Our problem becomes one not so much of how to live with it on a feather-bedding basis, but how to control it so that the proper objectives of education may be served and the human being remain central in the process.

In relating the concept of negative entropy to education, he gave us Finn's Law which states: *The thrust and energy of technology will force a greater organization upon us at every point at which it is applied to instruction.*

Finn saw organization as the base for programming—the piling up of energy. He stated that: "He who controls the programming heartland controls the educational system. Who is it to be? Will it be done on human terms?"

Prior to closing the article with his philosophy of adventure, he redefined audiovisual education as learning technology. He then restated his position that the profession of education as a whole must be made to sense the powerful movement to instructional technology and be made ready to seize the great opportunity it offers to make all teachers highly professional.

Automation and Education
1. General Aspects

In the year 1869, Thomas Henry Huxley addressed the Liverpool Philomathic Society at a dinner meeting. His subject was scientific education. During the course of his speech, he said, "As industry attains higher stages of its development, as its processes become more complicated and refined, and competition more keen, the sciences are dragged in, one by one, to take their share in the fray; and he who can best avail himself of their help is the man who will come out uppermost in that struggle for existence. . ." (Huxley, 1896, p. 114).

Huxley was a mighty prophet in many ways, but never was his crystal ball more clear than that night slightly less than 100 years ago. Over the century since he spoke, the sciences and their offspring technologies have, indeed, one-by-one and two-by-two, entered the arena of industry (and of military science) until practitioners of the most unlikely sciences can be found busy on Madison Avenue, in the General Motors Research Institute, or in the Pentagon Building.

Somewhere along the line, the First Industrial Revolution—which was the product of inventors—became the Second Industrial Revolution—which is the product of scientists and technicians. We are told on all sides by competent observers that we are just beginning to reap the harvest of the Second Industrial Revolution; that we have, in fact, just entered the era; that atomic power, rocket engines, and automation, for example, are in their technological infancies.

Of all of these phenomena which have resulted from the application of science to industrial and military problems, automation is, or should be, of greatest interest to the professional educator. The problems and promises inherent in automation are directly related to education. Educators must re-examine science and technology in the second half of the 20th century, keeping in mind Huxley's phrase that ". . . he who can best avail himself of their help is the man who will come out uppermost in that struggle for existence. . . ."

What is needed in 1957 and for many years to come is (a) a thorough assessment of the nature of the new technological world and its implications for education; (b) a revision of educational organization, procedures and content, which will close the gap between the rocket-like progress of technology and the snail's pace of education; and (c) the application of

Reprinted from *Audio-Visual Communication Review*, Vol. 5, No. 1, Winter 1957, pp. 343-360.

1. General Aspects

useful concepts and processes from the new technology to the educational enterprise itself in order to close this ever-widening gap.

Such a procedure will be a serious and difficult matter. It is not here suggested that educators take up a new fad. If automation and its related concepts are grasped by eager curriculum workers, school administrators, and professors in teacher education institutions, at the level of understanding and performance that some of the recent fads in education have been handled, we shall be worse off than before. We emerged from the "group dynamics" fad witnessing hundreds of people at various levels in education talking about a process they didn't understand and, fooled by their acquisition of the vocabulary, attempting to do things with group processes and with the results of group dynamics research that couldn't be done—as well as not doing other important things that should have been done.

We are still gripped today by a "three Rs" and a "gifted" fad which, presumably, will be dropped at the introduction of some other new idea that will make "experts" of its first practitioners long before they have an inkling of what they are either doing or talking about. Automation cannot be treated this way. To repeat, it is a serious and difficult business and offers no haven for the educator looking for something with which to impress the Teachers' Institute next fall.

Inasmuch as the audiovisual approach to communication is, strictly speaking, an attempt to solve a difficult human problem with technology, and, inasmuch as ideas have recently been advanced in several quarters that audiovisual technology might well become the means by which the process of education itself might become "automated,"[1] it follows that students of audiovisual communication must lead in the examination of the relationships of automation and education. The writer proposes to do this under three general headings, this paper covering the first. The three headings are:

1. General Aspects.
2. Automatizing the Classroom.
3. Built-In Dangers.

Future papers will deal with the other two topics.

In considering the topic of automation and education from the audiovisual point of view, there is a temptation to go directly to the central question continually raised by Dr. Eurich, namely: Can we, with audiovisual materials and devices such as film or TV, extend the range of the master teacher and replace the mediocre or nonexistent teacher? However, because of the complex nature of the automation concept and because it will have other effects on education which, in turn, will affect the audiovisual movement, it is necessary to begin the study by analyzing the more general relationships of automation with education. This background study is the theme of this article.

The idea that the Second Industrial Revolution and automation, one of

---

[1]See, for example, numerous speeches and statements by Alvin C. Eurich, Executive Vice-President of the Fund for the Advancement of Education, including his 1956 Statement to the Association for Higher Education.

its main products, will have a profound effect on education is not exactly new. Statements abound to this effect, including one of the author's (Finn, 1955). However, the existence of ideas and their immediate or even consequent effect on educational practice is very difficult to establish, fads excepted. There is, as is well known, generally thought to be a 50-year lag between idea and practice. With the speed of the new technology, symbolized by the space satellite, this 50-year wait simply cannot be tolerated. We are living in an era of fantastic change and everyone in education, from the superintendent to the janitor, must recognize this fact and be prepared to act upon it.

Dean Irving R. Melbo, in addressing a conference of school administrators, cited a need at the administrative level for educational statesmen. He then went on to define the characteristics of such a statesman, including the ability to recognize issues. In speaking to this point, he emphasized this modern problem of change in education:

> The ability to recognize issues is all too often lacking among our educational administrators. But perhaps the fault lies with the institutions which trained them, because no man can recognize issues if he has not learned to think in terms of issues.
>
> The simple truth is that great economic and social forces flow with a tidal sweep over communities that are only half conscious of that which is befalling them. Wise statesmen are those who foresee what time is thus bringing, and endeavor to shape institutions and to mold men's thought and purpose in accordance with the change that is silently surrounding them. (Melbo, 1954, p. 2)

And, as the Second Industrial Revolution with its automatic controls so sweeps silently into our communities, all schoolmen would do well to remember the words of Whitehead, written over 20 years ago:

> Our sociological theories, our political philosophy, our practical maxims of business, our political economy, and our doctrines of education, are derived from an unbroken tradition of great thinkers and of practical examples from the age of Plato . . . to the end of the last century. The whole of this tradition is warped by the vicious assumption that each generation will substantially live amid the conditions governing the lives of its fathers and will transmit those conditions to mould with equal force the lives of its children. We are living in the first period of human history for which this assumption is false. (Whitehead, 1933, p. 116)

Automation is one of the reasons why the immediate future will bring conditions that cannot be accounted for in the experience of mankind. Automation will radically alter our system of education. It will be well, if we are to recognize and study these issues, to attempt to define this phenomenon.

Some Definitions

Automation is one of those concepts capable of several definitions. This is, no doubt, due to the fact that its meaning has not yet been fully developed. At present, automation is a term better described than precisely defined; automation represents a territory that can be mapped, but not surveyed to the last decimal point. Drucker, in his very illuminating discussion, states that "Automation can be defined simply though superficially as the use of machines to run machines" (Drucker, 1955, p. 41).

Macmillan indicates the range of meaning inherent in the term. He speaks of:

> . . . a great extension in the use of automatic devices, to take over the simpler kinds of work previously done by the men in charge of machines. The word "automation" has been applied to this process and widely adopted. It was first used several years ago by engineers of the Ford Motor Company in the United States to describe their methods for automatically conveying workpieces between successive machines; but since then it has come to have a much wider significance, implying any process in which the lower functions of a human operator—both physical and mental—are taken over by self-acting devices. Interpreted in this way, the word "automation" clearly means more than mere mechanization. . . . (Macmillan, 1956, p. 2)

The CIO Committee on Economic Policy refers to a three-level description developed by Baldwin and Shultz of M.I.T. The three levels are:

1. The level of *integration* or "*Detroit automation*," which means the linking together of conventional machines so that the work proceeds from one to the other without human intervention (referred to by Macmillan in the statement about Ford engineers above).

2. The level of "feedback" control devices, which means the introduction of servo-mechanisms or other devices which operate machines without any human control. Adjustments of work output, etc., are governed by an automatic comparison with some standard with the information so derived fed back to the operating device to further govern its work.

3. The level of information control, which means the introduction of computing machines into the system. The computing machines record and store information, make computations and feed the results back into the system, establishing complete control of the whole operation (CIO Committee on Economic Policy, n.d., pp. 3-4).

The United Automobile Workers, reflecting the concern of their president, Mr. Walter Reuther, have conducted extensive studies of automation. In a special report to the union, the UAW Education Department referred to automation by its characteristics, stating that automation involved:

1. *Information* that is fed to a

2. *controller*, which operates (and supervises to make sure the operation is done correctly)

3. *a servomechanism* which in turn operates the

4. *machine*, which can be a drill, a milling machine, a lathe, a typewriter, or anything else which serves a purpose, and all this happens in a self-correcting system which is known as a

5. *feedback circuit* (UAW-CIO Education Department, 1955, pp. 30-31).

Examples of Automation

As will be shown later, the descriptions above represent only the hard-core, mechanical attributes of automation, and indicate that the concept, as it is developing today, has abstract, ideational properties which may be even more important. However, these abstract properties are not possible without the solid grounding in industrial and scientific developments. To increase understanding, then, of the basis for these descriptive definitions, some examples of automatic production may be helpful.

The following examples, for the most part, are taken from the CIO report. Others could be cited. The Cleveland plant turning out Ford engines under the "Detroit automation" system links a battery of machines on a line 1,500 feet long with automatic machine tools performing over 500 operations. The plant processes 154 engine blocks an hour with 41 workers. The same operation, under older mass-production methods, would take 117 workers.

The *Wall Street Journal* is the authority for the statement that Raytheon's television and radio division plant will be able to turn out 1,000 radios a day with only two workers on the line, as compared with a former need of 200 workers. Office automation (this involves computers and other machines, but also structured planning) is shown in the John Plain mail order house in Chicago, where a computer handles 90,000 tallies a day in keeping track of an 8,000-item catalog. Insurance companies now going to computer operation are represented by one cited which makes 7,000 daily billings, 20,000 monthly calculations for dividends, and 130,000 calculations for agency commissions at a saving of from 175 to 225 clerical personnel. Finally, it is reported that General Mills has produced an automatic machine which will produce electronic equipment. This machine, it is claimed, will require only two workers and a supervisor and has a capacity of over 200,000 assemblies a month.

While these are examples of automation in office and factory, it should be emphasized that they are, as yet, examples. Outside of modern oil refineries, which are almost fully automatic, there are practically no fully automatic plants yet in operation. As John I. Snyder Jr., president of U.S. Industries, points out, many industries are coming closer to it and have already automatized one or more *parts* of the production process (Snyder, 1955). His own U.S. Industries was, in 1955, operating the only fully automatic factory—a government-owned, 155mm shell plant in Rockford, Illinois. However, a recent study by the Research Institute of America of 1,000 major companies found that 16 percent saw automation as a factor in their industry at the present time and that an additional 23 percent predicted extensive automation in their industry within a decade. Only one-third of the respondents felt that automation would never affect them (Bund, 1955).

Characteristics of Automation

When automatic operations of various kinds are looked at together, certain characteristics seem to emerge. These characteristics are not mechanical or technical in a machine sense. Rather, they are conceptual, and, as such, have educational significance.

The first, and probably the most important, characteristic of the new industrial revolution is the concept of *systems*. While this concept has

1. General Aspects

been discussed before in connection with audiovisual administration, its pervasiveness in business and industry—in any large-scale organization—has yet to be realized. The systems idea is best grasped in connection with an office which is undergoing automation. While many processes—billing, posting, etc.—will be done by automatic machinery in such a set-up, the important elements are the planning necessary for the whole procedure, the location and function of necessary personnel, the types of forms used, etc. A system has to be created which gears all of these elements into a smooth flow, with controls feeding back information throughout the process. The system is much more important than the computer which calculates interest rates, for example. A representative of a firm which installs automatic processing in offices was heard to remark recently that it would take his firm eight years to install a system in a large company.

The problem of systems and systems analysis is complex and cannot be treated here at any length. Much automatic machinery can do certain routine work infinitely better and faster than man. Aircraft warning devices are an example of this. But somewhere in any operation, there must be men. Men, machines and information have all to be coordinated into a system which may begin when a prospect is furnished an insurance salesman or a blip appears upon a radarscope, and ends with the widow receiving a check or the refueling of an interceptor plane after its mission has been accomplished. No longer can the in-between operations be considered piecemeal. Automation means system integrated with system.

Secondly, such systems and such potentials for rapid production need control. The old fairy story of the salt-making machine that went on and on grinding salt could easily be translated into a modern engineer's nightmare. There are many elements here, and at different levels. At the level of the mechanics of the production line, information is needed constantly as to the condition of the machines and the work; at the marketing level, information is needed as to style, inventory, price, etc. If a machine gets too hot, information must come back to a control point and orders issued for lubrication, reduction of speed, or some other remedy; if too many of the wrong color widgets are popping off the end of the line, information again must go to a control point so that fewer pink widgets and more blue ones are produced. The constant here, and a general characteristic of automatic operation, is the flow of information and the control of the system by that information.

The third general characteristic of automation springs logically from the two which precede it. The systems and control characteristics of automation demand *thorough analysis and long-range planning*. The huge capital involved, for example, in automatizing an operation[2] means that the operation cannot be carried on by caprice. Once an automatic operation is underway, it is very costly to stop it for extensive changes. Hence, accurate planning is required. This planning is, obviously, one of the main factors in the eight-year installation problem described above in

---

[2]According to a story in the *Los Angeles Times*, the Farmers Insurance Group in Los Angeles recently installed an IBM 705 computer to control automatic operations. The cost was quoted as a rental of $36,000 per month.

1. General Aspects

connection with another insurance firm.

Planning, however, cannot proceed without analysis. Neither can a system function indefinitely without analysis. This is the reason that the terms "operations analysis" or "operational research" are closely allied to the development of automation, and, for that matter, of all modern technology. The term "operational research" is usually used to refer to the study of management problems from the point of view of the mathematician, the physical scientist, and the biological scientist. The operational research staff, for example, in a military headquarters, attempts to strip a problem down to its bare essentials, create a mathematical model of it, and solve for the best answer. As Dr. John Abrams of the Royal Canadian Air Force pointed out, in military operational research use is made of such things as probability statistics, the theory of games, search problems, and the newer theories of value and decision (Abrams, 1956). These, of course, are all in the realm of mathematics. Methodologies from other sciences are also used.

While there is no doubt that the armed forces of several countries have pioneered in the area of operational research, the term is now being heard in reference to similar problems in industry. The very existence of computers as part of the automation process makes possible the solution of complex mathematical problems relating to operational research in a very short time. For example, in a recent issue of the *Harvard Business Review*, Gaumnitz and Brownless review the mathematics that should be required of managers (Gaumnitz and Brownless, 1956). Decision making in all forms of automation processes and complex systems requires long-range planning and operational analysis.

The mention of decision making in connection with automation highlights the fourth characteristic of the process which is *decision making*. Drucker points out that, in order to control such processes and systems and to make and carry out the necessary plans, many more better educated managers will be needed in the future (Drucker, 1955). Decisions dealing with a complex system will have complex consequences. The implications here for education are tremendous.

Finally, although it should be obvious, it is necessary to point out that automation and its clusters of systems is a high-order product of technology. That is, it is a *highly technical* business. While the operations themselves may be reduced to watching an occasional gauge or checking the accuracy of a bill for an insurance premium that has just come through a machine, the machines themselves and the systems within which they operate are exceedingly complex. Maintenance, for example, will require very able technicians.[3] The engineering behind some of the equipment involved in automation also presents great problems of design.

These, then, are some of the characteristics of automation and its associated processes and systems: (a) the concept of systems, (b) the flow and control of information, (c) scientific analysis and long-range planning,

---

[3]For anyone who has not experienced this reality, let him see an ordinary automobile mechanic attempt to solve the problem of making a modern automatic transmission function when it has been jammed or otherwise damaged.

(d) an increase in the need for wise decision making, and (e) a high-level technology.

Effects on Education

Those who sense that all social and economic forces affect education in general, as well as specific ways, will realize that a list of the effects of automation on education could be almost endless. Obviously, there are concerns about technological unemployment, even if it may be temporary, and the age structure of our population for the next 20 years will bring less people into the labor force than we need. The need for maximum productivity for prosperity placed against a reduction in the labor force will hurry, rather than stay, the development of automation. A shorter work week seems to be inevitable. Industry will be interested, as never before, in a predictable expanding stable market—the risks are too great in capital, etc. to break down an automated system. This would imply that industry may have a stake in some kind of an annual wage. The initial high costs of capitalization for completely automatic production may operate to further change our economy in the direction of more big business. One authority has suggested that, with a great sameness of product possible from automatic factories, the day of the artisan and hand-worker may once again bloom (RCA, 1956).

All of these and many more possible social and economic effects have educational implications. It is too early to say what many of them will actually turn out to be. However, there are a number of more direct effects that can be identified and these effects will be pronounced at all levels of education. Consider carefully the following quotations:

. . . Automation's most important impact will not be on employment but on the qualifications and functions of employees.

And this increase both in the numbers of managers and in the demands made on them may well be the largest of all the social impacts of Automation.

One large manufacturing company (now employing 150,000) figures that it will need *seven thousand* college graduates a year, once it is automated, just to keep going; today it hires three hundred annually.

But the need is above all qualitative—for *better educated* people. The "trained barbarian," the man who has acquired high gadgeteering skill, will not do. Even in routine jobs, Automation will require ability to think, a trained imagination, and good judgment, plus some skill in logical methods, some mathematical understanding, and some ability well above the elementary level to read and write—in a word, the normal equipment of educated people. Under Automation, a school could do a student no greater disservice than to prepare him, as so many do today, for his first job. If there is one thing certain under Automation, it is that the job—even the bottom job—will change radically and often. (Drucker, 1955, pp. 44-45)

In the conditions of modern life, the rule is absolute. The race which does not value trained intelligence is doomed. Not all your heroism, not all your social charm, not all your wit, not all your

1. General Aspects

victories at land or at sea, can move back the finger of fate. Today we maintain ourselves. Tomorrow science will have moved forward yet one more step, and there will be no appeal from the judgment which will then be pronounced on the uneducated. (California Teachers Association, 1955)

Information theory is the science underlying all exchange of information, and is of especial importance in the development of telecommunications and automatic control systems. The treatment adopted by the author has enabled him to give *a clear account of recent developments without requiring a knowledge of mathematics in excess of that which may reasonably be expected from the professional engineer or physicist.* (From a book jacket—italics mine.)

The First Industrial Revolution, the revolution of the "dark satanic mills," was the devaluation of the human arm by the competition of machinery. There is no rate of pay at which a United States pick-and-shovel laborer can live which is low enough to compete with the work of a steam shovel as an excavator. The modern industrial revolution is similarly bound to devalue the human brain, at least in its simpler and more routine decisions. Of course, just as the skilled carpenter, the skilled mechanic, the skilled dressmaker have in some degree survived the First Industrial Revolution, so the skilled scientist and the skilled administrator may survive the Second. *However, taking the Second Revolution as accomplished, the average human being of mediocre attainments or less has nothing to sell that is worth anyone's money to buy.* (Quoted in *Saturday Review*, January 22, 1955, p. 13—italics mine.)

These quotations are enough to give any thoughtful educator pause. Some of their possible implications and the implications in general of automation may be discussed under four headings:

1. The curriculum and program of studies.
2. Administration.
3. The possibilities in automatizing the process of education itself.
4. Changes required in theory.

These will be briefly delineated.

The first problem under the program of studies category is the role and the placement of subject matter. It can be suggested that those educators who relegate "subject matter" to limbo or who suggest that "we only work through subject matter to something else" are echoing the clichés of the 1920s without meaning in 1957. For certain groups of students, subject matter *per se* will have a new importance. This is not to suggest that all the gains in dealing with individual persons, in attempting to change behavior through problem solving, etc., need be thrown out. It does suggest that a technological society needs technical information in the possession of a rather large group of its members.

Specifically, mathematics will continue to play a greater role than ever in our lives. A thorough revision of the classical mathematics curriculum at the secondary level is called for, in addition to a strengthening of the program at the elementary level. Solid geometry, for example, is, according to many modern mathemeticians, an interesting but obsolete subject and should be dropped. The high school mathematics program can be revised (and this is being done in a few places) so that, for example, the elements of calculus will be taught to *some* students in high school.

The science phases of the elementary curriculum need revision and strengthening. In many elementary schools in the United States, science instruction is a joke. If so, this may be a pretty high price for our society to pay for a few laughs. Actually, at least for some children, the self-contained classroom in the upper-elementary level with its extreme emphasis on social living should be due for a thorough going over. It may be that for some children, we will want some form of departmentalization with a pushing downward of some of the elements of the present high school curriculum. This will mean an earlier transition from a developmental curriculum than most people in elementary education are now willing to make.

Another concept that must be re-examined—and at the very highest levels of education—is the concept of general education. The last 20 years have brought great pressures for general education—for an education so that we may appreciate our common heritage, etc., etc., etc. Survey courses, Great Books, the Harvard essentials, "exploration" courses, and work-experience programs are all examples of this sort. There has been a great deal more worry among the Mark Van Dorens, Mortimer Adlers, and A. Whitney Griswolds about whether or not our technicians, engineers and scientists could talk to each other than whether or not they were good technicians, engineers or scientists (I include the social sciences here). The spokesmen for general education have, actually, attempted to create a *specialist in general education* under the guise of dealing with the well-known ignorance of specialists.

It is a moot point whether or not a highly technical society can survive and maintain itself without highly specialized technological know-how. It takes time to acquire this know-how. Without decrying general education, it would seem that this extreme emphasis upon it should be re-examined with a view to reducing the time devoted to it. It is recognized that this is in opposition to Drucker's position that the managerial talent of the future will have to have a general education. It is my feeling that Mr. Drucker has become enmeshed in his own net in this instance. Management is, or will become, a highly skilled profession. Mr. Drucker does not want general education; he wants general professional education, and the distinction is important.

The implication of a few paragraphs above—that not all students should have the same education—can be stated in the proposition that education for an age of automation will require homogeneous grouping of students in the public schools and recognition of different patterns of preparation. This is not necessarily to say that an intellectual elite will be created after Grade 4, although this is possible.

The well-known facts of psychology relating to the extreme variations in individuals and the fact that individuals may be homogeneous in one respect and not in others will require a re-examination of even the principle of homogeneous grouping. But it is still reasonable to expect that not all high school students will be able or find it necessary to take the ele-

1. General Aspects

ments of calculus. The great problem here is to find ways and means of shortening and improving the general education periods so that all students can function as citizens, parents, etc., and, at the same time, meet the demands of a technological society for specialists (Gardner, 1957). Perhaps the greatest educational dilemma of our time may be stated as follows: *If these two objectives come into complete conflict, a technological society, in order to survive, has no choice but to pick technology.* We may not like the world we so create (and some possibilities in that direction will be discussed in part three of this series), but we must recognize it for what it is.

Computers, cyclotrons and magnetrons—all paraphernalia of the age of automation—point up another great fact that is powerful enough to shake the structure of the whole curriculum from the kindergarten to the graduate school. I refer here to the almost unbelievably rapid growth of knowledge. This is a problem which most curriculum people have ignored by the simple process of ducking their heads in the sand of "attitude formation" and "problem-solving skills." Incidentally, it has also been completely ignored by their opposite members of the *Saturday Review* set who still think that Aristotle had no need to count the number of teeth in his wife's mouth. The poets, the Greek plays, the socio-drama, and the resource unit do not necessarily have the answer to this problem. The solution must lie in thorough revision of content and the development of means to teach more to more people in less time.

The systems concept and related matters emphasize the fact that, while group dynamics has not yet got beyond the stage of a primitive religion in most curriculum circles, it has the key to many problems that will beset an automatized world. Human relations, group relations, even international relations become proper subjects of study and necessary areas in which to develop real skill and proper attitudes. Again, here, the problem may be methodological. The solution will not be found in the technique used by a professor of education in holding a "buzz session" or in having committee reports made in class—any more than it will be found in some of the so-called "workshops" run by school districts under the euphemism of an in-service training program. We need to develop content and method which yield measurable results in this field.

There are many other implications for a curriculum content in an age of automation. To create decision makers means that the thought processes of students must have experienced conflicts and made decisions; to analyze, to plan, to create systems, and to live within these patterns implies an ability to think in patterns, to take the long-range view, to be creative. Those information-mongers, those fact-purveyors who felt that the position stated above with regard to subject matter was their meat had better keep this fundamental in mind. No mountain of facts will do more than provide the background for this process. Our new curriculum must do so, but not in any sloppy way, based on the assumption that if the children learn how to solve the problem of waste paper in the cafeteria over a semester's period of time, they have demonstrated outstanding ability in pattern thinking.

The problems of additional leisure time, the problems of vocational training in a world where vocations will change and buttons or magnetic tapes take their place all need consideration for the new curriculum. The best that we can say is that we have had curriculum challenges before, but never of the intensity and urgency that the coming age of automation presents us.

The second category of possible effects is that of administration. As I have discussed some of these problems elsewhere (Finn, 1956b, 1956-57, 1956a), I will merely enumerate them here. The new management concepts—systems, operational research, the use of the theory of games, etc.— all have much to offer school administrators. The theory of school administration—in my opinion already somewhat creaky—needs a thorough re-examination in the light of these new developments. It is not overstating the case to say that education for an age of automation needs an administrative theory that is at least able to keep up with it. This is a great challenge to theorists in the field of administration and to practicing administrators. If administrators are to lead, it would be well to begin by streamlining their own operations to fit the demands of a new world.

Third, much needs to be done in considering how far we may go in automatizing the process of education itself. Here, as was mentioned before, lies the direct relationship between the audiovisual field and the concepts of automation. This topic will be the subject of the second of these studies. (See part two of this series). Can the somewhat cold process of closed-circuit television, for example, increase educational efficiency without losing something in the process?

Finally, when all this is added up, what changes can be expected in our intellectual structure—in the context of "education" itself? Certain things have been mentioned—a re-examination of curriculum theory, especially as it applies to the program of studies and of administrative theory. It is to be expected that new bodies of such theory will be created and studies conducted on hypotheses based upon these theories.

Actually education, as a subject of study, has stood on a plateau for about 20 years. The audiovisual field is one exception to this statement. The curriculum books still say much the same thing; the speakers at institutes and meetings are still reworking the old material. Education, as a discipline, is in the position it was around 1900 when a new graduate content was created almost overnight by Thorndike, Dewey, Monroe, and a few others.

We are on the edge of a breakthrough. It will, in fact, be forced upon us by circumstances if we do not make it ourselves. A new content for education as a discipline needs to be made. Already, beginnings can be detected. For example, most of our psychology—and certainly all of it that reaches the teacher-training levels—is individual learning psychology with overtones of growth and development. Social psychology—the study of the behavior of people in groups, their responses to mass communications, etc.—has emerged as a separate discipline. Much can be gained by working pertinent material from this important field into teacher education programs. (Again, the audiovisual field has tapped social psychology more for its data than any field in education.)

New philosophies of education need to be created. The recent NSSE

1. General Aspects

Yearbook (1955) is rich in suggestions in this matter. All of the remaining social sciences—anthropology, political science, international relations, etc.—can contribute much. Experiments in the biology of the mentally retarded, new advances in psychiatry, and the new stress theory of disease can all give a little to the new content in education. We need now to work with experts in these disciplines and get the material flowing into our thinning lifeblood that still carries the weak corpuscles of "life-adjustment" and "purposive activity." The age of automation will require it.

One final word as to the content of education as a discipline. Much has been heard in recent years from critics of education, such as Arthur Bestor, that "there ain't no such thing" as a content in education. Bestor's position, of course, is untenable. However, it has, no doubt, been true that some courses in education have been sadly lacking in the content that was there for the asking. The new content that the age of automation suggests will be much more demanding intellectually than that we already have. The problems of teacher education will become, therefore, more acute. No true profession in the age of automation will be able to slough off its intellectual obligations. This includes teachers now who consider a course difficult when it has two 20-minute examinations.

Summary

An attempt has been made in this paper to describe automation, both as an industrial process involving flow and control and as an intellectual system involving planning and decision making. General effects on education in the areas of curriculum and program of studies, administration, general education, organization for instruction, the necessity for teaching new patterns of thinking, vocational and technical education were discussed. The concept of the possibility of automatizing the instructional process itself was introduced, and the outlines of a new content for education as a discipline were suggested. The following paper will deal with the problem of automatizing instruction through audiovisual means, and the third and last paper in the series will consider the dangers and difficulties involved in such a process.

References

Abrams, John W. "Military Application of Operational Research," Paper delivered to the Tenth Annual Convention of the American Society for Quality Control, Montreal, Canada, June 1956, pp. 107-113.

Automation and Technological Change. Report of the Joint Committee on the Economic Report, U.S. Congress, U.S. Government Printing Office, 1956.

Bowden, B. V. (Ed.). *Faster Than Thought: A Symposium on Digital Computing Machines.* London: Sir Isaac Pitman & Sons, Ltd. 1953.

Bund, Henry. "I.R. II: Yes or No?", *Saturday Review*, January 22, 1955, p. 21+.

California Teachers Association *Report Card*, November 1955.

CIO Committee on Economic Policy. Automation. Washington, D.C.: The Union, n.d.

1. General Aspects

Cleator, P. E. *The Robot Era.* New York: Thomas Y. Crowell, 1955.

The Coming Revolution in Industrial Relations: 1955-1975. New York: Industrial Relations News, 1955.

Davies, Daniel R. "The Impending Breakthrough," *Phi Delta Kappan,* Vol. 37, No. 7, April 1956, pp. 275-281.

Drucker, Peter F. "The Promise of Automation: America's Next Twenty Years, Part II," *Harper's Magazine,* Vol. 210, No. 1259, April 1955, pp. 41-47.

Dyer, Henry S.; Robert Kalin and Frederic M. Lord. *Problems in Mathematical Education.* Princeton, N.J.: Educational Testing Service, 1956.

Eurich, Alvin C. "Better Instruction With Fewer Teachers," *Current Issues in Higher Education, 1956.* Washington, D.C.: Association for Higher Education, NEA, pp. 10-16.

Finn, James D. "A Look at the Future of AV Communication," *Audio-Visual Communication Review,* Vol. 3, No. 4, Fall 1955, pp. 244-256.

Finn, James D. "What is Educational Efficiency?", *Teaching Tools,* Vol. 3, Summer 1956, pp. 113-114. (a)

Finn, James D. "AV Development and the Concept of Systems," *Teaching Tools,* Vol. 3, Fall 1956, pp. 163-164. (b)

Finn, James D. "Teacher Productivity," *Teaching Tools,* Vol. 4, Winter 1956-1957, pp. 7-9.

Gardner, John W. "The Great Hunt for Educated Talent," *Harper's Magazine,* Vol. 214, No. 1280, January 1957, pp. 48-53.

Gaumnitz, R. K. and O. H. Brownlee. "Mathematics for Decision Makers," *Harvard Business Review,* May-June 1956, pp. 48-56.

Hurwicz, Leonid. "Game Theory and Decisions," *Scientific American,* Vol. 192, No. 2., February 1955, pp. 78-83.

Huxley, Thomas H. *Science and Education.* New York: D. Appleton and Company, 1896.

Macmillan, R. H. *Automation: Friend or Foe?* London: Cambridge University Press, 1956.

Melbo, Irving R. "Wanted—Educational Statesmen." Speech delivered before the Summer School Administration Conference, University of Southern California, July 9, 1954. (Mimeo.)

National Society for the Study of Education. *Modern Philosophies and Education.* Fifty-Fourth Yearbook, Part I. Chicago: The University of Chicago Press, 1955.

Radio Corporation of America. "Automation May Bring Broad Revival of Skilled Artisans, Dr. Ewing Says," News release by the Corporation, September 20, 1956.

Saturday Review. Special issue on Atoms and Automation, January 22, 1955.

Scientific American Editors. *Automatic Control.* New York: Simon and Schuster, 1955.

Shultz, George P. and George B. Baldwin. *Automation: A New Dimension To Old Problems.* Annals of American Economics. Washington, D.C.: Public Affairs Press, 1955.

Snyder Jr., John I. "The American Factory and Automation," *Saturday Review,* January 22, 1955, pp. 16-17+.

1. General Aspects

UAW-CIO Education Department. *Automation.* Detroit: The Union, 1955.

Whitehead, Alfred North. *Adventures of Ideas.* New York: The Macmillan Company, 1933.

Wiener, Norbert. *The Human Use of Human Beings.* New York: Doubleday, Anchor, 1954.

Wiener, Norbert. "Eight Years of Cybernetics and the Electronic Brain," *Pocketbook Magazine No. 2,* New York, 1955.

Automation and Education
2. Automatizing the Classroom—
Background of the Effort

The first paper in this series developed the background of that phase of the Second Industrial Revolution known as automation. It also attempted to explore generally the current and future relationships that exist or will exist between education as an institution and automation as both a complex of industrial processes and a concept of wide application to human affairs. This general study attempted to supply background so that a closer look might be taken at the application of the concept of automation to education itself.

As indicated in the first paper, the student of audiovisual communication has a special interest in any proposals that the educational process become "automated" or "automatized." The majority of these proposals have suggested that audiovisual devices, particularly television, are the means by which many pressing problems of education can be solved. In general, these proposals have been made by persons or agencies outside of the mainstream of the audiovisual movement as it has developed in this century. Much of the major work now being done in exploring this field is being carried on by groups that have not been closely associated with previous audiovisual developments. Floyde Brooker, executive secretary of the Department of Audio-Visual Instruction, NEA, has recently pointed out that " . . . Many superintendents these days are talking about educational automation (films, television etc.) seemingly without realizing that this *is* audio-visual education" (Brooker, 1957, p. 64).

It will be the purpose of this paper to present to the audiovisual field a statement of the rationale surrounding the proposals that audiovisual communication devices and techniques can bring automation into the educational process and thereby solve many of the most perplexing of educational problems now and in the future. This paper will discuss educational (audiovisual) automation under two general headings:

1. The institutional background.
2. The theoretical position.

The Institutional Background

In the decade between 1930 and 1940, two institutions can be said to have given the audiovisual movement great impetus. These institutions were the Payne Fund of New York and the Rockefeller Foundation, prin-

Reprinted from *Audio-Visual Communication Review*, Vol. 5, No. 2, Spring 1957, pp. 451-467.

cipally through one of its funds known as the General Education Board. The so-called Payne Fund studies of the effects of the theatrical motion picture on children and youth and the subsequent motion picture appreciation movement are now classics, not only for students in the audiovisual field, but in the field of social psychology as well.

The work of the Rockefeller Foundation is, perhaps, not as well known, but was tremendously influential. Probably the most important project created by the Rockefeller grants was the Motion Picture Project of the American Council on Education. Here, Charles Hoban, Floyde Brooker, and Francis Noel, to name but a few, discovered and perfected techniques that later almost totally governed the audiovisual programs of the Navy (Noel), the Army (Hoban), and the U.S. Office of Education (Brooker) during World War II. Those who have ascribed the startling developments in audiovisual education post-1945 to "the Armed Services" would do well to review the history of the four years immediately preceding the war when the American Council Project was in full swing.

Other projects financed by the General Education Board and the Rockefeller Foundation included the Evaluation of School Broadcasts project at Ohio State University under the direction of I. Keith Tyler, the Minnesota motion picture production project under Robert Kissack, the Wisconsin Radio Study under Lester W. Parker, the human relations film excerpting project under Alice Keliher (with an assist from the Motion Picture Producers Association), a project with recordings, a project with film laboratories which later resulted in EFLA, the Rocky Mountain Radio Council, and several more. In addition, the Foundation trained personnel by providing fellowships for qualified people—permitting them to work in these projects, in the radio networks, and even to study broadcasting in England and in Europe.

This bit of history, not too well known by many people presently engaged in the audiovisual movement, is extremely important in that it established the influence of outside institutions on what was, until the time of these projects, a halting and almost ineffectual effort to bring audiovisual devices into American education.[1] It can be said categorically that for 15 crucial years (1930-1945) the audiovisual movement was influenced critically by two institutions—the Payne Fund and the Rockefeller Foundation. One additional point should be made clear, however. The Rockefeller Foundation, in particular, had no *specific* program for American education. The influence it provided was simply that it made money available for research, development and the education of personnel. The result of all this activity—much good, some indifferent, a little useless—was the responsibility of the people and institutions who received the money. *The result, how-*

---

[1]The Carnegie Foundation's Committee on Scientific Aids to Learning should not be forgotten in this account, and the contributions of earlier philanthropic grants such as the Commonwealth study (Freeman) and the Eastman Kodak study (Wood and Freeman) have to be recognized. It is not the writer's purpose, however, to write an entire history of grants for audiovisual research and development. Rather, it is to point out the tremendous influence on the development of the movement resulting from the Rockefeller and Payne contributions.

ever, was that the audiovisual field developed and grew as it never had before.

We come now to the present. The impetus toward this new idea of educational automation via the audiovisual route is also the product of an institution. That institution is the Ford Foundation. Much of the impetus has come from the Fund for the Advancement of Education (now merged back with the parent Ford Foundation) and its able executive vice-president, Alvin C. Eurich. However, the Fund for Adult Education (another subsidiary) has also contributed with its support of educational television and educational television stations. Most of the experimental programs in closed-circuit television have been sponsored by the Fund for the Advancement of Education (in some cases, as at Hagerstown, Maryland, in cooperation with other groups).

There are two differences between the foundation support of the 1930-1940 period and the present. The first has to do with the amount of money expended. Evening out differences in the value of the dollar, the Ford group obviously has invested much more money in its current audiovisual projects than did the Rockefeller group. Secondly, this money has been made available in a shorter time.

However, a still greater difference exists between the two types of foundation influence, and this difference is qualitative and not quantitative. The Ford Foundation operates on a different philosophy than most of the foundations that were created in an earlier day. The Ford Foundation develops a program and sets out to implement it with grants. In the case under consideration—automatizing the classrooms of the nation—the Ford Foundation has set this as a desirable goal and has made grants for both research and development programs in order that this goal may be reached.

This implementation of a goal with money distributed to institutions in locations as far apart as Pittsburgh and San Francisco would seem to be a more efficient way of exerting influence. While the older foundations may have agreed that an area of endeavor was worth support, groups or institutions went to them with proposals, and grants were made pretty much in terms of the proposals *per se*, not in terms of whether or not the proposal fitted a program the foundation had already set up as desirable.

It may seem, upon cursory consideration, that this excursion into foundation philosophy and procedure may be an unwarranted digression in a discussion of the audiovisual automation of education. *Actually, this is probably the most important point that could be considered.* It has been established that the earlier—should we say *laissez faire*—grants had a large influence on the audiovisual movement. It must be soberly realized that, in the present case, we have in operation a tremendously wealthy and influential institution with a specific program for changing educational procedures, methods and organization in the direction of classroom automation with audiovisual communication devices. *It is entirely reasonable to predict that this influence will be greater and will make itself felt in a much shorter time than anything heretofore experienced, including World War II. In studying the concept of automation in the classroom, this is the primal fact.*

The Ford Foundation, similar to other foundations, for the most part does not conduct its activities itself. Leaving out, for the moment, grants to educational institutions, school systems, and such, it has created several other institutions which now become part of the institutional background of the movement toward automatizing the classroom. It would be well, in order to round out this institutional background, to list these institutions and define the areas in which they are working.

According to the 1956 *Annual Report* of the Foundation, more than $11 million had been allocated, prior to the year covered in the report, to educational television in its various phases. During the year covered by the 1956 *Report*, an additional $8 million was added to this amount. Organizationally, the money was allocated as follows:

Education Television and Radio Center,	
Ann Arbor, Michigan	
General Program	$6,263,340
Public Information	90,500
National Association of Educational	
Broadcasters	94,000
Joint Council on Educational Television	140,000
Committee on Television, American Council	
on Education	12,000
Total organizational grants	$6,599,840

(Ford Foundation Annual Report, 1956, pp. 37-39)

The Foundation also has allocated $1.5 million from which to make grants to colleges and universities so that they can release time of faculty members for participation on educational telecasts. This grant is in the form of a matching grant, requiring each individual institution participating to partially ($22,500) match a grant of $37,500. The telecasts involved are apparently both closed-circuit and broadcast.

Mention should also be made of the underwriting of the Omnibus program which, although in a sense commercial television, is obviously educational. The productions of Omnibus by the Ford Foundation's TV-Radio Workshop also pay extra educational dividends. Some 10,000 requests for kinescopes have been received from schools and colleges (*Ford Foundation Annual Report*, 1956, p. 44). Films compiled from past programs are being distributed to additional television stations, and some have been placed on loan with the Educational Television and Radio Center. During the year covered by the 1956 *Report*, the TV-Radio Workshop was underwritten with $1,733,887. Sponsors furnished $1,058,017 in income, leaving a deficit of $675,870 covered by Foundation funds (*Ford Foundation Annual Report*, 1956, p. 193).

Excluding direct grants (mostly through the Fund for the Advancement of Education) to schools and colleges for closed-circuit and special developments, including grants as well as paid-out monies for the items noted above, and evening out the TV-Radio Workshop figures, the Ford Foundation has put approximately $8,700,000 in educational television in a one-year period. Over $7 million of this amount has been placed through organizations, and these organizations now become the dominating influence of this movement. Briefly, the organizations receiving over $6.5 million carry out the following functions:

Educational Television and Radio Center

By all odds, the dominant force in the educational television movement in the United States today, the Center develops and distributes programs to the affiliated ETV stations (six hours a week, kinescope and film, to be increased to ten hours a week); it provides information on ETV to the general public and to interested organizations, consulting services, speakers, etc. Many of the series produced under its subsidies are available in film form for use as audiovisual aids or on closed-circuit installations through a distribution arrangement with the University of Indiana. The Center has a distinguished Board of Directors, a large staff under the direction of Harry K. Newburn, former president of the University of Oregon, and has a program of bringing in, for a year at a time, specialists in film-making or television in the capacity of program associates. With the possible exception of Robert B. Hudson, Program Coordinator of the Center, not one member of the Board or of the staff has had anything significant to do with the development of audiovisual education in the United States prior to the establishment of the Center.

Joint Council on Educational Television

The JCET essentially is concerned with the preservation and utilization of educational television channels. It maintains legal representation in Washington, appears before the FCC, and gives advice to communities and institutions seeking aid in applying for and constructing ETV stations. In general, the members represent the higher institutions and public school systems of the United States and the operation of the JCET is in some way related to that of the American Council on Education, although this relationship is not clear from the literature (The American Council many times serves as the voice of education, particularly higher education, in Washington). The members of the JCET are all distinguished educators, none of whom has ever been associated with the audiovisual movement.

The National Association of Educational Broadcasters

The NAEB was a relatively small association of college and university radio stations until the advent of television and foundation subsidy. Its membership now includes educational institutions that operate either TV or radio stations, institutions which produce programs for such stations, and has recently, by way of a merger, acquired the membership of the old Association for Education by Radio and Television (AERT). It has an imposing list of services to its members—technical and program consultation—and has a wide range of very valuable publications. The NAEB also operates a radio-tape network among educational institutions. Among the officers, names long associated with educational radio seem to be dominant. As of January 1957, Burton Paulu, of the University of Minnesota, was listed as president, and Harry J. Skornia as executive director.

2. Automatizing the Classroom

Committee on Television, American Council on Education

Compared to the three agencies above, the Committee on Television is a small operation. According to the 1956 *Annual Report* of the Ford Foundation, the Committee exists to maintain liaison with "those educators not formally engaged in the educational television movement" (p. 39). Eight of the 14 members are college presidents or other high administrative officers of colleges and universities, two are public school superintendents and the other four represent agencies, associations, and like organizations. The Committee has held several conferences and produced some publications.[2]

The grants to and for these organizations and their programs have had the effect of forwarding the entire educational television effort without necessarily pinpointing the direct classroom, automation idea. The grants for such purposes have been made mainly through a subsidiary fund of the Ford Foundation, the Fund for the Advancement of Education, as noted above. The majority of these grants has been made since the 1952-1954 *Report* of the Fund. In that report, only one grant, $29,300, is reported as being awarded to the Montclair (New Jersey) State Teachers College to investigate the possible uses of television as a classroom aid (FAE *Report*, 1954, p. 90).

Figures are not easily available since the 1952-1954 *Report* and not all accounts of the projects developed report the amount of money involved. Lombard refers to the Fund grant to San Francisco State College as " . . . over a quarter million dollars" (Lombard, 1956, p. 19). Carpenter and Greenhill mention that their first proposal to the Fund " . . . would cost over one hundred thousand dollars" (Carpenter and Greenhill, 1955, p. 11). A search of all major libraries in the Los Angeles area revealed no report from the Fund for the Advancement of Education since the 1952-1954 *Report.* However, an approximate figure was turned up in the *NET News* for May-June 1956—a figure which can be considered reasonably accurate, as the *NET News* is published by an agency subsidized by the Foundation. The quotation reads: "In the past two and a half years, the Fund has awarded 21 grants totalling $1,590,286 to six colleges, eight universities, and four public school systems for experiments in closed- and open-circuit television teaching" (*NET News*, 1956, p. 3).

Accepting these figures as accurate, the total investment of the Ford Foundation and its subsidiaries in educational television in the space of less than four years would amount to over $20 million. This is a very large amount of money, indeed, and would probably amount to almost 20 times as much as was invested by the Rockefeller Foundation and the Payne

---

[2]The situation described above, i.e., that the entire television field is dominated by agencies outside the DAVI, is a serious indictment of the officers of the DAVI (including the writer) and of the national staff. Space does not permit an analysis of the situation from the professional organization point of view and to raise questions why the Ford Foundation and its agencies have decided to ignore or shortcut the existing national educational organization which, for over 20 years, has been concerned in these matters. Such an analysis should become the number one item on the agenda for the DAVI Board of Directors.

2. Automatizing the Classroom

Fund in the earlier decade. The influence of such an investment will, if history is any guide, be very marked. The amount already expended by the Fund for the Advancement of Education alone can be expected to exert greater influence on American educational procedures than the sum total of all the efforts of audiovisual professionals or of the DAVI since its post-war reorganization.[3]

There are, of course, compelling reasons behind this interest and financial support of educational television and of studies and projects designed to explore the possibility of automatizing the classroom. The next section will describe the theoretical position (and certain practical factors) which has guided the efforts of these organizations and groups as they attempt to reorganize, re-orient and otherwise rebuild American educational procedures in the direction of educational automation.

The Theoretical Position

Before outlining the theoretical position upon which all of the effort toward introducing television (and, to a lesser extent, motion pictures) into the educational process is based, it might be well to state explicitly just what is meant by automatizing the classroom. Essentially, the proposal is to introduce entire systems of audiovisual experiences into the classroom *in lieu* of the teacher. Thus, a fifth-grade arithmetic lesson may be taught to large numbers of 10-year-olds in a single school, a school system, or an entire county by one master teacher using closed-circuit television as the medium. In a recent project undertaken by Encyclopaedia Britannica Films in connection with the ETV station in Pittsburgh, an entire course in high school physics was filmed, using a master teacher. These films may be broadcast, sent out over closed-circuit, or projected in individual classrooms. In any case, the films (162 30-minute subjects) would be used to convey the content of the course. Any school district may obtain the services of this expert film teacher for approximately $13,000.

The fundamental basis for this idea is the continuing teacher shortage coupled with the tremendous increase in enrollments at all levels of education from kindergarten through college. This is basic in all statements, speeches and articles appearing on the subject. Nowhere, I suppose, is the proposition more baldly stated than by Harry Skornia, of the NAEB, as quoted by Richard Hull:

---

[3]An interesting numbers game can be played in this connection. If you take Stoddard's figures as to current enrollments in elementary and secondary education as 37 million (Stoddard, 1957, p. 15) and multiply them by the median amount per pupil spent on audiovisual education in the United States in 1953-1954 according to the NEA Research Division ($0.65) (Research Division, 1955, p. 118), the answer comes out about $24.75 million—not too much different than the $20 million figure for the Ford grants. Actually, the figure developed from the NEA per-pupil factor is probably high, since the study received replies from only about one-third of the school districts in the United States. The nonresponding districts were mainly in the smallest categories where no audiovisual programs might be expected. It is probably pretty close to the truth to say that the Ford grants for educational television have been equal to or have exceeded the entire amount of money spent on audiovisual education in the public schools of the United States in a year.

2. Automatizing the Classroom

As NAEB Secretary Harry Skornia says, education's crisis is the firmest basis for ETV's growth. We (the NAEB) were prepared for educational television. We were able to lead the way.

Hull then goes on to expand this point of view:

For better or worse, the rough outlines of a national framework for radio and television education have now been sketched out in a time of population pressure and an increasing array of problems for all levels of education. . . . The question is not whether to use educational television or educational radio in the educational crisis—the question is whether or not the personnel and the facilities available will be *sufficient* to meet the need in the future. (Hull, 1956, p. 34)

Other evidence may be adduced to show that the crisis in teaching is basic to all thinking on the subject. Carpenter and Greenhill, in their report of the research at Pennsylvania State University on closed-circuit television, indicate that they approached the Fund for the Advancement of Education with a proposal for what might be described as "pure" research in the field of television methodology using only segments of some courses for experimental purposes. The reply they received is described as follows:

Officers of the Fund reacted to this proposal by stating the following points either explicitly or by implication:

1. *The Fund's principal concern was with the impending shortage of teachers.* (Italics mine.)

2. Closed-circuit television seemed to have possibilities for making it feasible for a few good instructors to reach and influence large numbers of students. (Carpenter and Greenhill, 1955, p. 11)

Carpenter and Greenhill went on to say that the Fund wanted whole courses taught by television, the feasibility of the medium studied, and the acceptability by faculty members explored.

Readers of this paper are no doubt familiar with the general outline of the facts surrounding the existing and anticipated teacher shortage. However, to construct the rationale for proposed classroom automatization, it would be well to review them briefly. Two major trends are involved:

1. The rise in enrollments at all levels of schooling.
2. The subsequent shortage of teachers.

Two publications of the Fund for the Advancement of Education (1954, 1955) make the enrollment trend crystal clear. Stoddard points out that there are " . . . in excess of 29,000,000 pupils in our elementary schools (by 1965, 36,000,000), over 8,000,000 in our secondary schools (by 1965, 12,000,000), and over 3,000,000 in our colleges (by 1965, 4,500,000)" (Stoddard, 1957, pp. 15-16). These startling figures are compounded by the fact that, in addition to the rise in total enrollment, the *rate* (percentage of those in any given age group) attending school is also rising. For example, 62 percent of the 14-21 age group are now in school, compared to only eight percent in 1900. By 1960, it is expected that elementary school enrollments will be 68 percent above 1946, and by 1969 secondary school enrollments will be 70 percent above the 1954 level (FAE, 1955, p. 12).

College enrollments are somewhat harder to predict. The Ford study used two projections, one anticipating no rise in the *rate* of those attending college, the other taking a rise in rate into account. The report, however, points out that whatever projection is taken, " . . . it is evident that college enrollments are likely to reach double their 1954 level sometime between 1966 and 1971" (FAE, 1955, p. 14).

The tremendous increase in enrollments obviously has a direct relationship to the number of teachers needed and the number that can be obtained. "In order to provide for replacements, expansion and the maintenance of present pupil-teacher ratios, the schools of the nation must find 16 new teachers between now and 1965 for every 10 teachers now on the job. This is the equivalent of replacing all the teachers we now have and finding 60 percent in addition—all within 10 years" (FAE, 1955, p. 19). With reference to college teachers, the figures are somewhere between 16 and 25 new teachers needed by 1970 for every 10 existing today.

The matter of quality of teaching also enters at this point. Where can that many qualified teachers be obtained? And, for that matter, how may they be obtained? The Ford study states: "About one-fifth of all 1954 graduates of four-year colleges entered school teaching. But during the next ten years one-half of all college graduates . . . would have to enter school teaching. . . . Nothing approaching this proportion . . . can be expected to enter teaching" (FAE, 1955, p. 23). Stoddard estimates the shortage of fully qualified teachers at 250,000 in 1965 (Stoddard, 1957, p. 17). At the college level, where the Ph.D. is considered (rightly or wrongly) the mark of qualification, it is estimated that the number of *qualified* college teachers will decline 50 percent by 1970.[4]

Coupled with the teacher shortage is a shortage of buildings and physical plants. While not so directly concerned with the problem, a consideration of physical plant is certainly germane because the nature of the methods used to solve the teacher shortage (class size, television, etc.) would have an effect on both the amount and nature of the plant constructed. The Twentieth Century Fund, in a special study, estimates the need under present instructional conditions for new physical plants in the public schools through 1960 to be in the neighborhood of $30 billion (TCF, 1955, p. 4). Less than half of that amount is now being spent. Capital needs for *public* higher institutions (including dormitories, etc.) were estimated by McDonald at about $15 billion by 1975 (McDonald, 1956, p. 256). As an example from one state, Semans and Holy early this year advised the California State legislature that the capital outlay needed to expand existing units of the University of California and the California State Colleges and to add new units as necessary to meet swelling enrollments would cost approximately $1 billion by 1970 (Semans and Holy, 1957, p. 79, p. 97). Junior colleges were not considered in this estimate.

These are, as the philosophers say, the brute facts of the situation. The word "crisis" has been overworked, but the shortage of teachers and lack

---

[4]This estimate is probably too conservative. It assumes that almost the same proportion of Ph.D.s will enter college teaching as did in the 1930s. The technical requirements of industry and government have, however, changed radically. One of the characteristics of an age of automation is, apparently, a shortage of educated talent. The percentage of Ph.D.s likely to enter college teaching will probably decline against the competition of industry and government.

of plants facing this country in the next 20 years *is* a crisis of the first magnitude. When considered in the context of the cold war, the atomic age, the demands of automation, and the expansion of knowledge, the question can well be asked whether or not a technological society such as we are attempting to build can, in fact, provide its own means for *survival*, not to mention advancement.

The crisis has not gone unnoticed and some efforts at solutions have been made. The picture, however, is mixed. At this writing, federal aid for school construction is bogged down in Congressional disputes; the Federal Housing loan program for colleges and universities is in danger. Some local communities are becoming reluctant to vote more taxes for schools. Gifts by industry to higher education—much of which is in the form of scholarships—are on the increase. Many states and many communities are making valiant efforts to increase school support. On the other hand, the United States Chamber of Commerce and certain widely read columnists, such as Raymond Moley, stoutly maintain that no crisis exists and the whole thing is an invention of the NEA.

Into this scene of crisis has flowed money and counsel from the Ford Foundation and its several funds, from organizations, committees and conferences which have been helped or stimulated by Ford money and ideas. Leading the effort has been the Fund for the Advancement of Education, and the latest statement of the position of the Fund in this matter is an extremely important booklet by Stoddard referred to above (Stoddard, 1957). Implicit in this booklet, in public statements by Alvin C. Eurich, and in the monetary grants made for various projects is that a large share of the solution to the crisis in education must come from automatizing the process through the means of audiovisual devices and techniques.

Basically, the problem has been attacked by scrutinizing the accepted pupil-teacher ratios at all levels and questioning the whole concept of class size. Eurich and Stoddard both insist that in order to maintain quality of instruction (instruction by able teachers), class size must be increased because the number of able teachers will be smaller. Eurich points out that present pupil-teacher ratios are not sacred. Historically, the idea of one teacher to 25 students, he says, " . . . goes back at least to the middle of the 3rd Century. . . . the rule was established by Rabbi Raba . . . '25 students are to be enrolled in one class. If there are from 25 to 40 an assistant must be obtained. Above 40, two teachers are to be engaged'" (Eurich, 1956, p. 11). This ruling probably reflects earlier procedures and, of course, existed long before the invention of printing.[5]

Dr. Eurich further makes the point that research has never proved conclusively that a small pupil-teacher ratio results in more learning (Eurich, 1956, p. 12). This being the case, he and the rather large group of American educators who agree with him feel that class size must be increased in

---

[5]It is interesting to note that the idea that a device (in this case, printing) ought to change methodology is not new with Dr. Eurich. Twenty-five years ago, Nicholas Murray Butler said, "The lecture system as a means of communicating facts should have been dispensed with when the art of printing was invented. . ." (Knight, 1940, p. 261).

order to rescue teaching quality and properly to reward teaching both economically and socially. For example, O. Meredith Wilson, the president of the University of Oregon, stated:

But others may ask teachers to accept more students in return for higher salary on the theory that it is better to be 40 feet from distinction than ten feet from mediocrity. As Charles Johnson remarked . . . if we are forced to employ mediocre teachers in order to keep our classes small, we may only insure that the teacher communicate his mediocrity in an atmosphere of intimacy. (Wilson, 1956, p. 172)

The solutions suggested for increasing class size and the effectiveness of a given teacher include more than mere audiovisual automation of the classroom. The *aide* idea has been given much publicity. This concept is based on the comparison of a teacher's job with that of other professional workers, such as a doctor's job. The doctor now has various assistants who relieve him of routine, subprofessional tasks. Studies have shown that much work in the teaching situation is of a subprofessional nature (Stoddard, 1957, p. 21). Hence, it is felt that teachers can handle larger groups of students for the *professional* phases of their work, allowing aides to carry out routine and clerical functions.[6] The aide program has been widely discussed in the professional and popular literature and, no doubt, most readers are familiar with it. The aide idea might be considered a program separate from that of audiovisual automation except that any consideration of the use of closed-circuit television, for example, presupposes personnel handling the classes at the receiving end. In some of the college experiments, these people have been graduate assistants or other people at the lower instructional levels—aides, really. Further, Mr. Eurich himself has suggested that, part of the time, instruction must be under local people or aides, and makes reference to the fact that "Clusters of people and clusters of jobs must be related to clusters of objectives" (Eurich, 1956, p. 16).

The aide idea, however, is only an adjunct to the concept of the automatic classroom. The automatization of the educational process can occur only through the mass use of audiovisual devices and techniques. This was suggested by Dr. Eurich when he said:

In addition to making better use of teachers, education must also learn, as it has not yet learned, to make the best possible use of the numerous aids of this kind [audiovisual aids] produced and perfected by man's ingenuity. The phonograph, radio and motion pictures have long been familiar to most American households; but they remain relative strangers in the field of education, where their potentialities are vast. In television, we now have available an almost perfect educational instrument. (Eurich, 1955, p. 10)

---

[6]It should be emphasized, in this discussion of the efforts to meet the crisis of expanding enrollments and teacher shortage, that the Fund for the Advancement of Education has also spent a great deal of money in efforts to tap heretofore unavailable sources of teachers—liberal arts graduates, housewives who have left teaching, and others—with special training programs which the Fund has subsidized. However, these efforts are not in the main line of the automation movement.

2. Automatizing the Classroom

The use of television in automatizing the classroom is stated in the following propositions by Stoddard:

. . . Let there be experimentation to determine, first, whether there is a practical way to incorporate television into the school program in such manner as to substitute for and lessen what the regular teacher must do so she can do the remainder better and also possibly to teach some things even more efficiently than she could do otherwise; and second, to determine which phases and types of learning experience lend themselves best to the television medium of presentation; and third, whether a high level of teaching efficiency can be attained with fewer trained teachers than would be involved in the usual school organization. (Stoddard, 1957, p. 43)

When summarized, the thinking represented by Eurich, Stoddard, Faust, and many other associated with various television and film projects may be said to involve:

1. The introduction of mass audiovisual experiences—by picture, tape, television, or film—into the classroom. This would mean, for example, a whole course in chemistry or French or all of the content of fourth grade social studies taught by master teachers through one or more of these media.

2. Leaving much (perhaps all) of the *systematic* (logical, content, etc.) aspects of teaching to transmission over one or more of these audiovisual media and the *developmental* (personal, social and growth) aspects to be handled by other persons in the classroom.

3. Large classes as audiences for the transmissions at least part of the school day. (Stoddard proposes a plan for the organization of an elementary school which includes a viewing group in a "resources room" of no less than 150 students while another group of 75 is having some kind of a mass experience in an auditorium. At other periods, students work in small groups (Stoddard, 1957, p. 45).)

4. Developing a group of master teachers who (with assistants) will prepare and deliver the lessons or lectures in a form for audiovisual transmission. These master teachers will draw higher pay equivalent to their ability to operate effectively in the mass media.

5. Carefully planned, highly systematized lessons and lectures equivalent, perhaps, to the careful planning and programming necessary in other forms of automation described in the first paper.

6. Fewer regular classrooms, resulting in a saving of money and time and fewer positions that need to be filled by *trained* teachers, also resulting in a saving of money as well as making a dent in the teacher shortage problem. These savings could be used for the audiovisual equipment (closed-circuit television, etc.) required and to raise the salaries of qualified teachers.

7. Raising the quality of teaching (at least in its systematic aspects) because the master teachers will be experts, both in content and presentation, and possess specialized knowledge not available to the line teacher. Thus, no longer will Johnny or Jane have to suffer through general chemistry with the typing teacher because the school had no one qualified to teach chemistry. The chemistry teacher now comes every day or every other

day via film or television. Exercises can still be supervised by the typing teacher, but the presentation or content is not her responsibility.

This, then, is the basic theoretical position for automatizing the classroom. There are, in addition, certain factors implicit in this concept that should be stated. These factors have not been directly stated in the literature. They are:

1. A larger number of audiovisual technicians than audiovisual enthusiasts ever imagined in the wildest dreams of post-World War II will have to be found and made available for such a program. Many will be used in preparing materials for the filmed or television presentations; others, presumably, would fit into the "aide" category and render the same service to local teachers.

2. It must be recognized that the classroom automation proposal will attempt to short-cut what might be called normal audiovisual development which, granting Mr. Eurich's point, has really never achieved its potentiality in most schools and colleges of the nation. What, however, will happen to existing audiovisual libraries and collections of equipment?

3. Such an automatization as visualized really has no limits of extension, at least until national boundaries are reached. Dr. Eurich has stated in two places (Eurich, 1955, p. 13; Eurich, 1956, p. 15) that some of these television presentations might well be on a *national* basis. Further, the possible tie-ins, through the Educational Television and Radio Center, permit almost indefinite extension of these presentations through television stations and film circulation. In fact, considering the movement as a whole, while most of the activities of the ETRC and other organizations are concerned with *telecasting* as opposed to closed-circuit transmission, the boundaries between the two areas become almost nonexistent at times. Stoddard's statement of ETV achievements, for example, jumps from a description of the St. Louis experiment which deals with *broadcasting* to the Hagerstown study, which is *closed-circuit* (Stoddard, 1957, p. 34). Since the difference between the two is only a technical one concerned with the use or nonuse of carrier waves, there is reason to believe that some automatization might originate with broadcasting stations, some with closed-circuit, and some using both at the same time.

4. The entire idea has brought the existing audiovisual movement to a point of serious crisis with no particular evidence that those who have been concerned with the movement over the last two decades are doing or can do much about it. Here and there, as, for example, at San Jose State College in California, the audiovisual people have undertaken the developments in television. For the most part, however, the power of control over this new movement for automatization rests with the organizations described in the early part of this paper.

Summary

This paper has attempted to deal with two aspects of the problem of automatizing the classroom through audiovisual means. These aspects were: (a) the institutional background in which the earlier foundation support of the audiovisual movement by the Payne Fund and the Rockefeller Foundation was compared to the present, much more extensive support of the mass audiovisual approach to teaching supported by the Ford Foundation, its several funds and the organizations which it underwrites; and (b) the theoretical position requiring audiovisual automation which was shown to be based on the teacher and plant shortage facing this country in the next two decades of swelling school and college enrollments and, therefore, necessitating technological solutions to problems of class size and building space.

For a complete picture, some case histories and research results, as well as a technical analysis of the entire position, would prove useful. It is hoped to supply these three aspects of the study in the next paper, to be followed, if thought desirable, by a fourth discussion which would subject the whole idea to critical scrutiny.

References

Brooker, Floyde E. "DAVI Convention: Market Place of Ideas," *Audio-Visual Instruction*, Vol. 2, Issue 2, February 1957, p. 64.

Carpenter, C.R., L.P. Greenhill, et al. *An Investigation of Closed-Circuit Television for Teaching University Courses.* University Park, Penn.: The Pennsylvania State University, 1955.

Eurich, Alvin C. "The Teacher's Dilemma." Unpublished paper, June 1955.

Eurich, Alvin C. "Better Instruction with Fewer Teachers," *Current Issues in Higher Education, 1956.* Washington, D.C.: Association for Higher Education, NEA, pp. 10-16.

Finn, James D. "Automation and Education: 1. General Aspects," *Audio-Visual Communication Review*, Vol. 5, No. 1, Winter 1957, pp. 343-360.

The Ford Foundation Annual Report. New York: The Foundation, 1956.

The Fund for the Advancement of Education. *A Report for 1952-1954.* New York: The Fund, 1954.

The Fund for the Advancement of Education. *Teachers for Tomorrow.* New York: The Fund, 1955.

Hull, Richard B. "Consider Basic Problems," *AERT Journal*, Vol. 16, No. 3, pp.5-9, 34-37. December 1956.

Knight, Edgar W. *What College Presidents Say.* Chapel Hill: The University of North Carolina Press, 1940.

Lombard, Edwin H. "Experimental Teaching by Television," *Proceedings, Western College Association* (fall meeting, 1956), Fresno, Calif.: The Association, pp. 16-26.

McDonald, Ralph W. "Financing Higher Education: Successful Capital Expansion Programs," *Current Issues in Higher Education, 1956.* Washington, D.C.: Association for Higher Education, NEA, pp. 256-262.

NET News, May-June 1956.

Research Division, National Education Association. "Audio-Visual Education in Urban School Districts, 1953-54," *Research Bulletin*, Vol. 33, No. 3, October 1955.

Semans, H.H., T.C. Holy, et al. *A Study of the Need for Additional Centers of Public Higher Education in California.* Sacramento: California

2. Automatizing the Classroom

State Department of Education, 1957.

Stoddard, Alexander J. *Schools for Tomorrow: An Educator's Blueprint.* New York: The Fund for the Advancement of Education, 1957.

The Twentieth Century Fund. News release by the Fund, September 1955.

Wilson, O. Meredith. "Use of Technical Aids and Assistants: Ingenuity in Improving Quality of Instruction while Accomodating Larger Numbers of Students," *Current Issues in Higher Education, 1956.* Washington, D.C.: Association for Higher Education, NEA, pp. 170-173.

Automation and Education
3. Technology and the Instructional Process

As an instrument of work engaging human energies in a manner far surpassing the lure of war, as a social dissolvent and readjuster, and as a philosophy of action, technology must be brought into the main stream of history, if the course of history is to be surveyed correctly and "the dark imminence of the unknown future" is to be in any way penetrated. —Charles A. Beard, 1955, p. xxiii

If we can grasp what John Stuart Mill called the "principia media" of a society, if we can grasp its major trends; in brief, if we can understand the structural transformation of our epoch, we might have "a basis for prediction." —C. Wright Mills, 1959

Recently, by way of the mass media, the American and, presumably, the Russian people were treated to a vigorous sporadic and, at times, heated debate between Vice-President Nixon and Premier Khrushchev. The occasion, as is well known, was the opening of the American Exhibition in Moscow. The strange thing about this debate was its subject. Although the discussion ranged superficially over many cold war problems, the central subject of the debate was *technology*—both directly and by implication. Mr. Nixon and Mr. Khrushchev were debating the values of American *versus* Russian technology. Mr. Khrushchev's later visit to the United States merely continued the international colloquy.

Nothing could have emphasized the interests and commitments of the modern world more clearly than this clash between statesmen of the great opposing powers. They were singing, in effect, a gigantic and possibly terrifying version of the musical comedy favorite of some years ago—"Anything you can do, I can do better." And what is to be done involves the technology of washing machines, missiles, atomic power, mass production of housing, frozen orange juice, and lunar satellites, to mention only a few items. Technology, unlimited; technology, international; technology, totally competitive—in weapons, in commerce, even in the arts.

In addition, science, technology and invention on a large scale have become such a handmaiden of industry in the United States that research and development are now regarded as the basic keys to progress in our economy. Research lengthens our lives and, until very recently, our

Reprinted from *Audiovisual Communication Review*, Vol. 8, No. 1, Winter 1960, pp. 5-26.

automobiles; it filters our water and our tobacco smoke; it weaves miracle fabrics out of chemicals and television commercials out of our suppressed desires; it shoots missiles at Venus and attempts to construct a Venus out of every American woman with the aid of brassieres designed by engineers acquainted with the latest in stress technology.

Modern historians of technology, such as Oliver (1956), emphasize the fact that American civiization is fundamentally a technological civilization. It is not surprising, then, to find that at a time of international crisis occurring simultaneously with the period in history remarkable for what is now generally called the "explosion" of knowledge, Americans would take the view that further development of science and technology is our number one national objective.

Technology, however, is more than invention—more than machines. It is a process and a way of thinking. And, as Mumford reminds us, "Every technical process tends, in its perfection, to eliminate the active worker from participation and to produce an effective substitute: the automaton" (Mumford, 1955, p. 52). And so in the midst of great national pressure to increase the number of scientists and engineers, to change the curricula of schools and colleges in the direction of more abundant and rigorous offerings in science and mathematics, to invent and apply new machines and systems in the armed forces, in government, in industry and commerce, it is also not surprising to see emerging powerful social pressures that are opposed to unrestrained technological development. The great strike of 1959 in the steel industry was, so the analysts tell us, at bottom an issue over automation, displaced jobs, and the new role of the worker. The active worker has resisted, since the beginnings of the several industrial revolutions, his elimination by an automaton—whether loom, sewing machine, computer, television camera, or automatic steel puddler.

These two forces—the national (and private enterprise) drive for technological superiority and the resistance by special groups to the introduction of specific technologies into various segments of our society—are not new, but have certainly increased in degree until they are different in kind from similar actions in former times.[1] In the long view, however, we do not see much visible effect upon the onward march of research, development, invention, and application.

The most powerful movement today is, without question, that associated with the *increase* of technology. Admiral Rickover, for example, is not an isolated phenomenon. While no scientist, he is a technical administrator and, to the public he speaks with the voice of SCIENCE as he demands an increasingly technological education for all. Even 20 years ago the thought that an admiral in the U.S. Navy would become, almost in an official capacity, a spokesman for reform of the public school system of the United States would have been unthinkable. Today, many sectors of the general public expect that he should.

The result of all this pressure for increase is a sort of Parkinson's law of technology. One technician—be he pure scientist, applied scientist, or

---

[1] A brief statement identifying several approaches to the study of the social effects of technology appears in the Notes at the end of this paper.

3. Technology and the Instructional Process

engineer—enters a new sector of development and four others automatically follow in his footsteps. And four more for each of the first four. Before long, a new territory has succumbed.

Technology, in this sense, absolutely refuses to be confined. There are few areas of human interest that are sacred from invasion. Religion, perhaps, but not love, nor poetry, nor the composition of symphonies, nor the teaching of the young. With this continual extension, there must be great temptation for the technologist, literally, to "take over." As an example of this phenomenon close to the hearts of educators, it might be well to examine the activities of the National Science Foundation and the National Academy of Sciences. It will be found, I suspect, that the two of them have already taken over some functions of the U.S. Office of Education, the NEA and its departments, and other possible "nonscientific" groups of educators,[2] and are, in fact, busily spawning a "scientific" educational bureaucracy. Such a bureaucracy is but the beginning—and there are many others in other fields—of a movement I have dubbed "Neo-Technocracy." The Neo-Technocrats have, in fact, moved farther into positions of power and control in a few short years than Howard Scott ever imagined when he was promoting the original Technocracy in the days of the Depression.

Turning now to education, it becomes apparent under this national and international drive for technological superiority that: (a) those concerned professionally with education have not developed a well-conceived point of view and a position and/or positions concerning technology and education,[3] (b) because of this lack of a point of view and because of certain cultural lag factors naturally associated with education, the acceleration of technological development has tended to bypass the entire educational enterprise until very recently, (c) professionals in education are not prepared now to deal with the tremendous impact that technology is beginning to have on the instructional process itself as, by the technological process of extension, technology begins to invade education in full force, and (d) the absence of understanding and a point of view in the profession creates a situation where the Neo-Technocrats not only can, but are beginning to move into the field of instruction.[4]

---

[2]As an amusing sidelight, the NSF has a brochure offering certain kinds of rewards to personnel in the various sciences and the "scientific" side of certain social sciences, for example, mathematical economics. Included, with approval, in the "scientific" studies is the subject of philosophy of science! To be fair, the NSF itself is probably not to blame. Congress and the conditions surrounding the use of funds have set these boundaries. The point, however, is still valid.

[3]During the Depression, Harold Rugg, George Counts, and others attempted this, but an examination of their work shows that almost the entire focus was on the economic effects of the increasing technology and the kind of education necessary for economic reorganization and survival. Other aspects were either slighted or missed entirely. See, for example, Rugg's *The Great Technology*.

[4]An immediate example—the so-called MIT Physics Group—comes to mind. Professor Zacharias has gone into the educational filmmaking business.

3. Technology and the Instructional Process

Statement of Rationale

The statement of the rationale of this paper is best accomplished by presenting four propositions, as follows:

1. For purposes of anaylsis, the American educational system taken as a whole may be considered as a *society*.

2. The introduction, on a large scale, of technology into this (educational) society will be accompanied by the same types of stresses and strains, initiate the same types of problems, accomplish the same classes of objectives, and force similar reorganizations as the introduction of large-scale technology into any society, subject of course, to modifications by the unique factors inherent in the specific (educational) society.

3. Because professional educators for the most part have not, up to this time, viewed current technological problems from this orientation, but rather, if at all, in specific contexts (for example, problems associated with the sudden introduction of television into education), it will be necessary not only to review thoroughly current educational theories from this more general point of view, but to create new theories encompassing and anticipating the new technological directions.

4. This new orientation toward technology and education will require rigorous analysis of the nature and possible roles of specific technological developments, the relation of these to educational purposes, and the prediction of effects on the society.

The remainder of this paper will attempt to delineate some of the elements of this technological study, to relate and analyze current and predicted technological developments, and to suggest some educational implications that might bear examination by the profession.

Definition of Technology

In sophisticated quarters of the intellectual world, definition is usually left to popularizing speech makers, as definitions are often assumed—mistakenly, perhaps—in professional discourse. However, when considering technology in its relations with education, definition cannot be left to chance. Many curriculum specialists, philosophers, administrators, and subject matter experts are inclined to think of technology exclusively as *machines*. Nothing could be further from the truth. Machines are, of course, an indispensable part of technology. Even the dictionary definition, however, goes beyond machinery. The *Funk and Wagnalls College Dictionary* (the one I have closest at hand) defines technology as: (a) "Theoretical knowledge of industry and the industrial arts," and (b) "The application of science to the arts."

A much more useful concept for our purposes is contained in a little-known discussion of technology by Charles A. Beard. Beard says, in part:

> What then is this technology which constitutes the supreme instrument of modern progress? Although the term is freely employed in current writings, its meaning as actuality and potentiality has never been explored and defined. Indeed, so wide-reaching are its ramifications that the task is difficult and hazardous. Narrowly viewed, technology consists of the totality of existing laboratories, machines, and processes already developed, mastered, and in operation. But it

is far more than mere objective realities.

Intimately linked in its origin and operation with pure science, even its most remote mathematical speculations, technology has a philosophy of nature and a method—an attitude toward materials and work—and hence is a subjective force of high tension. It embraces within its scope great constellations of ideas, some explored to apparent limits and others in the form of posed problems and emergent issues dimly understood. (Beard, 1955, pp. xxii-xxiii)

The educationist, in considering the effect of technology on the instructional process must remember that, in addition to machinery, technology includes processes, systems, management, and control mechanisms both human and nonhuman, and above all, the attitude discussed by Beard —a way of looking at problems as to their interest and difficulty (broadly considered) of those solutions. This is the context in which the educator must study technology.[5]

The Possible Relations of Technology with Education

Technology relates to education in at least three major ways. First, in a society in which science and technology are primary, such as America, the society requires that the educational system insure an adequate supply of scientists and associated technicians. This requirement sets a curriculum problem, an organization problem, and many other problems associated with the screening, selection and education of young people as potential additions to the nation's technical manpower.

Second, as a society becomes more and more technologically oriented and controlled, the question of the general education of all citizens is raised. The survival and management of the whole society theoretically requires more general education in the sciences and technology for all. Pressures arise for more mathematics and science to be taught to the entire population. Again, curriculum problems, organization problems, and a host of nagging, persistent general education problems arise.

Third, because of the tendency for technology to have no limits and constantly to extend into new areas, it is inevitable that, in an advanced technical society, technology should begin to extend into the instructional process itself. As will be shown, this is particularly true when education has been, for a century or more, one of the areas of American society which has walled itself off from technological advances and, consequently, has created a technological vacuum. That vacuum is now rapidly being filled.

It is with this third relationship—the application of technology to the educational, or to be more precise, to the instructional process—that the balance of this paper is concerned. The three relationships just discussed—

---

[5]For purposes of this argument, no distinction is made betweeen "science" and "technology" nor between "pure" and "applied" science. The distinction between "basic" and "applied" research is perhaps more meaningful, but these two activities are still regarded here as positions on a continuum. From a societal point of view, it matters little whether the society is hit with new ideas or new processes that stem from the ideas, or, as is more likely, a combination of the two.

development of technicians, general education in technology, and the application of technology to the instructional process—cannot eventually be completely separated. However, the third relationship is sufficiently different to merit thorough analysis.

Again, for purposes of this paper, a finer distinction will now be drawn. Within the educational process itself, there are three general areas in which technology can or is being applied. These are: (a) general administration, (b) testing, and (c) instruction. The uses of technical management systems, modern equipment, etc., represent the fairly obvious applications to the field of general administration. While it probably can easily be shown that this area of management is, taken as a whole, two or three decades behind its counterpart in industry, the problems associated with technology and school management are not as difficult as some of the others and will not, at this time, be considered. This is not to say that administration is not highly interrelated with the other two areas of testing and instruction, but the position can be taken that the problems arising from the latter two should guide the technical solutions in administration, not the other way around.

The second category, testing, is in many respects the most highly developed technology at present existing in American education. This is true both from a machine and from a systems standpoint. Further, the close relationships between psychological, achievement and other types of testing and the instructional process are so well defined as to need no comment. However, with some exceptions, the technology of the instructional process can, for purposes of analysis, be isolated from testing. This arbitrary decision is made here in order to further the remainder of the discussion. We are left, then, with the instructional process by itself and can now turn to the impact of technology upon that process.

The Development of Instructional Technology

The development of a technology of the instructional process is relatively new. For guidance as to this development, the reader may refer to the chart which roughly compares the development of industrial technology with instructional technology. In the pre-industrial phases of both education and industry while industry was principally at the handwork, artisan level, the instructional process relied upon such devices as the slate, the hornbook, the blackboard. chalk, and limited single textbooks with few illustrations. Although attention will be confined principally to some of the symbols of technology—equipment and machines—it should be emphasized that other factors—organization, etc.—were in the same state. As is well known, for example, the graded school is a late development.

At the beginning of the 19th century we note the famous changes in industry—the invention of a group of related machines for power, spinning, weaving, etc.—which made possible the factory system. Based on the work of Toynbee, the elder, the term "Industrial Revolution" is applied to this period.[6] However, in education the same revolution did *not* occur and instructional technology (with some exceptions here and there) re-

---

[6]Now called the First Industrial Revolution.

3. Technology and the Instructional Process

Industry	Instruction
Pre-Industrial Production Technology	**Pre-Industrial, Instructional Technology**
. . . artisans, handwork, etc. . . . invention of specific machines and technics	. . . hornbook, slate, blackboard and chalk, limited single textbooks, few illustrative and graphic materials
TO 1800	
Mass Production Technology	. . . some signs of American leadership . . . maps, globes, blackboards, etc. (1875)
. . . establishment of factories, gradual development of machines . . . assembly lines . . . modern capitalism . . . management concepts . . . new power sources, etc.	TO 1900
	. . . addition of laboratories, project methods, libraries, very small amounts of audiovisual materials . . .
	Still Pre-Industrial in Concept and Execution
TO 1950	TO 1950
Pre-Automation Period	**Potential Mass Production Technology** (Unrealized)
. . . "Detroit automation," working computers . . . development of systems and operational concepts	. . . Legacy of WWII for AV materials . . . ETV possible . . . "saturation concept" . . . all not applied significantly
TO 1955	TO 1955
Automation Period Dawns	**Eurich-Stoddard Mass Production Period Dawns**
. . . the automatic factory . . . the space age.	. . . the pressure for ETV and mass instruction with master teachers . . .
TO 1960	TO 1960
	Potential for Educational Automation Appears
	. . . the era of teaching machines . . . stratovision . . . the automatic classroom

By 1959, in actuality, instructional technology had by no means reached the mass production stage on any scale.

mained at the pre-industrial level. During the last quarter of the century, there was some indication of a change. Oliver notes that an exhibit from an American school "with maps, charts, textbooks, and other equipment" won admiration at the International Exposition in Vienna in 1873. He adds that the American school display at Paris in 1878 was even more outstanding (Oliver, 1956, pp. 298-299). Significantly, Oliver's last mention of international attention to American instructional technology was at Melbourne in 1880.

By 1900, on the other hand, industry had established factories and was moving into assembly-line operations; had begun to apply, in a crude way, research and development concepts; had introduced the beginnings of modern management; and had developed a sophisticated financial system. Developments had reached such a state that the perceptive Henry Adams, the true prophet and seer of technology, foretold the coming of the Age of the Atom and the problems it would bring (Adams, 1958, p. 2). In the years to 1950, aided by the acceleration of two wars, technology burgeoned and developed, piled machine upon machine, system upon system, added fantastically to power, and invented the method of invention. Technology transformed American society, philosophy and art.

Technology, however, during the period from 1900 to 1950 only washed lightly upon the shores of instruction. During this same time span, when high-speed printing techniques, radio, sound motion pictures, television, and other pieces of communication technology were invented, developed and exploited, American education failed to apply these devices in quantity to the instructional process and failed to develop the appropriate technological systems necessary for this application. There were always rumblings, to be sure, as evidenced by the statement attributed to Edison in 1916 that the motion picture would replace the teacher. However, looked at from the vantage point of 1960, laboratories, project methods, libraries, and minute arrangements for audiovisual materials—the provisions to 1950—constituted what was still a pre-industrial technology for instruction.

By 1950, industry had entered what may be called the pre-automation period, which heralded the beginning of a movement toward true automation. At this time, much of the basic work on computers and other control mechanisms had been done. At this time also, mechanical systems of transporting work goods between machines had integrated some production lines into a system of mechanical or "Detroit automation." The period of automation was dawning, and I have arbitrarily set 1955 as the date at which this electronic-mechanical-systems analysis procedure began to be significantly applied. It is no accident that man began to push toward space at about the same time.

By 1950, American education had the potentiality, to carry on with the analogy, of a mass production technology. The hardware—projectors, recorders, television—and the materials were present. The systems concepts—saturation with audiovisual materials at the point of an instructional problem, for example, a concept derived from the military experience of World War II—were developed and known in a few audiovisual circles. A certain amount of incentive and public acceptance, also derived from World War II experience, could be drawn upon. The cake of custom, however, proved to be too tough and the mass production stage, at least 100 years behind industry, was not entered except here and there on isolated little islands.

At this point, approximately 1955, the god came out of the machine in the form of the Ford Foundation and Dr. Alvin C. Eurich (and a little later, his associate, Dr. Alexander J. Stoddard) to give the instructional processes of American education a sharp push into a mass production technology. The time was ripe. There was a shortage of teachers; education and educationists were under fire from all sides; Neo-Technocracy was turning its attention to education; the race with Russia was underway; the natives were restless indeed.

There are several interesting facets to this shove into mass instructional technology. First, television, both closed-circuit and broadcast, was chosen as the prime hardware. This was due, I think, to the unconsciously assumed basic concept of *mass production*, usually stated in terms of a shortage of teachers, large classes, and quality instruction. Second, the Ford impetus made provision, for the first time, for a technology of systems to go along with the hardware. Stoddard's own work in designing the school of tomorrow (Stoddard, 1957) and the work of Trump and his associates in drawing the instructional images of the future for the American secondary school (Trump, 1959) are, essentially, proposed systems.

Third, the Ford group made use of technology in creating the impetus in the first place. The technology of social psychology and public relations was drawn upon; educational decision makers were the prime targets, and the means used, including high-priced public relations counsel, were the best available. This accounts for the fact that the teaching profession in general and audiovisual specialists in particular—really the only technologists of the profession—were for the most part left out of this move into technology.

Instead, the Foundation went to administrators and board members and influential opinionmakers of all types. New organizations were approached—the National Education Television and Radio Center (now, I believe, related to something called the Learning Resources Institute), the revitalized National Association of Educational Broadcasters, and the Educational Facilities Laboratories, Inc. Others were created. As a sign of this policy to short-circuit the profession, it is significant that, except for a small grant to promote a seminar or two, the Department of Audiovisual Instruction of the NEA was left out of the largesse.

The Ford people assumed, I think, that if the cake of custom was to be broken the priesthood was not the agency to break it. This may mean, in part and as far as audiovisual and curriculum people are concerned, that to an outsider their technological orientation is the level of the water wheel and the hand-loom, not the computer and the rocket. It is something to think about. On the other hand, it may also mean in part that the Ford Foundation is merely another instrument of Neo-Technocracy and that these actions forecast even more loss of control of education by the existing pre-technological profession.

3. Technology and the Instructional Process

The Impact of Present Trends

With this background of development, we can now come to grips with the impact of present technology. Here, our first principle should be recalled. If we consider education as a societal universe, what happens when we introduce energy into that universe suddenly and on a large scale? That is apparently what is happening. Two general effects, based on two analogous concepts can be predicted. If the concept of entropy holds good in this connection, the educational system will become more highly organized and less random in nature (negative entropy will increase); if the observed historical-anthropological law of the introduction of a system of technology into a culture holds good, the distortions, stresses and nonpredictable effects will increase until the culture becomes unrecognizable. Two historical events serve as examples—one taken from Mumford, the introduction of the clock to regulate the habits of monks (Mumford, 1959) and one from Muller, the invention of writing in Egypt for the purpose of keeping accounts (Muller, 1952). Both changed the world several times over and the effects of neither were, of course, predictable.

What, then, are the present trends? If you don't count the continuous, grinding, sweating struggle carried on by audiovisual specialists, dealers, manufacturers, and producers to introduce, one item at a time, film libraries, projectors, recorders, etc. into the educational system in the slow process of conversion from a pre-industrial technology, two major trends can be identified. These two trends at the moment, lead in opposite directions.

The first is the trend toward a mass instructional technology and is governed by machines and systems suitable for that purpose. Foremost, of course, is television, of which there are four instructional types:

1. Broadcast on an educational channel.
2. Broadcast on a commercial channel.
3. Closed-circuit of the Hagerstown-Penn State type in which live instructors are used either to supplement instruction or to provide direct instruction exclusive of classroom teachers.
4. The Compton type in which filmed lectures are distributed via the closed-circuit medium as replacement for classroom teachers.

In all cases, the desire is to reach more students with fewer teachers or to obtain "quality" instruction.

Mass instruction technology in another form includes, of course, the massed film systems. The prime example is the EBF series in physics and chemistry, amounting to over 300 half-hour motion pictures intended to be used where there are no science teachers, where there are teachers not considered qualified, as audiovisual aids to more qualified teachers, and with very large groups of students. These films, of course, may theoretically be used over television. The exploitation of the overhead transparency projector is another example of mass instructional technology as shown in the Newton (Massachusetts) experiment in English grammar and composition.

In opposition to this trend of mass instruction is a growing technology for individual instruction. This trend is the audiovisual wave of the future (for the moment). The most dramatic development here is that class of systems and instruments known as teaching machines. Actually, I would class all teaching equipment designed for individual or near-individual operation as being in the category of teaching machines. At present, then, there are approximately five types, listed here on an ascending scale of sophistication:

1. Individual reading pacers and similar devices.
2. Individual viewing and listening equipment for existing slides, film-strips, motion pictures, and recordings.
3. Language laboratories of all types.
4. Specifically programed printed materials such as scrambled textbooks, the proposed Lumsdaine notebook, etc.
5. True teaching machines of the Skinner or Pressey type containing carefully worked out verbal or pictorial programs with various ingenious mechanical or electronic arrangements to test student reaction, inform him of his progress, errors, etc.

Extrapolation I

These two major trends toward technologies of mass and of individual instruction are not science fiction; they are with us now. Assessment of this impact is our most pressing and difficult problem. It does not take much of a crystal ball to see that a combination of these two technologies is the next immediate step. For example, let us take Maurice Mitchell's new series for his chemistry course on 150-plus films. Let us further assume that Mr. Mitchell makes a deal with the Rheem Califone Corporation to program their multiple-choice Didak teaching machine with material to go along with the chemistry films (Mr. Eurich might have to put up the money for the initial effort). Let us continue by saying that we will project the films hourly on the Compton College closed-circuit television film distribution system and place the machines in proper cubbyholes for Compton College students to work at specified times. We could get, almost immediately, to what can be called total educational automation.

Or, in another context, let us suppose that the new Ford creation, the Learning Resources Institute, initiates another Continental Classroom in mathematics over national broadcast television; and that the new audiovisual teaching machines now being developed by Hughes Aircraft are programmed to relate to this new series; and that these machines are placed by the Foundation in every public library in America; and that, at the conclusion of the instructional period, students report to designated centers where the Educational Testing Service examines them and certifies them to the colleges of their choice. Think, for a moment, about that one. It is now possible not only to eliminate the teacher, but also the school system.

These may be considered extreme applications, but such applications, as least experimentally, are inevitable. In fact, one such study, testing the combination of these two technologies, is now underway. The point is that both the mass instruction systems and the technology of individual instruction—teaching machines—are getting terrific momentum. These technologies are going to hit education with a million-pound thrust. What will be the effect upon our educational society? What is the role of the teacher, the audiovisual specialist, the curriculum director? *Quo vadis* the curriculum itself? This last question is particularly pertinent if the medium or image governs the message, as McLuhan maintains.

Extrapolation II

Technology never sits still. The extrapolations given in the last section are immediate. What of those that can be made for a time slightly farther into the future? I see several developments. First, it is reasonable to expect that programming for teaching machines will move from the verbal-Socratic-Skinner type to the audiovisual-branching type. That is, the machines will present, based upon student pre-tests, conceptual content using films, slides, filmstrips, tapes, and/or videotape as the medium. The presentation sequences will be longer and the student will be given an opportunity to select additional sequences for further explanation if the machine, through testing, informs him that he needs it. Records, of course, will be maintained instantaneously by miniaturized computers.

Secondly, the work in England of Ashby, Pask and others with teaching machines based on the design concepts of biological computers will affect the technology of the present teaching machine much as transactional psychology affects stimulus-response psychology. Pask has already produced a machine which, in training key punch operators, actually senses the characteristics of the student as he works and automatically adjusts the program of the machine to the student's individual needs. This is a pure transaction, and Pask maintains he can develop such a machine to teach decision making with present hardware and know-how[7] (Pask, 1958).

A second development of the future begins next year with the stratovision experiment of the Ford Foundation over a city in Indiana. News of this experiment is gradually filtering into educational circles and has appeared here and there in the press. One or more airplanes, carrying multiple television transmitters circling overhead can receive multiple signals from the ground and broadcast them over much greater distances than conventional tower transmission. The programs will go into several states and, because of the location, into both urban and rural areas and into innumerable heretofore independent school districts. Effects of this on local control of curricula, curriculum planning procedures, images of master teachers, patterns of school organization, etc. have yet to be speculated about, let alone explored. Nothing technical would prevent, by the way, a national network of such aircraft as an addition to our other national means of transmission.

A third predicted development is based on information that it is technically possible, although not yet economicaly feasible, to produce a motion picture film with characteristics similar to the Polaroid film that is instantaneously developable. What this would do to videotape; to production techniques, and to the possibilities of local production of materials is

---

[7]A machine produced by the Western Design Division of U.S. Industries was presented to the press in New York on October 15, 1959. The machine "interacts" with the student, but, if press information can be relied upon, not in the manner of a biological computer (*Santa Barbara Evening News*, October 14, 1959).

of great interest to the audiovisual field.

The fourth development that can be anticipated is the most spectacular. Essentially, it is based on a systems concept. It is theoretically possible now to design an automatic classroom under the control of the teacher. Most of the elements are present or can be designed. Such a classroom would have total light and air control, automatic projection and television systems, technical provision for the best possible discussion environments, display situations, etc. which could be changed at will. By planned programming the classroom could be made to function for major presentations, small group discussions, individual work at teaching machines, creative periods, etc. All of this *could be* under the control of the teacher. The classroom then would become the teaching machine. Adrian TerLouw at Eastman Kodak has done some work in this area and it is, of course, implied in the provocative article by Simon Ramo which is so often cited in discussions of teaching machines (Ramo, 1957). An acquaintance of mine refers to this concept as the "mad scientist" classroom.[8]

If we look at this series of possibilities, we can sense a change in trend. Of the four cited, only one—the stratovision experiment—is clearly on the side of mass production. The other three lean toward the technology of individual instruction. The automatic classroom is a combination of both, but one in which the human element still plays the central part with the machines being the slave of man, not the other way around.[9]

Implications

To assess the implications of this current and predicted impact of technology on the instructional process is not easy. As with any such speculation, it is also hazardous. It is, however, a job we must attempt.

We have to start, I think, with the proposition that, based upon the historical development of instructional technology, our educational society is in the position of a backward or underdeveloped culture suddenly assailed by the 20th-century engineer. In addition, reactions in some professional circles to the advent of television are similar to those factory workers of the 19th century who attempted to destroy the machines that were replacing their jobs. In the long run, the machines remained; instructional technology is, no doubt, here to stay. Our problem becomes one, not so much of how to live with it on some kind of feather-bedding basis, but how to control it so that the proper objectives of education may be served and the human being remain central in the process.

---

[8]The multiple projection system developed by Teleprompter and a somewhat similar unit announced by Harwald would seem to make it possible to *simulate* the "mad scientist" classroom, although admittedly, this is reaching a little. The new motion picture projector developed by MIT Physics Group and the proper programming of a device called the "Perceptascope" could accomplish similar effects.

[9]Since writing this, I have seen documents describing, in some detail, the objectives of the Ford Foundation's Learning Resources Institute. The LRI is to be a gigantic source of mass instructional technology. The stated ambitions are so great, the funding so large that the direction postulated in Extrapolation II might well be reversed. The potential dangers of such a development are also very great.

3. Technology and the Instructional Process

Certain things are obvious. I think the concept of negative entropy[10] will hold. The thrust and energy of technology will force a greater organization upon us *at every point at which it is applied to instruction*. Such an arrangement as the stratovision experiment will require a tighter and different organization in those five or six states and in those many school districts of the Midwest. The closed-circuit experiment just beginning in Anaheim, California, is already, it seems to me, making requirements for organization—of scope and sequence in the curriculum, of the district itself, and of the nature of the television presentations—the rigor of which has never been felt before in Anaheim.

Programming, in its machine, computer and automation sense, is a matter of extreme organization—the piling up of energy.[11] Most of the devices and systems of the new technology require, one way or another, this type of programming—television, teaching machines, mass films, language laboratories. This means that programmers are needed. Who is to do this? This is essentially, in the old sense, both a curriculum and an audiovisual problem. It is also a social problem. The heartland is programming. He who controls the programming heartland controls the educational system. Will it be a Foundation, a committee of scientists, textbook publishers and film producers, the NEA, the school superintendent, the board of education, the students, or the general public?

Too, how many of us will go overboard and sink with the old concepts that will be absorbed or outmoded and tossed to the sharks by the new technology? Take the concept of instructional materials which some curriculum specialists love so well and which caused quite a controversy in DAVI circles a few years ago. The fight was futile, as is the love. The concept of programming and the systems and systems analysis it implies completely absorbs the idea of materials. Instructional materials becomes an outmoded atomistic, pre-technological concept useful mainly to the historians of education.[12]

The concept of audiovisual education may accompany instructional materials down the drain, or it may not, depending on whether or not it can be redefined acceptably.[13] The Skinner- and Pressey-type teaching machines and their descendants are, for example, primarily verbal devices, and yet their management, programming, etc. as technical electronic devices belongs somewhere. At a practical level, within a school situation, someone is going to have to worry about them. It is my position that the audiovisual field is in the easiest position to help integrate these mechanisms properly into the instructional process. They are not primarily audiovisual; they are primarily technological. The AV field, I think, must now sudden-

---

[10]In all the tests I have been able to think of, this concept holds. I am almost convinced that this idea can achieve some status as, forgive me, "Finn's Law."

[11]The role of energy in relation to AV communication theory needs much study. For example, the energy involved in the negative entropy concept should be examined in much the same way that Cottrell studied energy in general in *Energy and Society*.

[12]McLuhan maintains the same thing about audiovisual aids (McLuhan, 1959).

[13]Redefinition is the subject of a paper now at press.

ly grow up. We, the audiovisual specialists, are, of all educational personnel, the closest to technology now; we have, I think, to become specialists in *learning technology*—and that's how I would redefine audiovisual education.

What of the curriculum people? Their work is just beginning. It can become infinitely more rigorous and, at the same time, more satisfying. I suspect some of the emotional Rousseauians will have to get out of the curriculum business. The rest can step in and face this problem of programming a learning technology in human terms and find great excitement and reward.

Finally, we must remember the stresses and strains of introducing technology into our pre-industrial culture and the unpredictable consequences of any technological advance. If the profession continues to react like the California Teachers Association did toward television, we'll end up with a new profession. Not that I approve of the so-called Compton concept which shocked the CTA, because emphatically I do not. The great mistake of the CTA statement was that, at a point where precision was an absolute necessity, the CTA chose to lump all educational television—broadcast and closed-circuit, general and in-school—into one pile and then indicate that the CTA was against sin and in favor of home and mother. The statement was unrealistic, imprecise and failed to deal with the issues. We have a right to expect better from a large professional association of teachers. The profession as a whole must be made to sense this powerful movement to instructional technology and be made ready to seize the great opportunity it offers to make *all* teachers highly professional.

The Philosophy of Adventure

In closing, I should like to return to the position I took in my Los Angeles address to the DAVI National Convention in 1955. Whitehead has said that it is the business of the future to be dangerous. Technology is now making this aphorism into the outstanding fact of our time. Technology is now making the future of instruction capricious and hazardous. But in doing so it has presented us with more opportunity and more choices than ever before. If the future is an adventure, it is an adventure *because* of technology. The cost of civilization is the fact that we can make wrong choices because of the alternatives technology presents. The reward of civilization is the freedom provided by technology and the opportunity to make the right choices. This cost and this reward we now face with the technology of the instructional process. We *must* look forward to the adventure and not present what Herbert Muller noted as "the curious spectacle of civilized man forever marching with his face turned backward—as no doubt the cave-man looked back to the good old days when men were free to roam instead of being stuck in a damn hole in the ground" (Muller, 1952, p. 65). The "good old days" are gone; approached with intelligence and zest, the days of the future will be better.

Notes

It is not the intention of these notes to start another paper, or for that matter, a book-length overview of man's efforts to consider technology and its relation with society. Students of technology have approached the subject with various emphases. There are generalists, too, as shown by the great four-volume work of Lewis Mumford (*Technics and Civilization, The Culture of Cities, The Condition of Man*, and *The Conduct of Life*). However, Mumford himself pointed out that no one chose to follow his technique of study in depth (Mumford, 1959). Bertrand Russell is another generalist returning to the subject in many of his works and considering it directly in *The Impact of Science on Society*.

At least four identifiable approaches to the problem of technology have been taken by critics, philosophers and theorists. By far, the largest volume of literature has been concentrated on the relation of technology to economics, organization patterns, and general social problems. The classical criticisms of industrial developments in England and Europe, such as those presented by Owen and Marx, are in this vein. Thorstein Veblen, with his *Instinct of Workmanship, Theory of the Leisure Class*, and *The Engineers and the Price System*, critical analyses of the economic problems of a technical society, is an ancestor, perhaps, of Stuart Chase and Lewis Mumford, writing some time later. These are but samples; there were the technocrats, the social novelists, and college presidents like Nicholas Murray Butler, all having something to say pro or con about the growing American industrial society.

At the present time this general concern continues. We have William Whyte worrying about the Organization Man, David Riesman is concerned with the other-direction supplied by a technological society; John Kenneth Galbraith suggests new economic directions for a society that has solved the problems of production; and C. Wright Mills studies the new elites and the growing technocracy in social science thinking.

Secondly, we have a small and quite vocal minority that views technology in terms of its potential force for destruction, both of the race of man and of the human personality. Descending perhaps from H. G. Wells, such thinkers would have to include Aldous Huxley, Erich Fromm, and the small number of rebellious scientists who are worried about THE BOMB, fallout, and eventual wearing out of all our natural resources in the face of population growth. Included here, too, would be certain religious groups, writers, and some government officials.

In a third category is a group—almost a cult—devoted to intellectual resistance to the flow of technology. This group would include many members of the literary set in this and other countries, devotees of the New Criticism, the Humanities, novelists, critics, etc. Their literary world is being shaken to its roots by science and technology. Joseph Wood Krutch is one American spokesman for the literary wing of the anti-technologists. His *Human Nature and the Human Condition* is the latest in this view. One would also have to include here people who hold to Existentialist philosophy and, probably, the lowly and much-maligned beatniks.

The fourth approach to the analysis of technological effects stems from those who give thought to the problems of art and beauty; here the concern is with technology and aesthetics. Some of the controversies in modern art no doubt stem from this origin. There is much variation in thinking here. Ortega y Gasset, for example, speaks of the "dehumanization of art."

3. Technology and the Instructional Process

Mumford, on the other hand, now states that the great contributions of technology to art are fully realized, but that technology is, in some cases, retrogressing in this regard (Mumford, 1959). Critics and philosophers such as Robert Shayon, Gilbert Seldes, and Marshall McLuhan deal with the artistic possibilities and symbolic problems associated with the mass media, the most ubiquitous children of the new technology.

References
This bibliography is divided into two parts. The first part deals with general comment on the relations of technology to society in a variety of ways; the second covers directly certain aspects of the technology of the instructional process.

Part I—Technology and Society

The references that follow are, of course, incomplete, as the subject is overwhelming. An attempt has been made to include most of the classic references from which information as to other works may be obtained. A reference or two from each of several important areas (the future of technology, the role of the scientist, etc.) have also been included.

Adams, Henry. *The Education of Henry Adams.* New York: The Modern Library, 1931 (First published in 1918).

Adams, Henry. *The Degradation of the Democratic Dogma.* New York: Capricorn Books, 1958 (First printed by Macmillan in 1919).

Beard, Charles A. "Introduction," in J. B. Bury *The Idea of Progress.* New York: Dover Publications, 1955 (First published as the American edition by Macmillan in 1932).

Bogart, Ernest L. "Industrial Revolution," *Encyclopedia Americana.* New York: Americana Corporation, 15:93-95, 1958.

Brickman, William W. "Scientific Genius and Educational Wisdom," *School and Society*, Vol. 87, No. 2160, 411-12, October 24, 1959.

Brown, Harrison, James Bonner, and John Weir. *The Next Hundred Years.* New York: Viking Press, 1957.

Chase, Stuart. *Men and Machines.* New York: The Macmillan Company, 1929.

Cleator, P. E. *The Robot Era.* New York: Thomas Y. Crowell, Inc., 1955.

Cottrell, William Frederick. *Energy and Society; the relation between energy, social change and economic development.* Toronto: McGraw Hill, 1960.

Counts, George S. *Education and American Civilization.* New York: Bureau of Publications, Teachers College, Columbia University, 1952.

Diebold, John. *Automation: The Advent of the Automatic Factory.* New York: D. Van Nostrand Company, Inc. 1952.

Einzig, Paul. *The Economic Consequences of Automation.* New York: W. W. Norton & Company, Inc. 1956.

Fromm, Erich. *Escape from Freedom.* New York. Rinehart and Company, 1941.

Fromm, Erich. *The Sane Society.* New York: Rinehart and Company, 1955.

Galbraith, John Kenneth. *The Affluent Society*. Boston: Houghton-Mifflin Company, 1958.

Grigson, Geoffrey and Charles Harvard Gibbs-Smith (General Eds.). *Things*. New York: Hawthorn Books (in the series which also includes *People*, *Places*, and *Ideas*). n.d.

Huxley, Aldous. *Science, Liberty, and Peace*. New York: Harper, 1946.

Huxley, Aldous. *Brave New World Revisited*. New York: Harper and Brothers, 1958.

Mead, Margaret. "Closing the Gap Between the Scientists and the Others," *Daedalus*, Vol. 88, Winter 1959, pp. 139-46.

Mills, C. Wright. *The Power Elite*. New York: Oxford University Press, 1957.

Mills, C. Wright. *The Sociological Imagination*. New York: Oxford University Press, 1959.

Muller, Herbert J. *The Uses of the Past: Profiles of Former Societies*. New York: Oxford University Press, 1952.

Mumford, Lewis. *Technics and Civilization*. New York: Harcourt, Brace and Company, 1934 (Part of a four-volume series generally referred to as *The Renewal of Life*, completed in 1951).

Mumford, Lewis. *The Human Prospect*. Boston: The Beacon Press, 1955.

Mumford, Lewis. *In the Name of Sanity*. New York: Harcourt, Brace and Company, 1954.

Mumford, Lewis. "An Appraisal of Lewis Mumford's *Technics and Civilization* (1934), *Daedalus*, Vol. 88, Summer 1959, pp. 527-36.

Oliver, John W. *History of American Technology*. New York: The Ronald Press Company, 1956.

Oppenheimer, J. Robert. *Science and the Common Understanding*. New York: Simon and Schuster, 1953.

Rugg, Harold. *The Great Technology: Social Chaos and the Public Mind*. New York: The John Day Company, 1933.

Santa Barbara Evening News. October 14, 1959.

Toynbee, Arnold. *The Industrial Revolution*. Boston: The Beacon Press, 1956. (First published in 1884 as *Lectures on the Industrial Revolution in England*.)

Part II—The Technology of the Instructional Process

These references, again, are highly selective. The emphasis has been placed on teaching machines because it is assumed that readers are familiar with the basic audiovisual literature and with the large and growing literature on educational television. Enough of the latter has been included to provide both a sample and a working base.

Automated Teaching Bulletin, Vol. 1, No. 1, September 1959, South Gate, Calif.: Rheem Califone Corporation.

Bissex, Henry S. *For Instance Number 1*. Holyoke, Mass.: The Tecnifax Corporation, 1958.

Burns, John L. *Communications and Education*. Camden, N.J.: Radio Corporation of America, 1958 (Reprint of a speech delivered before the National School Boards Association, 1958).

3. Technology and the Instructional Process

Corrigan, Robert E. *Automated Teaching Methods: A Solution to Our Educational Problems* (Speech published by the Rheem Califone Corporation, South Gate, Calif., October 1959).

Earnest, Ernest. "Must the TV Technicians Take Over?", *American Association of University Professors Bulletin*, 44: 3, 582-88 (September 1958).

Eurich, Alvin C. "Better Instruction with Fewer Teachers," *Current Issues in Higher Education, 1956*. Washington, D.C.: Association for Higher Education, NEA, pp. 10-16.

Finn, James D. "A Look at the Future of AV Communication," *Audio-Visual Communication Review*, Vol. 3, No. 4, 244-56 (Fall 1955).

Finn, James D. "Automation and Education: 1. General Aspects," *Audio-Visual Communication Review*, Vol. 5, No. 1, 343-60 (Winter 1957).

Finn, James D. "Automation and Education: 2. Automatizing the Classroom—Background of the Effort," *Audio-Visual Communication Review*, Vol. 5, No. 2, 451-67 (Spring 1957).

Finn, James D. "From Slate to Automation," *Audiovisual Instruction*, Vol. 4, No. 3, 84-85+ (March 1959).

Galanter, Eugene. *Automatic Teaching: The State of the Art*. New York: John Wiley & Sons, 1959.

Henning, Gordon. *Meeting the Teacher Shortage*. (Paper published by the Technical Military Planning Operation, General Electric Company, Santa Barbara, Calif., February 1957).

Lumsdaine, A. A. "Teaching Machines and Self-Instructional Materials," *Audio-Visual Communication Review*, Vol. 7, No. 3, 163-81 (Summer 1959).

McLuhan, Marshall. "What Fundamental Changes Are Foreshadowed in the Prevailing Patterns of Educational Organization and Methods of Instruction by the Revolution in Electronics?", *Current Issues in Higher Education, 1959*. Washington, D.C.: Association for Higher Education, NEA, pp. 176-82.

Miller, Neal E. and others. *Graphic Communication and the Crisis in Education*. Washington, D.C.: Department of Audio-Visual Instruction, NEA (Published as Vol. 5, No. 3 of the *Audio-Visual Communication Review*), 1957.

Mitchell, Maurice B. *Education—A New Era Begins* (Speech published by the Encyclopaedia Britannica Films, Chicago, Ill., May 1958).

National Education Association, Department of Audio-Visual Instruction, *Television in Instruction: An Appraisal*. Washington, D.C.: DAVI, 1958.

Pask, Gordon. "A Teaching Machine for Radar Training," *Automation Progress*. May 1957, pp. 214-17.

Pask, Gordon. "Electronic Keyboard Teaching Machines," *Education and Commerce*, 24: 16-26 (July 1958).

Pask, Gordon. "The Teaching Machine," *The Overseas Engineer*, February 1959, pp. 231-32.

Pask, Gordon. "Tomorrow's Control Systems Can Learn from Experience," *Automation Progress*, February 1959, pp. 44-46 and 57.

Porter, Douglas. "A Critical Review of a Portion of the Literature on

Teaching Devices," *Harvard Educational Review*, Vol. 26, No. 2, 126-47 (Spring 1957).

Ramo, Simon. "A New Technique of Education," *Engineering and Science*, 21: 17-22 (Spring 1957).

Raymond, R. C. *The Need for Labor-Saving Machinery in Education* (Speech published by the Technical Military Planning Operation, General Electric Company, Santa Barbara, Calif., February 1957).

Skinner, B. F. "Teaching Machines," *Science*, Vol. 128, No. 3330, pp. 969-77 (October 24, 1958).

Stoddard, Alexander J. *Schools for Tomorrow*. New York: The Fund for the Advancement of Education, 1957.

Teaching by Television. New York: The Ford Foundation and the Fund for the Advancement of Education, 1959.

Trump, J. Lloyd. *Images of the Future: A New Approach to the Secondary School*. Washington, D.C.: National Association of Secondary-School Principals, NEA, 1959.

Section III

From Audiovisual
to
Instructional Technology

From Audiovisual to Instructional Technology

Faced with the choice of craft or profession, James D. Finn chose to be a professional. In an age where quality education must be a universal goal, he saw no alternative for the audiovisual educator and educators in general but to move toward becoming professional.

The necessity and vehicle for this move he saw as technology:

1. A technology which is more than invention, more than machines.
2. A technology which is a process and a way of thinking.
3. A technology which increasingly frees man to have greater power to make choices which will control his own destiny.

As a framework for converting a movement into a profession, Finn sets out six characteristics of a profession in "Professionalizing the Audio Visual Field" (1953):

1. An intellectual technique.
2. An application of that technique to the practical affairs of man.
3. A period of long training necessary before entering the profession.
4. An association of the members of the profession into a closely knit group with a high quality of communication between members.
5. A series of standards and an enforced statement of ethics.
6. An organized body of intellectual theory constantly expanding by research.

The sixth characteristic he saw as the most fundamental and most important. After examining the status of audiovisual education in terms of each criterion, Finn found it lacking in depth and direction, or "flying blind." His argument then describes steps which need to be taken before audiovisual education can claim the status of a profession.

By opening "A Look at the Future of AV Communication" (1955) with an attack on the "hole in the ground" attitude of the "literary set," Finn focuses on the need for an appropriate theory of history which he first outlines and applies to three problems facing society: the problem of knowledge; the problem of the second industrial revolution; and the problem of public philosophy. In examining technology, he looks at the negative and the positive, the unacceptable and the acceptable, the inherent dangers and the challenges. "The future presents us with the picture of an incredible load of knowledge, a radically new social organization, and a necessity to communicate the public philosophy."

The statement, "We need a new audiovisual *systems* theory; we need it NOW," concludes the article "AV Development and the Concept of Systems" (1956). This predates by about 10 years the surge in the mid-1960s toward the systems approach in education.

Finn briefly describes the "systems concept" and notes its application in industry and the armed forces. He holds that the general theory of educational administration is at least two generations behind the times. This lag has extremely negative effects on the audiovisual field—a field which he sees as geared to the technological world of the future with interlocking, complicated systems of men and machines.

Originally made as the president's address to a Department of Audio-Visual Instruction summer meeting in 1960, "Technological Innovation in Education" (1960) begins with a reference to King Canute and Ethelred, the Unready. Finn relates these two concepts to DAVI's role in innovation: "We sit on the rising curve of swift technological change in education with some hope of giving it direction."

Finn adds a seventh criterion to the six basic criteria he had earlier discussed for a profession—a profession must have the ability to exercise its own leadership. He sees the seventh criterion as based essentially on the sixth, "an organized body of theory constantly expanding by research."

In an examination of the needs and difficulties of innovation and leadership, Finn draws attention to the problem of the "90-day pioneer" who tends after partial success to become orthodox and change resistant.

What does all this mean to the audiovisual profession? As Finn states:

It means, first, that we have to understand and live with accelerating technological change in our own business; it means, second, that we must learn to master these changes and help integrate them into the educational process so that their benefits may be useful and rewarding; it means, third, that we cannot let the practical sociology of the situation deter us or we will lose control of the movement; it means, fourth, that we must lead an intellectually demanding professional life with the forces of technology with which we daily work.

Writing to educators in general through the *NEA Journal* in 1960, Finn titled his article "Teaching Machines: Auto-Instructional Devices for the Teacher" (1960). He stated: "It would help everyone in the teaching profession desiring to assess the teaching machine movement if it were clear exactly what was being talked about, what has been claimed for these devices, what their advantages and limitations are, and what they will do and will not do for teachers and students."

The succinct and descriptive article which follows describes teaching machines, theories of programming, types of machines, and the state of the art in 1960. The relevance and implications of the technology of individual instruction are outlined in an attempt to help teachers understand and manage instructional technology to further human ends and build teaching into the most human of all professions.

"New Techniques of Teaching for the Sixties" (1961) was written principally for those in the field of teacher education. By describing technology as involving systems, control mechanisms, patterns of organization, and a way of approaching problems, Finn creates a new perspective for teacher education in the article.

A brief historical overview of developments is provided and the double-pronged growth of "individual instructional technology" and "mass instructional technology" is outlined. Finn introduces the concept of instructional systems diagrammatically as a combination of individual, mass and conventional instruction.

Four patterns of teacher education in the audiovisual area are identified and discussed. Teachers must consider the implications of technology in education and the need for becoming higher level professionals in order to remain in control of education. "A technology of instruction forces us back to our basic mission—methodology."

"Instructional Technology" (1965) was written to support Finn's proposal that DAVI be renamed. The article identifies the high level plateau in the territory the author had spent many professional years mapping and describing so that the profession could reassess, rename and redirect itself toward the future as an integral part of the technological society. "The DAVI that has been my professional life for 25 years must now connect with the future and announce to the world that its members are the technologists of the educational profession."

Seeing the only alternative to looking at the future as looking back at the past, he saw no realistic choice at all. He did not regard the concept of adventurers nor the concept of instructional technology as comforting but as essential if we are to keep in touch with the accelerating times and a world of man-machine systems.

"Properly constructed," he wrote, "the concept of instructional or educational technology is totally integrative. It provides a common ground for all professionals, no matter in what aspect of the field they are working; it permits the rational development and integration of new devices, materials, methods as they come along." Finn saw the concept he proposed as a viable alternative to professional splintering and foundering and as a vehicle for professional unity and direction.

In "The Marginal Media Man" (1965), Finn seems to be saying that the field is like a supersonic aircraft which has soared up into unexpected air currents and now has spun into control problems for which the crew was not prepared. Lots of potential, lots of success, but lacking direction—control was being lost.

"For the plain truth of the matter is that our business, field, profession—whatever you want to call it—is in serious trouble." Finn re-emphasizes the need to change traditional concepts about the professionalization of the educational media field. He saw the media man as always being marginal to many fields.

Finn states the paradox of success and failure, briefly reviews the historical development of DAVI toward professionalism, and examines the current growth and leadership situation. He itemizes publications, conventions, conferences, and councils on the positive side together with the emergence of new technologies, organizations and events such as federal funding. On the negative side of the paradox he makes his case for retrogression of professional development, including the downgrading of the educational media field by the U.S. Office of Education and the White House Conference on Education that didn't include media.

"A Possible Model for Considering the Use of Media in Higher Education (1967) is an edited extract from a memorandum prepared with others for Harvard University. The model represents a way of thinking which emerged over a period of time while Finn was working at the national level with professors from several disciplines. He noted how, according to their experience and the structure of their disciplines, the instructors' approach to new instructional processes differed.

Finn summarized that the newer educational media can be applied at several levels—as individual instructional tools, as data storage, for behavior control-type instruction, to build meaning, as research tools, as the core of instructional systems, and as a means to increase the distribution efficiency of learning experiences of all kinds.

In "Educational Technology, Innovation, and Title III" (1966), Finn records an assessment of Title III NDEA funding toward innovation in education. A considerable portion of the original paper, devoted to providing a general context and historical overview, is not included because it overlaps other articles in this section (see "The Emerging Technology of Education"). The part used is subtitled "Educational Innovation and Title III" and deals with proposals, confusion of proposal writers, objectives, nature of innovation, hurdles, and procedures. Finn states that at federal and state levels there is an attempt through Title III to institutionalize educational innovation through bureaucracies. He notes: "It is still a very moot point whether or not bureaucracy is the exact opposite of innovation."

Other items discussed include scholarly support, dissemination problems, and weaknesses or gaps. The article concludes with 11 recommendations.

The closing article in this section is "What is the Business of Educational Technology? Some Immodest Comments on Mr. Muller's Paper" (1968). Finn summarizes Muller's argument that educational technology had been oversold at every point along the way; that it had had little effect on the educational enterprise which fundamentally depended upon people, not technology; and that the computer can make contributions to instruction. Agreeing in part regarding the contributions of the computer, Finn then thrusts with the disarming style of an authority at Muller's one-sided stance. In refuting Muller's argument, Finn aligns him with the overcautious and pessimistic on the literary side of the Snowian dichotomy.

Citing statistics from data which had been gathered while he was director of the NEA's Technological Development Project, Finn casts doubt on Muller's assertion that technology has had little effect. Emphasizing that the educational system is still highly undeveloped technologically, Finn maintains that we are much further ahead than Muller assumes.

After responding to Muller's categorical statements, Finn goes on to stress the need for instructional technology in today's tri-revolution setting. He raises some questions yet to be resolved regarding the role of the teacher in the classroom, the administrator in the board room, and the educator in society. Admitting that educational technology could be used to condition young people to live in a world of standardization, conformity and alienation, Finn stands on the side of mankind and says we must move toward freedom, creativity and worthwhile sense of self.

Professionalizing the Audiovisual Field

Specialization of occupation is a growing social factor in modern life. This factor is as applicable to education as to any other field. Where once there were only teachers, there are now administrators, psychologists, curriculum consultants, counselors, and many other educational specialists. Each of the specialties is developing into a profession within the general profession of education. Educators whose main responsibility lies in the preparation, distribution and use of audiovisual materials represent another group of specialized personnel newly developed and integrated into the field of education.

In addition to the fact that people working with audiovisual materials are devoting the major share of their time to a specialized phase of education and are developing special interests, techniques, etc., there is also the fact that the audiovisual field itself is somewhat unique in that it embraces all branches of the communication arts and technology and brings new disciplines to bear upon the problems of education. This second fact makes the audiovisual field even more of a specialized educational activity than, say, the teaching of reading.

In recent years audiovisual workers have become sensitive to the professional problems of their specialty. Questions have been raised as to the possible degree of professionalization of the movement; as to what, if any, certification requirement should be set up for audiovisual directors; and as to the long-range professional objectives of associations such as the Department of Audio-Visual Instruction (DAVI) of the NEA. DAVI has set up a Committee on Professional Education to study the general problem of professionalization.

It is the purpose of a series of papers, of which this is the first, to present a study of the problem of professionalization to the membership of DAVI from the Committee on Professional Education. These papers will analyze the present status of the field to determine, if possible, the degree of professionalization that has been developed, to review the historical development of this status, and to suggest some problems that must be met and some possible solutions that might be developed in order to move the field further in the direction of a true profession.

It is hoped that these studies will stimulate the membership of DAVI and other people working in the field to undertake appropriate action. It

Reprinted from *Audio-Visual Communication Review*, Vol. 1, No. 1, Winter 1953, pp. 6-17.

is very significant to the Committee on Professional Education that this series of papers is inaugurated in the first issue of the new professional magazine of the Department of Audio-Visual Instruction.

Tools of a Profession

In considering the audiovisual field as a possible area of professionalism, a good place to begin is with the question: What are the characteristics of a profession?[1] A profession has, at least, these characteristics: (a) an intellectual technique, (b) an application of that technique to the practical affairs of man, (c) a period of long training necessary before entering into the profession, (d) an association of the members of the profession into a closely knit group with a high quality of communication between members, (e) a series of standards and a statement of ethics which is enforced, and (f) an organized body of intellectual theory constantly expanding by research.

The statements identifying these characteristics need little comment. That a profession is primarily intellectual in character can be readily seen by viewing the activities of any profession; a doctor who did not reflectively think before prescribing is inconceivable. That a profession applies its knowledge directly to the benefit of man is also obvious.

The long periods of training necessary to develop specialists such as design engineers or oral surgeons are common examples of the third characteristic. Professional associations which began their evolution in the Middle Ages are a part of every civilized society. They identify the members who have successfully passed through the long training stage and, in fact, even control to a great degree the nature of that training. Communication between members of the profession is carried on by meetings, journals of high quality, consultations, and other means.

Architects, actuaries, engineers all have their codes of conduct or statements of ethics and various forms of standards. Coupled with this ethical formulation is a means of enforcing it in the more highly organized professions. Sometimes this enforcement responsibility rests with the professional association, sometimes with the state as a licensing body, and sometimes with both. Although there is much criticism of many professions at this point and some evidence (Landis, 1927) that many codes are window dressing to protect the profession from public interference and are not enforced except to the advantage of the profession as against the public, the fact remains that the idea of an ethic with the power of enforcement places a personal responsibility on each member of a profession not as-

---

[1]The best quick source on the nature and development of the professions with special reference to the teaching profession may be found in Smith et al. (1951). Part Four of this volume, "The Nature and Status of the Teaching Profession," contains pertinent articles by A. M. Carr-Saunders, Abraham Flexner, Alfred North Whitehead, and William O. Stanley. Many of Flexner's other works also consider this problem. See also articles relating to the professions and professionalization in the *Encyclopaedia of Social Sciences*. A good idea of the development of a profession to a status closely resembling medicine may be obtained by studying the last four or five years of the *American Psychologist*. In this journal reports of committees on standards, ethics, training, etc. are particularly revealing.

sociated with other types of occupations.

Finally, the most fundamental and most important characteristic of a profession is that the skills involved are founded upon a body of intellectual theory and research. Furthermore, this systematic theory is constantly being expanded by research and thinking within the profession. As Whitehead says, "...the practice of a profession cannot be disjoined from its theoretical understanding, and *vice versa*....The antithesis to a profession is an avocation based upon customary activities and modified by the trial and error of individual practice. Such an avocation is a Craft..." (Smith et al., 1951, p. 557). The difference between the bricklayer and the architect lies right here.

Professional Status of Audiovisual Education

We can now examine the present status of audiovisual education when measured by these six tests of a profession. Are audiovisual personnel, in fact, professionals? By audiovisual personnel is meant, for the moment, those individuals who spend 50 percent or more of their time working with audiovisual programs in schools and colleges as directors, supervisors, producers, consultants, etc., or those who engage in in-service and pre-service teacher training or research in this area.

An intellectual technique. First, the audiovisual worker does possess an intellectual technique. He has to think reflectively in such varied areas as the critical evaluation of materials, the visualization of abstract concepts, the improvement of instruction, and in many aspects of planning and administration. Audiovisual personnel, as a group, meet this criterion fairly well.

Practical application of the technique. Second, audiovisual techniques and materials justify their existence only as they become operative in classroom communication. Hence the test of practical application is completely met. Here the personnel of the field is at its best. The practical problems of classroom design, equipment and materials are the meat and drink of most audiovisual people. As will be indicated below, there is, perhaps, even an overemphasis on this point.

Long period of training. The test of a high degree of professionalization of the audiovisual field, however, breaks down completely against the third criterion—a long period of rigorous training for the members of a profession. Most professions not only require this long period of training but are also in substantial agreement as to the nature of this training. This results in the professional associations specifying the nature of the training either through state regulation of some sort or through a system of accrediting training institutions.

The teaching profession as a whole does maintain training standards. But specific training for audiovisual directors and other personnel, with few exceptions, is still in the thinking stage. Although there have been directors of programs since before World War I, McClusky's bibliography lists only 15 articles in the literature which discuss the requirements for audiovisual personnel (McClusky, 1950, pp. 135-136). An examination of these articles reveals that only four are pertinent (Finn, 1941; Frazier, 1949; Lewin, 1947; and Shreve, 1950). The others are devoted to administrative relationships and duties of principals, building coordinators, students, and miscellaneous problems. There has been practically no thoughtful consideration of this problem by audiovisual people and no attempt to develop standards.

The history of all professions reveals that the lengthy and rigorous training programs came after a long period of evolution. So it is not surprising to find that the audiovisual field has not made an organized effort as yet to develop such a program. The audiovisual field has developed rapidly and has surmounted many professional problems without showing all the required characteristics of a profession. Now, in 1953, the field is really, for the first time, in a position to take a good look at the problem of professional training. The State of Indiana has already taken action, and proposals have been published in other states as to the training necessary for an audiovisual director and pointing to some form of certification. The Committee on Professional Education of DAVI has this as one of its direct concerns.

The nature and content of professional education for audiovisual directors and other workers present many problems that must be solved before audiovisual education can claim the status of a profession. The system of apprenticeship training that has been in operation is no longer adequate. Trained audiovisual personnel will not stay in their present jobs forever, and there is no longer the reservoir of service-experienced people to draw upon. Obviously, a graduate program that can provide the competencies generated by service and industrial experience coupled with a better theoretical background is required immediately. The audiovisual field cannot be upgraded into a profession until this occurs. Other unsolved problems include the nature of certification standards, admission standards and practices, and placement.

Association and communication between members. The fourth criterion of a profession—a closely knit association with a high quality of communication between members—is another point at which the audiovisual field does not measure up. Considering first professional association, the best that can be said at present is that a professional association is in the process of *becoming* and will someday emerge.

For many years DAVI was a comparatively weak organization held together by a small group of stalwarts. DAVI went through several reorganizations and managed to survive a depression and a war, but only in the last two or three years has the organization shown anything like the potential it can develop. The present arrangement which ties in the organization with the NEA through its executive secretary, with working national committees dealing with important problems, and with an increasing and interested membership promises much for the future.

The audiovisual field has also suffered from too many organizations. It is a moot question whether the organizations which represent special applications of the field such as The Association for Education by Radio-Television (AERT), the Educational Film Library Association (EFLA), and the Film Council of America (FCA) should remain outside of the mainstream of the DAVI or become divisions within it in order to develop the best possible organization for the profession. The men and women who founded and carried on these organizations deserve nothing but commendation for their continual struggle and achievements, but the field as a profession would probably benefit more by merger than by continued separatism. At least this possibility should be thoroughly explored.

At the state and local levels, the structure of audiovisual organization has not yet even approached the professional. There are some fine state units, to be sure. The Audio-Visual Education Association of California, one of the oldest and strongest, is a professional organization in every sense of the word. AVID of Indiana has achieved national recognition, and AVDO of Ohio is rapidly growing in strength. And there are others. But much work remains to be done on the state and local levels.

It is at the other half of the concept of association—the idea of a high quality of communication between members—that the audiovisual movement as a whole had failed until the decision was made to publish the journal in which this paper appears. With the exception of Edgar Dale's *Newsletter*, all of the journals serving the field had difficulty presenting professional content. This was true of *Educational Screen, Audio-Visual World, See and Hear, Audio-Visual Guide, Business Screen, The Journal of the AERT, Film News*, and all the rest. Most of the time these magazines were not able to print thorough and scholarly papers on the theoretical bases of audiovisual education; research studies for the most part were ignored and left to journals outside the field. When compared to the *Psychological Review, The American Journal of Sociology*, or a hundred other professional periodicals, the audiovisual magazines have simply not measured up professionally. There were good and sufficient reasons for this, but the fact remains.

This is not to say that these other audiovisual journals have failed to contribute as the audiovisual field struggled through its infancy. They have done their share in developing the field. In particular, Nelson Greene made a great contribution through the years with the *Educational Screen*. Greene was a scholar and had an intensive interest in professionalizing audiovisual education. Some examples of this interest were the publication of Krows' somewhat dull but important account of the development of the nontheatrical film, carried serially over two years; David Goodman's abortive column on research abstracts; and an attempt to carry a column which critically reviewed the literature of the field.

In general, the journals until now have made a contribution by carrying information on materials and equipment, occasionally publishing an article of professional merit, and everlastingly promoting and crusading for things audiovisual. This is a sign of the childhood and adolescence of the audiovisual movement. Audiovisual education is here to stay. Promotion and professionalization, while both are necessary, are not the same things. The time has come to add the dimension of professional content to the field's journals and it is hoped that the *Audio-Visual Communication Review* will fill the gap.

Professional communication is also carried on in meetings and conferences. The same criticisms leveled at the quality of the journals can apply to the quality of most audiovisual meetings. The meeting agenda seem to be of two types. One is a type designed to appeal to the practicing teacher and consists of a rehash of one or more chapters of Dale, Hoban or Kinder carried on for two or three days! To the audiovisual professional, this type of meeting is about as intellectually stimulating as a plateful of unsalted grits would be to Oscar of the Waldorf. The other type appeals to the ever-present gadgeteer in audiovisual circles and, while it may not be concerned any more with the "f-value" of lenses, the topics have merely changed to more efficient booking forms, the JAN projector, or the heat and pressure necessary to laminate a 20×24 print.

The writer is not arguing for the elimination of meetings designed primarily for teachers nor for the abolition of the technical problems of the audiovisual field from consideration. Certainly thousands of teachers need help with the elementary concepts of audiovisual instruction;[2] certainly the audiovisual field will always be plagued with technical problems which must be solved. But professional meetings are not professional meetings if they are limited to these two areas. The first can be best dealt with in regular gatherings of teachers rather than at audiovisual meetings, and the second area should be reduced to a section or two of professional audiovisual conferences to restore perspective.

Again, improvement in recent years has been noted. The agenda for the Boston and St. Louis meetings of DAVI showed many signs of professionalization not present at earlier meetings. Many state meetings have been improving programs. Nevertheless, the improvement of audiovisual conferences has a long way to go.

In summary, then, to achieve a real professional status, the audiovisual movement needs to develop a strong and creative association at the national and state and local levels; it needs to develop true professional journals; and it needs to improve conferences and meetings. When these things are done, audiovisual personnel can say they have met the requirements of the fourth criterion—the criterion of association and communication.

Code of ethics and standards. The fifth measuring point—ethics, standards and their enforcement—is a function of the fourth, a strong association. Statements of ethics and publications of standards are developed by professional associations. Audiovisual personnel, as members of the teacher profession, are subject to the ethics of the profession. As yet, nothing has been done to develop a separate code of ethics for the audiovisual movement.

In the field of standards, there are signs that professionalization is underway. The Committee on Buildings and Equipment of DAVI is studying standards in its field, and has produced an excellent publication (Planning Schools for Use of Audio-Visual Materials, No. 1, Classrooms, 1952).

However, the publication of codes and ethics and manuals of standards in itself guarantees nothing. Professionalization occurs when enforcement is possible and vigorous. Thus, the American Medical Association wages

---

[2]See, for example, the recording: Edgar Dale and James D. Finn, "The Improvement of Teaching Through Audio-Visual Materials," Educational Recording Services, Los Angeles, 1951, an aid for teachers' meetings, illustrating the position of the writer on this point.

war on quacks and malpractice; engineers and architects write building standards into the law; and the courts can disbar a lawyer for illegal practice.

Enforcement is closely tied in with admission to the profession by a licensing system (a function of criterion three—training), with placement which assures that licensed personnel are hired, and, most fundamentally, with an obligation on the part of each professional to the ethics and standards of his profession.

In view of the fact that the entire education profession has not met this criterion to the degree that the other professions have, it is questionable whether the audiovisual group will ever completely measure up to this point. And it is even questionable whether such a rigorous arrangement is either necessary or desirable. However, the audiovisual movement will at least have to reach the stage where it has a well-defined code of ethics, a series of standards based upon fundamental research, and a form of certification somewhat related to them. At the moment the field is not at this stage and does not meet criterion five.

Intellectual theory and research. As was indicated in the introductory phase of this paper, the most important characteristic of a profession was the sixth and last—that the technique of a profession is founded upon a body of systematic theory and research constantly being expanded by research and thinking within the profession. When the audiovisual field is measured against this characteristic, again the conclusion must be reached that professional status has not been attained.

Audiovisual workers have put a premium on "practicality" and have been criticized for this by colleagues within the field of education and by the literati from without. There is some merit to this criticism. For years, even at audiovisual meetings, someone has always been taking cracks at the "gadgeteers." As the writer has indicated above, it is his position that the audiovisual field is a result of the fruits of technology applied to the educational process and a certain amount of gadgeteering will always be necessary. Too much, however, reveals a poverty of thought.

The audiovisual field has never been too clear on the point that theory and practice must constantly interact in any intellectual activity of man. In line with some other mistaken educators, many audiovisual people have insisted that they want to be "practical" and not "theoretical," and that "experience" is the thing. This is, in part, an honest reaction against an older viewpoint that placed theory up somewhere near the Milky Way where it had no relation to practice except to cause aesthetic chills to chase up the spine of some professor.

This attitude, however, also represents a complete misunderstanding of the nature of reflective thinking, scientific progress, and the wellsprings of human behavior. As Dewey has said, "...we find that experience when it is experimental does not signify the absence of large and far-reaching ideas and purposes. It is dependent upon them at every point" (Dewey, 1960). Without these large and far-reaching ideas any field, and this is particularly true of the audiovisual field, can go only so far and then has to stop.

Many of the criticisms listed above (lack of content at meetings, journals with little intellectual meat, etc.) are merely symptoms of this greater trouble—lack of theoretical direction. Without a theory which produces hypotheses for research, there can be no expanding of knowledge and technique. And without a constant attempt to assess practice so that the theoretical implications may be teased out, there can be no assurance that we will ever have a theory or that our practice will make sense.

The audiovisual movement is new and growing, but it is in danger of becoming stunted if it is left to its present theoretical formulations. The present theory guiding the movement can be summed up in three references (Dale, 1946; Hoban et al., 1937; and Kinder, 1950). The basic concept around which all three have oriented is the notion of the concrete-abstract relationship in learning. This is perhaps most thoroughly explored in Hoban. Dale adds material on retention and forgetting with a brief historical section, and Kinder expands to a very short history of communication and deals slightly with perception and imagery. All also emphasize the gamut of materials approach to learning, the concept of utilization, the experience theory of learning, and the strengths and weaknesses of the various aids.

The remainder of audiovisual theory is scattered throughout the literature. McClusky (1949) has related audiovisual techniques to learning theory in a somewhat unique fashion; Brooker (1949) began a line of thinking of promise in his discussion of communication which remains to be explored; Exton's (1947) contribution of the concept of "optimum synthesis" has not received the attention it merits. All of these are but examples of the scattering of notions throughout audiovisual literature which, when brought together, might constitute a beginning of a fruitful theory.

To these examples, of course, much more would have to be added: most of the writings of Hoban, many of Dale's essays in the *Newsletter*, reports on the proceedings of conferences, generalizations derived from successful practice, generalizations derived from research, etc. The audiovisual field cannot rest its theoretical formulation on the contents of several textbooks designed for teacher training that do not include even all the useful theory to be found in audiovisual literature.

Because of the nature of the audiovisual field, however, useful theory is not confined to its own literature. As most workers realize, there is a literature of the film, of photography, of the museum, of dramatization, etc.; there is also the literature of educational method and curriculum; there is the literature of educational psychology, of social psychology, of social anthropology; there is the literature of art and design; finally and perhaps most important, there is the growing literature of communication. In fact, research and thinking in some parts of the physical sciences (neurology, physiology, acoustics, etc.), the social sciences (social communication and control, learning theory, etc.), and the humanities (art, music, etc.) are each pertinent to the field. We need to understand the filmic expression ideas of Slavko Vorkapich, the visual experiments of Samuel Renshaw, and the communication theory of Susanne Langer.

Viewed in this light, most people in the audiovisual field are still guided by a theory which is fragmentary; theory as now guiding the field (the generalizations held by many of the workers) is not even inclusive of the notions contained in audiovisual literature; it has never worked in most of the pertinent generalizations available from outside these narrow limits.

On the important test of theory the audiovisual field does not meet professional standards. Its workers are craftsmen, not professionals, in the majority of instances because they are operating, in Whitehead's words, on "customary activities modified by the trial and error of individual practice." Absolutely fundamental to the development of audiovisual education as a profession in the sense that DAVI is now using the term is the prior development of an all-inclusive body of theory upon which to assess and guide practice and base research. Once this is done, many of the other criticisms stated above will no longer be valid because their source will have disappeared.

The status of audiovisual research also reflects on professionalization. Not that research does not exist or that it is not being pursued. A recent bibliography (Dale, Finn and Hoban, 1948, pp. 284-293) lists 163 titles through 1946, and this is by no means all-inclusive. Because of the journal policy discussed above, research pertinent to audiovisual education is published throughout the literature of the social sciences and needs a staff of detectives to trace it down. Very little of it has been reported in audiovisual meetings. This means that many audiovisual workers must be "flying blind"—a black mark for professionalization.

The post-war years have brought an increase in research activities in audiovisual education and related areas. Much of this research is government sponsored and financed, but it is being published in pamphlet form, in psychological journals, or in other places more or less inaccessible to the practicing worker in the audiovisual field. A true profession, such as medicine, makes this information more easily available to its practitioners. Furthermore, outside of the volume by Hoban and van Ormer (1951), there is not much evidence that research is influencing the formulation of theory. Many of the hypotheses now being tested have been derived from learning theory and the "social perception" theories which have been developed by a number of social psychologists. *The audiovisual field is in the peculiar position of having much of its research carried on by workers in other disciplines using hypotheses unknown to many audiovisual workers, and reporting results in journals that audiovisual people do not read and at meetings that audiovisual people do not attend.* While the research is expanding the intellectual background of the profession, it seems to be having little effect. A tremendous amount of integration is necessary before this part of the criterion six can be met.

Summary

In summary, then, of the six criteria set forth in this paper: (a) intellectual technique, (b) application of technique to practice, (c) long training period, (d) association of members with a high quality of communication, (e) a series of standards and an enforced statement of ethics, and (f) an organized body of intellectual theory constantly expanding by research, audiovisual personnel meet only the first and second completely. The fourth and fifth are met to a degree which is not satisfactory but which is improving. And the third and sixth tests rate such low scores that failure is the only possible grade. This adds up, in the opinion of the writer, to the simply stated fact that *the audiovisual field is not yet a profession.*

References

American Psychologist. Published monthly by the American Psychological Association, Washington, D.C.

Brooker, Floyde E. "Communication in the Modern World," in *Audio-Visual Materials of Instruction*, 48th Yearbook, Part I of the National Society for the Study of Education. Chicago: The University of Chicago Press, 1949, pp. 4-27.

Dale, Edgar. *Audio-Visual Methods in Teaching.* New York: Dryden Press, 1946. Revised edition, Holt, Rinehart and Winston, 1954. *Audio-visual Methods in Teaching*, Third Edition, Holt, Rinehart and Winston, 1969.

Dale, Edgar, James D. Finn, and Charles F. Hoban Jr. "Research on Audio-Visual Materials," in *Audio-Visual Materials of Instruction*, 48th Year-book, Part I of the National Society for the Study of Education. Chicago: The University of Chicago Press, 1949, pp. 253-293.

Dale, Edgar and James D. Finn. "The Improvement of Teaching Through Audio-Visual Materials." Los Angeles: Educational Recording Services, 1951.

Dewey, John. *The Quest for Certainty: A Study of the Relation of Knowledge and Action.* New York: G. P. Putnam's Sons, 1960.

Encyclopaedia of Social Sciences. New York: The Macmillan Company, 1930.

Exton Jr., William. *Audiovisual Aids to Instruction.* New York: McGraw-Hill Book Company, Inc., 1947.

Finn, James D. "Adequate Training for a Director of Audio-Visual Education," *Education*, Vol. 61, No. 6, February 1941, pp. 337-343.

Frazier, Alexander. "How Much Does the Audio-Visual Director Need to Know?", *School Review*, Vol. 57, No. 8, October 1949, pp. 416-424.

Hoban, Charles F., Charles F. Hoban Jr., and Samuel B. Zisman. *Visualizing the Curriculum.* New York: The Cordon Company, 1937.

Hoban Jr., Charles F. and Edward B. van Ormer. *Instructional Film Research 1918-1950. (Rapid Mass Learning).* Port Washington, Long Island, N.Y.: Special Devices Center, Department of the Army and Department of the Navy, 1950. Technical Report No. SDC 269-7-19.

Kinder, J. D. *Audio-Visual Materials and Techniques.* New York: American Book Co., 1950. Second Edition, American Book Co., 1959.

Landis, Benson Y. *Professional Codes: A Sociological Analysis to Determine Applications to the Educational Profession.* Contributions to Education, No. 267. New York: Teachers College, Columbia University, 1927.

Lewin, W. (Ed.). "Duties of School Audio-Visual Coordinators," *Film and Radio Guide*, 13, 1947, pp. 10-11.

McClusky, F. Dean. *Audio-Visual Teaching Techniques.* Dubuque, Ia.: Wm. C. Brown Company, 1949.

McClusky, F. Dean. *The A-V Bibliography.* Dubuque, Ia.: Wm. C. Brown Co., 1950.

Planning Schools for Use of Audio-Visual Materials, No. 1, Classrooms. Washington, D.C.: Department of Audio-Visual Instruction, NEA, 1952.

Shreve, R. "The Superintendent Hires an A-V Supervisor," *See and Hear,* 1950, Vol. 5, p. 35.

Smith, B. Othanel; Kenneth D. Benne; William O. Stanley; Archibald W. Anderson. *Readings in the Social Aspects of Education.* Danville, Ill.: Interstate Printers and Publishers, Inc., 1951.

A Look at the Future of
AV Communication

My text comes from a rather famous speech by President Griswold of
Yale University on occasion of the National Book Awards dinner in 1952.
Dr. Griswold, an exceptionally literary man, for a college president, was
discussing his favorite topic—the decline of reading and thinking in the
United States. As he neared the climax of his speech, he shook the walls of
the Waldorf Astoria and later the pages of the *Saturday Review* and
Harpers with these machine-gun like statements: "What are we doing
under our forest of television masts . . .? We are succumbing one by one to
technological illiteracy. We have traded in the mind's eye for the eye's
mind. . . . Here and there . . . [reading] still hangs on in competition with
more efficient methods and processes, such as the extrasensory and the
audiovisual" (Griswold, 1954, p. 71).

Now, although I am a great reader of science fiction, I was not aware
that *extrasensory perception*—"thought transference" to those of you who
slept through general psychology—had reached the stage where it had become as efficient as we all know audiovisual communication to be. When I
was asked to examine the future of the audiovisual communication movement, it occurred to me that an exploration of President Griswold's thesis
might be a good starting point. If thought transference is possible in the
near future, we might as well disband now, go home, and buy a pack of
the famous Rhine cards.

Actually, of course, Dr. Griswold was speaking ironically. He was, however, speaking in a great tradition of both older and contemporary literary
figures, including William Wordsworth, Henry James, Clifton Fadiman,
and Joseph Wood Krutch, all of whom have attacked, in poem and essay,
the more efficient means of communication, be they illustrated books and
newspapers as in the case of Wordsworth, or all audiovisual materials and
devices as in the case of Krutch. Implicit in these criticisms of "efficient"
communication as compared to reading, there lies a theory of history. And
without a theory of history, it is not possible to assess the future.

Presumably, Dr. Griswold, Mr. Fadiman, and Joseph Wood Krutch
make use of running water, automobiles, and can openers and, presumably,
Wordsworth was willing to settle for the usefulness of Watt's steam

Reprinted from *Audio Visual Communication Review*, Vol. 3, No. 4, Fall 1955,
pp. 244-256. Based upon an address given by Finn before the DAVI Convention
in Los Angeles, April 19, 1955 entitled "Will ESP Replace AV?"

engine. It would seem, then, that this branch of the intellectual community decries certain kinds of material progress but not others. As the historian, Herbert Muller, has pointed out, Thoreau took civilized tools and materials with him into the woods when he wanted to survive *without* civilization; and, in another place Muller points this position up when he says, " . . . we have the curious spectacle of civilized man forever marching with his face turned backward—as no doubt the cave-man looked back to the good old days when men were free to roam instead of being stuck in a damn hole in the ground" (Muller, 1952, p. 65).

All forms of material progress, including the development of audiovisual techniques and devices for communication, admittedly have brought evil. Material progress and specialization—which is one definition of civilization—have also brought good. A theory of history which can encompass this contradiction in order to help us assess the future, even for the audiovisual field (which is almost a symbol of technical progress), is all-important. We cannot accept Griswold with his "hole in the ground" approach; we cannot accept his ironical suggestion that, in the immediate future, technology will be replaced by mental energy. What, then, can we say of our increasingly technical and specialized development in communication that will give us guideposts for the future?

Again, Muller seems to me to be pertinent. His position is that: " . . . in the long evolutionary view, reaching back to the cave-man and ape-man, there unquestionably has been progress—always granted the assumption that it is worth being a human being. Man has achieved greater mastery of his natural environment, greater freedom of action, and thereupon has discovered the finer possibilities of life implicit in his distinctive power of consciousness" (Muller, 1952, p. 68).

Progress, according to this view, is not inevitable. As man is increasingly freed by technology and civilization, he achieves a greater power to make choices which control his own destiny. These choices can be wise or foolish, but it is technology and civilization that give him the opportunity to make them. This view explains why man has, at times in his development, moved in the wrong direction. Today, we have more choices than ever before; we will probably make more mistakes. But, with Muller, I think that, "Our business as rational beings is not to argue for what is going to be but to strive for what ought to be, in the consciousness that it will never be all we would like it to be" (Muller, 1952, p. 70).

As Whitehead said, "It is the business of the future to be dangerous." Into this uncertainty, into this plural universe where we can make mistakes and intelligent choices, let us project the technological field of audiovisual communication and apply this theory to three great problems facing our society:

1. The problem of knowledge.
2. The problem of the second industrial revolution.
3. The problem of the public philosophy.

The Problem of Knowledge

Many dimensions of the problem of knowledge will plague us with communication problems in the future. One of the most difficult aspects of this problem is the fantastic and incredible *increase* in human knowledge. As the sociologists have told us, we have, indeed, invented a method of invention, and this method of invention continually spews forth knowledge and information day and night with a double shift working on weekends. For example, the *Quarterly Journal of Current Acquisitions* of the Library of Congress reported for August 1954 that the Library had received in the fields of science, technology, medicine, and agriculture alone approximately 30,000 journals, including 2,000 new titles; 25,000 research reports; 15,000 books and monographs; 15,000 manuscripts; 10,000 pamphlets; 5,000 prints, blueprints, microfilms, and the like; and 150,000 maps and charts.

Because President Griswold desires that we should read all the books in print in the Western World, I took pains to inquire as to the holdings of the Library of Yale University at this time. According to a story in the *Los Angeles Times* (1955), the Yale University Library now contains 4,245,583 volumes; it occupies second place of all university libraries in the United States (Harvard is number one, with almost 6,000,000 volumes).

The nature of the problem then becomes clear. It would take one person, reading 24 hours a day, approximately 1,000 years to get through the present Yale Library. In the meantime, at the present rate of increase of knowledge and information, at the end of that 1,000 years he would have not made a dent in the then-existing collection. In fact, at the present rate of increase, figuring two hours per book, it would take one person, reading 24 hours a day, approximately two years to read the 15,000 volumes received in the sciences and technology by the Library of Congress during a three-month period. Reading eight hours a day, it would take him approximately six years and that includes Sundays and holidays.

This log jam has to be broken if this knowledge is to be communicated and made useful. It obviously cannot be done by increasing reading speed, desirable as that may be. The World War II Navy slogan for audiovisual materials, "More Learning in Less Time," is one possible approach to this problem. Audiovisual specialists looking toward the future must realize that the next great development of the audiovisual field must be an organized and systematic attack on existing and nascent knowledge so that it may be communicated. This implies an expansion of materials, equipment and personnel that even the dreams of the audiovisual prophets that were so current right after World War II did not anticipate. This is a great social need; our society must meet it beginning now and projecting forever into the future. Money, vision and ingenuity must be continually applied to the communication of knowledge.

The skyrocket growth of the record of knowledge is not, however, confined to books. Each year our films, filmstrips, recordings, charts, and kinescopes increase. We are already beginning to meet, on a smaller scale, the problems faced by book libraries on an extensive scale. The general problem is that when such an extensive record of knowledge exists, how can we locate within this record an item of information we desire? If the existing collection of audiovisual materials is to serve us—if we are to get the right material to the right place at the right time, as so many of us try to do—we must begin now to do something about classification, cataloging and location.

The existing film catalogs of libraries; the existing methods used by producers to supply information on new materials; the existing general lists of films, filmstrips and records; the existing methods used by libraries to book and distribute materials belong, for the most part, as much in the middle of the 20th century as does the dugout canoe. There are today at least 50 companies that manufacture and install all sorts of ingenious data-handling equipment used by banks, insurance companies, and scientific institutions. Items can be scanned at the rate of 1,000 a second, digital computers can manipulate records, and photographic techniques may be applied to records in all sorts of ways. With the exception of a few IBM machines scattered here and there in some of the larger audiovisual departments, nothing has been done in the field of classification, cataloging and location of audiovisual materials that is worthy of this century. We are still publishing catalogs outmoded before they are printed, which grow thicker every year until eventually the post office will have special 10-ton trucks just for the delivery of audiovisual catalogs from which you will have to order 10 years in advance.

The techniques of the 20th century are now available to help us organize that part of knowledge and information that exists in the record of audiovisual materials. Where, then, are those who are working to code audiovisual materials so that they may be handled under some data processing system? The government—and, as Vannevar Bush has pointed out, such general coding is a legitimate task of government—cannot even modernize its postal system where modern data-handling methods could be applied immediately. The audiovisual industry, some of whose systems are roughly comparable to building a house in 1955 out of handhewn logs and tar paper, seems no better able to assist.

And the profession? The profession is always 50 years too late, although a case can be made that the nature of the educational enterprise in this country makes this inevitable. Again, looking to the future, we should set up projects, beginning tomorrow morning, designed to plan, code and introduce modern data-handling procedures into the audiovisual materials system. There is no technical reason, for example, why every teacher in the nation couldn't have almost instantaneous information on any given *scene* in a film, let alone the film.

We are also faced with the fact that knowledge, as it becomes more specialized, undergoes continual splintering—a fact which continually worries people concerned with general education. This splintering of knowledge into smaller and smaller compartments has reached the point where information and ideas are no longer being communicated generally among men of learning, not to mention the general public. The ear surgeon cannot speak to the nose surgeon and be understood; the sales tax lawyer can say no more than hello and goodbye to the income tax lawyer; the curriculum director and guidance director cannot even worship Rousseau at the same altar.

This situation, the creation of specialized ignoramuses, so prevalent in America today, is the legitimate concern of President Griswold and many others; for example, his somewhat older compatriot, Robert Maynard

A Look at the Future of AV Communication

Hutchins. Somewhere Mr. Hutchins has said that it would not be necessary to burn the books, only to leave them unread for a couple of generations.

Now the concern here is for ideas—great ideas. While I do not agree at all with Mr. Hutchins' basic philosophy, I certainly agree that a highly specialized illiterate is a danger to society. However, Mr. Hutchins and Mr. Griswold, it seems to me, are going at it "wrong end to." In order to be civilized, we must be specialized. At the same time, we need great generalizing ideas which can cement the fields of knowledge together and infuse them with value. And at the more practical level we need to communicate specialized findings between related specialists.

Although it is not often realized, this problem of communicating knowledge between specialists has held back science and thought even in times when the accumulation of knowledge was relatively small. Two predecessors of Charles Darwin, one William Charles Wells in 1813 and one Patrick Mathews in 1831, described evolution by the means of natural selection which Darwin, in ignorance, redescribed in 1859. Wells' paper was written when Darwin was four years old, and never communicated. Think what the situation must now be with our mountains of undigested, unlocated knowledge—knowledge held incommunicado—without communication.

It seems here that the audiovisual field of the future must play a crucial role. We cannot stop specialization for we will stop civilization, whether Mr. Hutchins thinks civilization lies in the great books or not. We must devote much of our future production to the preparation of materials which can generalize knowledge and convey great ideas efficiently, and we must devote other production to speeding up and making more efficient communication between groups of specialists. Generalizing ideas, value concepts will not necessarily die if we cannot read all the books. Audiovisual materials must be reoriented in part from the *Adventures of Bunny Rabbit* to such difficult generalizations as, for example, the role of science in society.

Several problems revolving around the concept of the precision of information are also pertinent to our general problem of audiovisual materials and knowledge. Two of these may be described as:

1. Simplification.
2. Information which cannot be coded either into language or pictorial means.

As knowledge increases and becomes more specialized, it also becomes more complex. At one time, for example, in the 19th century, Lombroso developed a system of criminology which classified all criminals by facial types and ascribed a biological basis for crime. Today, the causes of crime and the nature of the criminal are known to be extremely complex and possible only of simplification in the tabloid newspapers.

It is in the nature of audiovisual materials to oversimplify many complex ideas and chains of ideas. This is a very great danger as knowledge becomes more and more complex. A motion picture is a highly selective device that permits little qualification in its grammar. The users of audio visual materials in the future will have to have more subject matter knowledge, not less, in order to handle this difficult problem in the communication situation, or we will, indeed, produce a generation with little learning. Producers must face up to the same problem. And the consumer of audio-visual materials must learn to apply the techniques of critical thinking to this type of communication.

You are all familiar with Edgar Dale's "Cone of Experience" or some other concept of the ladder of abstraction of knowledge. All of these ladders put words or language at the top as the most abstract means of communication or formulation of knowledge. We have always assumed, in dealing with audiovisual materials, that, by using these various representations of experience—the film, filmstrip, recording, and the like—that we can help individuals better comprehend what they read by building meaning into abstractions (words).

This idea that words are the most abstract containers of knowledge is now a more or less false idea. Above words, exists a higher order of abstraction that comes more and more into use as this century progresses. I refer to mathematics. There are now many concepts impossible to describe except mathematically. Language will not do. One of the basic ideas of quantum physics—that light behaves both in terms of wave motion and in terms of bundles, quanta or corpuscles—cannot be adequately described in words. In fact, in much of modern physics, and, increasingly in other fields, mathematical formulations are the only means by which the knowledge can be handled. Information theory itself consists of the equations of thermodynamics as developed by Shannon. I recently read a paper describing the mathematical theories of history by a professor of mathematics at the University of Chicago.

Precisely what this development means, I am not at this time prepared to say. Certainly, it should make us more humble toward the possibility of communicating all of knowledge by audiovisual means in the future. While, for example, the concepts of entropy may be visualized, they may not be visualized in the mathematical sense and they cannot be stated in language. And here I am certain that even Dr. Griswold's ESP will not help us much. The problem of the formulation and communication of knowledge is vast; we must be prepared, in the future, to question a number of assumptions regarding audiovisual materials that are accepted without question today; basic here is the concrete-abstract relationship.

There is also a possibility inherent in the great ability of audiovisual materials to simplify and to communicate efficiently that can make the future of knowledge dark instead of bright, terrifying instead of hopeful. I refer here to the fact that audiovisual communication techniques can be used in the future to pervert knowledge and information to immoral ends. This is the last dimension of the problem of knowledge that I wish to consider.

In science fiction, as has been demonstrated over its long history, we can find great artistic insights into the possibilities of the future. Throughout modern Utopian science fiction, as handled by the greatest writers, runs a common, horrifying thread. This thread can be found in Huxley's *Brave New World*, in Orwell's *1984*, in Ray Bradbury's *Fahrenheit 451*, and even in such a short story as Henry Kuttner's incredible "Year Day." This thread is also present in the writings of philosophers. Bertrand Russell considers the possibilities of the scientific totalitarianism of the future

A Look at the Future of AV Communication

which may come into existence if we make the wrong choices.

The common thread of horror as seen by these insightful artists is that the mass communication media—audiovisual media for the most part—may in the future be the means by which society is conditioned both in school and out for life in a totalitarian regime. Audiovisual materials are the ideal means for this. With them, we can truly convince that black is white, that Big Brother is all good, that humanity is worthless. With these materials, we can make all of society into one vast Pavlov's dog, conditioned to accept whatever it sees and hears as true knowledge, and to salivate at the will of its director. It is entirely possible, with the right controls, to pervert knowledge and control information with audiovisual materials.

This, you may say, is not the concern of audiovisual people. You may say that our job is to take the materials as developed and see that they are used. This point of view I would call the transmission-belt concept of the audiovisual mission, and I will not accept it. If the job of audiovisual people is merely to transmit materials, if we turn our minds and energies to it, we can all be replaced by electronic computers in 10 years—computers that can do our jobs much more efficiently. (We might then reconvert to ESP specialists.)

I submit that it is our job to be concerned about the content and philosophy of the materials we use in the future. I agree with the anthropologist, Redfield, who, in recently discussing the scientific method with his colleagues, said, "Just who are you neutral for?" And he answered this question, " . . . I have placed myself squarely on the side of mankind, and have not shamed to wish mankind well" (Redfield, 1953, p. 141).

Already there are signs that pressure groups, government officials at various levels, and various people with axes to grind, desire to step in and control the audiovisual materials we use in the schools. The films of my good friend, Lester Beck, cannot be used in a school system not far from here, due to the efforts of an interesting combination of pressure groups. In the same system, great pressure is put on the audiovisual department to list the programs of a reactionary radio commentator in a radio-listening bulletin for schools and to eliminate naming any liberal commentators whatsoever. This is the beginning of 1984 in 1955. In the public arena, we have in existence phony television forum programs and allegedly impartial magazines which are elevating the perversion of information into a fine art.

If we are on the side of mankind; if we believe that the future must contain choices, we must, in the future, defend the freedom of audiovisual communication and see that the materials contain, insofar as it is possible, all knowledge, not a perversion of it.

As I have tried to show, then, the problem of knowledge, in terms of its scope, in terms of its organization, location and distribution, in terms of its increasing specialization, in terms of conveying it precisely, and in terms of its possible perversion are all intimately connected with the future of audiovisual communication. The problem of knowledge is one of the greatest our society must carry into the future; it is an absolute social necessity that we begin now to apply audiovisual techniques to aid in its solution, and to protect its integrity.

A Look at the Future of AV Communication

The Problem of the Second Industrial Revolution

Knowledge and technological civilization are two sides of the same coin. Technological civilization has been the product of the Industrial Revolution. Today we are told that we are in the beginning stages of the Second Industrial Revolution. Two factors are making this possible—atomic energy and automation.

Atomic energy, in its peacetime aspects, will help us do better the world's work. Automation, while ostensibly also helping us with the world's work, will apparently redesign the world in which we live. Automation is simply defined as the use of machines to run machines. Electronic computers and other forms of control mechanisms now make it possible to move materials to a machine, control the machine while it is in operation, change the setting of the machine, and keep track of its production. Feedback of information at all points keeps the system running.

Mr. Walter Reuther first called the attention of the nation to automation and to the specter of unemployment *and* promise it presents. Since then, we have heard much concerning it. A complete automobile engine can now be made, it is said, untouched by human hands in a sort of push-button factory controlled by electronic brains. Application of the process is seen in oil refineries, pipelines, and even in forestry.

However, as Peter Drucker has recently pointed out, automation is not a manless, machine-operated production. Its primary characteristic is a process—a way of thinking involving patterns and self-regulation. It is here that the educational implications are tremendous. Automation is not simply a problem of technological unemployment or employment. To quote Drucker: "...Automation's most important impact will not be on employment but on the qualifications and functions of employees" (Drucker, 1955, p. 44).

This means that very large numbers of workers of high technical skill will be needed to create, maintain and control the machines. More important, large numbers of managers will be required to operate the process— to think, to analyze, to decide. Some economists feel that another complete upgrading of all workers is absolutely necessary to run an automatized world. The peasant, with his wooden hoe, was once upgraded into a semi-skilled factory worker over a long, long period of time. The factories and offices of tomorrow will need higher-grade employees to be created immediately—employees working under a system which makes possible the human use of human beings.

In order to avoid technological unemployment, it is the best estimate of many familiar with the situation that much of our labor force will have to be retrained quickly for new jobs; that much of the future labor force will have to be provided with the kind of generalized education required to produce men capable of high-level analysis and synthesis; that the increased use of leisure time—the 30-hour week is not far off—will demand a change in all education.

Audiovisual materials may be the only answer to the retraining problem—audiovisual materials created and used jointly by the schools, industry and labor unions. The estimates of the number needing such training run into the millions. The speed needed approaches that of World War II.

If this hypothesis is true, the future demands on the audiovisual field will be fantastic.

The type and use of audiovisual materials in the schools of the future to educate young people to live and participate productively in the society created by the Second Industrial Revolution will have to be much different. The curricular implications are clear. Life Adjustment, a curriculum movement which is already being questioned in some quarters, in my opinion, will be dead as the dinosaur.

The creeping control of the emotional "Rousseauians" over all levels of education will be halted and restored to its proper perspective. We are on the edge of an era requiring an iron content in our education—whether the curriculum supervisors and some of their audiovisual worshippers are ready for it or not. This content will, by no means, be the same as the much maligned subject matter of yesteryear, and so will give scant comfort to Professor Bestor and other critics, but it is also certain that it will not consist of courses in how to paint your house. New audiovisual materials must be created and used to carry this content. It was not easy to move the peasant with his hoe to the level of a machine tender; it will be more difficult to make the next move, upgrading the entire intelligence and achievement of the American people. We will have to do it, and audiovisual materials are our best bet to help us.

The Problem of the Public Philosophy

Finally, I should like to say a word about the communication of the public philosophy. Walter Lippmann has recently produced what is probably his greatest work, *The Public Philosophy* (Lippmann, 1955). In this book, Lippmann traces the causes of the decline of the Western democracies to the disintegration of the public philosophy—our general traditions of government and civility. One reason for this decline, he suggests, is due to the fact that we have not remade this public philosophy—the glue that holds us all together—to fit the times; another reason is that, through a mistaken theory of education and a general surrender of the communication media to trivia, we have not communicated the public philosophy. His remedies lie in the reverse of this process—the remaking of our traditions of civility to fit the times and the rebuilding of an educational and communication system that will communicate them.

Mr. Lippmann's position in *The Public Philosophy* has been soundly criticized. I am inclined to agree with some of these criticisms, particularly those that point out that his analysis is principally based on European experience. At times, one wonders whether or not Mr. Lippmann really knows the United States at all. However, his thesis that a public philosophy adequate for the times needs to be communicated throughout our society is very sound.

This, it seems to me, is the final and great challenge for the audiovisual movement in the future. We can well disintegrate without blowing ourselves up in an atomic flash. We can become so immersed in trivia that a scientific dictatorship is inevitable. With an adequate and adequately communicated public philosophy made possible, in part, by the intelligent use of the audiovisual media of communication, we can survive to look forward to a freer and happier life.

This has been my theory of history as it frames the future of audiovisual communication. The future presents us with the picture of an incredible load of knowledge, a radically new social organization, and a necessity to communicate the public philosophy. ESP will not handle these problems at all; Dr. Griswold's books can make some contribution; audiovisual communication, I am convinced, must carry the major share of the load into the future—the future which will always contain the contradictory elements of good and evil—between which we must always make difficult and constantly more intelligent choices.

References

Bradbury, Ray. *Fahrenheit 451.* New York: Simon and Schuster, 1950.

Bush, Vannevar. From an address before the American Society of Mechanical Engineers, February 16, 1955.

Drucker, Peter F. "The Promise of Automation: America's Next Twenty Years, Part II," *Harper's Magazine*, Vol. 210, No. 1259, April 1955, pp. 41-47.

Griswold, Alfred Whitney. *Essays on Education.* New Haven, Conn.: Yale University Press, 1954.

Huxley, Aldous. *Brave New World.* Garden City, N.Y.: Doubleday, Doran & Co., Inc., 1932.

Kuttner, Henry. "Year Day," in *Ahead of Time* by Henry Kuttner. New York: Ballantine Books, 1953.

Lippmann, Walter. *Essays in the Public Philosophy.* Boston: Little, Brown & Co., 1955.

Los Angeles Times. March 10, 1955.

Muller, Herbert J. *The Uses of the Past: Profiles of Former Societies.* New York: Oxford University Press, 1952.

Orwell, George. *Nineteen Eighty-Four*, a novel. New York: Harcourt, Brace, 1949.

Quarterly Journal of Current Acquisitions. Washington, D.C.: Library of Congress, August 1959.

Redfield, Robert. *The Primitive World and Its Transformations.* Ithaca, N.Y.: Cornell University Press, 1953.

AV Development
and the
Concept of Systems

The other day an industrial film producer friend showed us a film he had made for a company which produces radio-controlled target airplanes for the Armed Forces. It was an interesting film; but while watching it, a sequence in the film set off a train of thought that had nothing to do with radioplanes—it had to do with the audiovisual movement and why it, unlike the planes in the film, never seems to get off the ground.

The manufacturer of radio target planes had started from scratch. As we all know, the audiovisual movement did likewise. Without pushing the comparison too far, the making of target airplanes is now big business, but we still struggle with lack of projectors; glass classrooms; films a thousand miles away not available until next February; record players suitable for Edison cylinders; and tape recorders, filmstrips, up-to-date map collections, and hundreds of other efficient tools still on dealers' shelves or in manufacturers' warehouses instead of in the hands of teachers in the classrooms of the nation. Why the difference? Why success on the one hand and crawling, halting, almost imperceptible movement on the other?

Many things probably explain the difference. Philosophers could knowingly talk about cultural values, war versus peace, etc. And they would probably be right in the large sense; but there seems to us to be more to this business of slow growth of the audiovisual movement in the United States than meets the eye. It cannot all be explained by the fact that the people of the United States consider safety or beer or a new model car more important than the efficient education of their children.

Part of the answer to this perplexing problem of the iceberg-like progress of the audiovisual movement lies within the subject of our last editorial on the concept of educational efficiency (Finn, 1956). It may interest our readers to know that over 3,000 reprints of that statement have been requested to date. Administrators and boards of education have not been thinking about the problem of classroom efficiency and the achievement of that efficiency through use of the proper tools.

Acquiring the concept of educational efficiency, however, is only part of the solution. Another part came to us with a sequence in the target plane film. This part, we believe, may be even more important than the idea of educational efficiency. We refer to the concept of *systems*. The plane manufacturer stated in the film that, as the company grew and the technical

Reprinted from *Teaching Tools*, Vol. 3, No. 4, Fall 1956, pp. 163-164.

problems grew, there evolved throughout the organization the concept of a *system*. By this he meant that the drone radioplane could not be considered separately. Everything associated with it had to be developed, studied or taken into account. This meant that the company designed launching platforms for special uses on land and sea; it manufactured recovery vehicles; it developed a system of repair and maintenance and spare parts service; it developed a training program and trained Armed Forces personnel; it gathered a staff of field men; it geared its research and development program to that of the Armed Forces.

Now, what the Army or Air Force was buying from this manufacturer was not a radio target plane. The Army or Air Force obtained, with the contract, a *system* of radioplane use from the means for getting the plane in the air to the means for recovering and repairing it. This *system* of radio target plane use was integrated, in each of the Armed Forces, into other systems to create a working, efficient organization.

Here is the lesson that school administrators and audiovisual directors must learn—the lesson of the modern concept of *systems*. A little inquiry will reveal that the *systems* concept is one of the hottest ideas now occupying the attention of top industrial management, industrial psychologists, the Armed Forces, and many others. The *systems* concept, while certainly not new, has evolved into a major corollary of the premises of automation. It is related to "operations analysis" or "operations research," as these terms are used to refer to management techniques now being studied and applied in the Armed Forces and in certain advanced segments of industry.

Essentially, the *systems* concept is an idea of organization. It is an idea of organization that includes what might be called the *gestalt* or *whole* function of a unit or organization. Thus, in advanced management research circles today, "men-machine systems" and "machine systems" are carefully set up and studied. When an aircraft—bomber or commercial—is in the air, it consists of an intricate system of men and machines made up of smaller unit systems of men and machines. To make that aircraft accomplish its objective—whether to deliver a bomb or a sack of mail—it is necessary that the *system* as a whole be managed. What is important is not the physical and psychological condition of the pilots, the electronic devices, the code used with the tower, *each taken separately*, but the *gestalt* or *field* of all these items and many more, *considered as they interact with each other in a system.*

We hold that the general theory of educational administration is years behind that of modern management. How many years, it is difficult to say. About two generations would probably be a good guess. The theory of educational administration is still occupied with bits—with atoms and pieces. The present theory of educational administration confines itself to such things as the problems of line and staff organization. For example, we know of a new junior college in the process of organization. The director of instruction set up a job on the chart for an audiovisual director—a professional person. But the college had a consultant—a well-known theorist in educational administration. When the consultant got through with the organization chart, the audiovisual director was gone and, in his place,

a great concession was made with the substitution of an audiovisual technician—a projector repairman, if you will. It can easily be shown that with
that one move, the audiovisual potential of the instructional program was
reduced 100 percent. Classroom efficiency went out the window in return
for what appeared to be an "efficient" organization chart. The audiovisual
program of the college was simply not considered as the *system* it is.

We do not imply here that this is an isolated case or that the administration theorist was acting either maliciously or with caprice. He was, as a
matter of fact, acting consistently within the best form of thinking abroad
in educational administrative circles today. The point is that this thinking
is at least two generations behind the times and it is for this reason, among
others, that the audiovisual movement never gets off the ground.

For an audiovisual program—and this is the heart of our argument—is a
clear-cut *system*. The *system* begins with the production of materials—
films, pre-recorded tapes, or even a classroom bulletin board—and ends
with the recovery or replacement of the materials. It is a man-machine system. Involved, within the school situation, are people—teachers, administrators, students, clerical, and technical help; materials, machines, other
systems (delivery, for example); and outside institutions—dealers, producers, distributors, to name some of the larger units.

What is the value of this concept of audiovisual administration as a *system*? It should change completely
tion. It should, for example, stop immediately the wasteful practice engaged in by some schools of putting the materials in one location under
the supervision of a "materials" person (the school librarian, for example)
and the equipment under the supervision of another (the science teacher or
audiovisual building coordinator, for example), leaving the teacher to do
the coordinating. This is no system; this can be chaos, although it looks
"efficient" on an organization chart—not so many "unproductive" personnel.

The concept of an audiovisual *system* would also reduce the ridiculous
lack of coordination which results in building classrooms with no light
control on the one hand and investing in projectors and materials on the
other. If there were no other fact in existence to establish that present-day
administrative theory and practice is an atomistic, old-fashioned, outmoded
business, this fact would do it. If, instead of considering buildings in the
category of "buildings" and audiovisual materials and devices in the category of "curriculum materials," administrators were, for five minutes, to
consider the audiovisual program as a *system*, obviously lighting, ventilation and even proper bulletin board and chalkboard space would be related
to the problems of teacher use of these materials. Buildings would be built
with the *system* in mind.

Equipment is another example. According to recently released National
Audio-Visual Association figures, the number of 16mm motion picture
projectors produced in the U.S. in 1954 was 10,000 less than the number
produced in 1947. Granted that in the early post-war years it was necessary
to fill a backlog of orders and granted that continuing production since
that time has put a lot of projectors in the field, the fact remains that this
is a bad sign for development of the audiovisual movement.

AV Development and the Concept of Systems

In the first place, it is highly doubtful whether the number of projectors in the field has even kept up with the widely publicized growth of the schools. Secondly, by no means all of these projectors go into the school market. The industrial market is very large; the expansion of television and television film activities has increased the market there; the churches are just beginning to wake up to the need for equipment. The schools are probably buying much less of the total product than in 1947. To say the school market is saturated, as some equipment people do, is to condemn educational administrators where they stand. The audiovisual program, considered as a system, would never operate on a one-or even two-projectors-to-a-school basis. This is atomistic thinking; this is almost no audiovisual program at all. And the same may be said for other types of equipment.

Materials reflect the same thing. Some years ago, seven book publishers in the U.S. made a study of the possibility of going into film production. Among other findings of the study was that schools in cities below 50,000 in population rent most of their films. This is probably just about as true today. And how schools obtain and use recordings, tapes, transparencies, filmstrips, etc. is one of the great mysteries of our time. This is a *system*? When a teacher or principal or AV director has to rent a film a thousand miles away and hopes it gets there within six months of the time needed, one of the major elements of a system is missing—and no system exists.

Professional audiovisual directors are also not without fault in this matter. In many cases, perhaps for very good reasons, but true nevertheless, the audiovisual director thinks and operates in an atomistic fashion as opposed to the fact that he should be managing a *system*. His *system* extends from the producer to teacher and class back to producer again. But he spends his time with booking forms or equipment repair or previewing committees—operating all the time in a piecemeal fashion.

The audiovisual movement is relatively young. It is also geared into the technological world of the future—a world of interlocking, complicated systems of men and machines. It cannot be administered under a theory useful for the production of buggy whips. We need a new audiovisual *systems* theory; we need it NOW.

References
Finn, James D. "What is Educational Efficiency?", *Teaching Tools*, Vol. 3, No. 3, Summer 1956, pp. 113-114.

Technological Innovation in Education

... when [King Canute] came home (as some write) he did grow greatlie into pride, insomuch that being néere to the Thames, or rather (as others write) vpon the sea strand, néere to Southhampton, and perceiuing the water to rise by reason of the tide, he cast off his gowne, and wrapping it round togither, threw it on the sands verie neere the increasing water, and sat him downe vpon it, speaking these or the like words of the sea: "Thou art (saith he) within the compasse of my dominion, and the ground whereon I sit is mine, and thou knowest that no wight dare disobeie my commandements; I therefore doo now command thée not to rise vpon my ground, nor to presume to wet anie part of thy souereigne lord and gouernour." But the sea kéeping hir course, rose still higher and higher, and ouerflowed not onelie the kings féet, but also flashed vp vnto his legs and knees. Wherewith the king started suddenlie vp, and withdrew from it, saieng withall to his nobles that were about him: "Behold you noble men, you call me king, which can not so much as staie by my commandement this small portion of water."　　　　　　　　　　　　　　　　　　　　　—Holinshed, 1807, p. 731

There were, in Holinshed's time, at least two different versions of the story of Canute and the sea. The heritage that has come down to us, however, is contained in the bare bones of the story—the futility of man commanding the sea to stand still. One does not have to be an historical determinist to accept the point. If old Canute had wanted to do something about the tide, he could at least have built a sea wall.

In checking into the story of Canute, I ran into an interesting sidelight. Canute wrested control of Britain from Ethelred, the King of Wessex. Some authorities called him Ethelred, the Unready; others, Ethelred, the Redeless—meaning a man who lacks counsel. Again, however, using Marshall McLuhan's concept of the bare bones of a code surviving subsequent changes in culture, the stark idea of ordering the sea to turn back and the equally stark idea of defeat because the leadership was unready appeal to me as a way to start thinking about the problem of technolgical innovation in education.

Reprinted from *Audiovisual Instruction*, Vol. 5, No. 7, September 1960, pp. 222-226. Originally an address to a DAVI summer meeting June 27, 1960.

Technological Innovation in Education

Actually, I am indebted to Francis Noel for the suggestion that this subject be examined in some detail. Last winter, at the state meeting of the Audio-Visual Education Association of California, Francis, in a brief speech, uttered a warning to the membership that they were running the risk of becoming fat and lazy on the job right in the middle of the greatest tide of technological change ever to hit the audiovisual movement—right at the time when they should be lean and hard and hungry. In such a context I would like to discuss DAVI, our own national organization. We must, I think, raise the question as to what is the role of such an organization in this era of swift change in our own business.

DAVI is now very important. As a group of people working with modern communication devices in education, we find suddenly that there is great national and international interest in what we do and say. We find many other groups either asking us for help or counsel, or invading our territory. The professional educator, the school board member, the PTA, the news media, politican and taxpayer groups, and even the critics of education have all suddenly discovered instructional technology and, consequently, DAVI. Incidentally, this sudden importance poses many internal problems which it is going to take us several years to solve. Our immediate problem, however, is to assess this importance—which, of course, we always felt we had—in more general terms.

The basic reasons for this interest are known to all of you. They are rooted in the twin explosions of population and knowledge, in international tension, in the teacher shortage, in the lack of facilities, and so on. Fundamental to all of these, however, is the fact that technological change—partly as an effort to solve some of these problems and partly by its very nature of constant expansion—has invaded all of education to the point where it is inescapable.

I am sure that most DAVI members are now well acquainted with the nature of these changes. Since I have discussed them at length elsewhere (1960), there seems to be no need at this point to re-examine the so-called leading edge of the audiovisual movement in detail. The list is long: television, teaching machines, language laboratories, and videotape are bringing us multiple problems; we are now faced with 8mm sound film, airborne TV, thermo-plastic recording, multiple projection units, and the development of whole instructional systems; along with this we are dealing with new school designs, new patterns of organization, new curricula, problems in the creative production of materials, problems of control, etc.

There are two elements that relate to this change that I should like to emphasize. The first is speed. Dr. Roy Simpson, state superintendent of public instruction in California, cited an interesting comparison in a recent speech. He referred to an official of the Defense Department who said:

> In Nero's time, or even 1500 years before that, man could travel only as fast as a horse could carry him. Thirty centuries later, when Columbus discovered America, the speed at which a man could travel was still limited to the speed of a horse. Two centuries later, when Paul Revere made Longfellow famous, he was still dependent on Dobbin.

Technological Innovation in Education

So, on our graph we would have a straight line from 1500 B.C. to about 1830. About then, after centuries of scientific stalemate, the Iron Horse broke the "oat barrier" and the technology of transportation was born. In 1910, our first military airplane had a speed of 42 miles per hour. Thirty-five years later, persistence had pushed speed up to 470 miles per hour.

Then in 1945, progress changed its pace again. It stampeded! In the decade from 1945 to 1955 man catapulted his speed from 470 miles per hour to more than 1,500 miles per hour. . . .

Our curve which ran dead level for more than 30 centuries, curved gently upward with an increasing slope from 1830 to 1945. Then it careened around a corner and shot skyward. The horizontal line has gone almost vertical. . . . (Simpson, 1960, pp. 4-5)

There is no doubt in my mind that we have broken through the "oat barrier" in education. This means that each successive development will take effect in a shorter time than the last one. Compare, for example, the rate of acceptance of the sound film with television. Ten years after the introduction of sound motion pictures into education there were film libraries that were not buying a single print of a sound motion picture, so convinced were they that silent film was as far as education ought to go.

Television has caught on much more quickly and with wider effect. I would predict that, other things being equal, teaching machines will outstrip television as to rate of integration into the educational system. The problem of programming, seen as a great deterrent just a few months ago, seems much closer to solution. One large educational film producer, for example, has five teams of programmers in the field turning out material with great rapidity. A school in New York that has experimented a great deal in this field is setting up a general programming service. Our curve is going up right beside the transportation curve. I hate to speculate on what might be possible with 8mm sound film at this point in time.

The second element of importance in technological change is that it is cumulative. Consider the automobile. C. L. McCuen of the General Motors Research Laboratories noted a few years ago that:

[The automobile] is the modern version of a road vehicle first built by the French engineer Cugnot in 1769. It is Otto's contribution of the four-cycle engine. It is the basic electrical discoveries of Henry, Faraday, Ampere, Volta, and Benjamin Franklin. It is Kettering's self-starter and Thomas Midgley's tetraethyl lead. (McCuen, 1952, pp. 12-13)

In this second element (the cumulative effect of technology), we find the explanation for the first—the speed of change. Cumulation has affected the audiovisual technological complex to such an extent that, today, we are, willy-nilly, in a new phase—as when water suddenly turns to steam, to borrow from Willard Gibbs by way of Henry Adams.

The nature of this change, then, presents the basic problems for people who are organized together to promote the technology of instruction. And I must insist that, going back to the Keystone slide sets and stereographs at the turn of the century, this is what the audiovisual field has always been about. We have been attempting to introduce and to manage instructional technology; this term of course, must be broadly defined; it includes the systems involved, the methods of use, the learning research which supports it, the measured reactions of social groups to media and many other elements.

This wave of technological change, like the waves of Canute, will not go away. One of the problems I have faced as president of DAVI is that, by the mere process of calling attention to these overwhelming facts, I have been shot at, verbally, even by some of our own members. It has been suggested, for example, that we should forget about teaching machines[1] on the grounds that they are primarily verbal devices—in the face of evidence that some of the better machines use audiovisual methods of presentation and of the more fundamental fact that these technological devices will obviously be treated within the school and college situation as are the other "teaching machines" such as the motion picture projector, the television receiver, and the tape recorder.

With the more general audiences of teachers and administrators, the unrest is even greater. I have been accused of inventing the vacuum tube and the transistor in order to replace teachers. I wish I had invented these devices; having missed that opportunity, I can only assure the threatened educators that they will not go away. Our problem remains as I have stated it before: ". . . not so much of how to live with [instructional technology] on some kind of feather-bedding basis, but how to control it so that the proper objectives of education may be served and the human being remain central in the process" (Finn, 1960, p. 21).

A Profession of Innovators

Within this pattern of unrest and change and desire of some people to order the waves to stand back, what is the role of DAVI? More generally, what is the role of an organization of educational innovators?

We have, I think, to begin with a realistic assessment of the nature of our membership and of our organization. Any organization such as ours is a conglomerate sum of a professional association, a learned society, a trade union, a trade association, and some kind of a social group within which we can carry on shop talk. Our members include full-time teachers who are personally interested in the use of audiovisual materials or who are in charge of a program for their department or building, college professors who do research in this field, administrators of audiovisual service programs large or small, producers of materials from graphics to television programs, psychologists, curriculum specialists, librarians, government information specialists, broadcasters, military, and religious audiovisual people—and many others. We are conglomerate in both purpose and membership and yet, as a whole, with this mutual interest in a field almost impossible to define, we sit on the rising curve of swift technological change in education with some hope of giving it direction.

---

[1] A new term now circulating among psychologists and others interested in teaching machines and likely to be adopted is "auto-instructional devices."

Technological Innovation in Education

If I were to sum up the meaning of this description of the membership of DAVI and its organization, I would have to say that we are moving in the direction of becoming a true professional organization of extreme importance to education and to America. We have a long way to go, but not as far as we did in 1952 when, as some of you know, I attempted to make an assessment of our status as a profession (Finn, 1953. See also chapter in this book.)

At that time, I examined the DAVI situation using six criteria usually accepted as measuring the degree of professionalization of a group of workers. I am interested here only in exploring further the sixth criterion which stated that a profession had "an organized body of intellectual theory constantly expanding by research" (Finn, 1953, p. 7). To this, I would like to add a seventh, namely, that a profession must have the ability to exercise its own leadership.

To some, the idea that a profession must exercise its own leadership may sound like a contradiction in terms. A little thought, however, will reveal that all professions, one way or another, tend to turn into priesthoods. And, long ago, William James said that a priesthood never initiates its own reforms. To some degree, this has happened with the audiovisual field. We must face the brute fact that many of the recent innovations giving us headaches today have been forced upon us from the outside. I would be the first to say that this has, by no means, been entirely the fault of the audiovisual profession. Some of this short-circuiting has been deliberate on the part of outside agencies for other purposes; some has been the result of technological accident—an invention or process just turns up and people outside of our group begin to work with it.

Other audiovisual innovations taken over by, offered to, or otherwise passed on to groups outside the mainstream of the audiovisual movement have ended up where they have because individual members of our profession have refused to see the relevance of these innovations in the pattern of instructional technology. A recent historic example— recent enough to have occurred within the last 20 years and yet far enough away in time so as not to be threatening—is illustrative Exactly 20 years ago at a DAVI meeting in connection with the NEA summer convention, a discussion was held as to whether or not we should change the name from the Department of Visual Instruction to the Department of Audio-Visual Instruction. It was suggested that the time was ripe as the people interested in educational radio and recordings were then meeting to discuss the formation of what was later called the Association for Education by Radio. (The descendant of this organization was absorbed into the NAEB a year or two ago.)

I'll never forget the remark of an audiovisual director (since passed on) who said, "What are we talking about radio and recordings for? I have enough to do just keeping up with films put out by Encyclopaedia Britannica and Coronet." There was other opposition, too, and we remained DVI for some years after that. The members of the organization refused to take the responsibility for their own leadership. True, we were small, weak and scattered. But perhaps that's why we remained so for much longer than necessary.

I should like to emphasize that this business of exerting leadership depends primarily upon the sixth criterion—an organized body of theory constantly expanding by research. No one, no profession, no organization can seize the present and bend the future to proper ends without it. We must know our own posture; we must know where we want to go and why. In an intellectual field, this becomes a demanding intellectual problem placing constantly higher requirements upon our members. The way of the innovator is hard.

The Practical Sociology of Innovation

If intellectual demands make the life of the innovator difficult, what I call the "practical sociology of innovation" makes it even more so. Actually, in the audiovisual movement we face two problems of innovation simultaneously. The first—and this is more of a problem than ever before— is resistance to innovation by our own members. Some of them are playing Canute and Ethelred, the Unready, at the same time; they want the waves to stand still and they're not ready to deal with the tide when it keeps rolling in.

Second, we always have been, to some degree, the innovators of the educational profession. This is because we have always been the technologists of learning and, in our society, the technologist is the innovator. Some of our members remember the time when they had to go out and sell the idea of using a motion picture in the classroom; and, in many places it was about as easy as selling Mr. Khrushchev on subscribing to a U-2 photographic service. As a matter of fact, this idea of using films extensively in teaching is by no means completely sold to the whole of American education to this day.

It should be obvious that we cannot develop new ideas and processes within all of American education until we first learn to live with change and innovation within our own professional segment. We are in danger, I think, of developing an audiovisual orthodoxy—which was what Francis Noel was worried about in Sacramento last winter. We have too many people who are what might be called 90-day audiovisual pioneers. That is, for a time they were out in front of education waving the banners for film libraries, tape recorders, central sound systems, or the concept that the concrete is better than the abstract. But pioneering is hard work, and many people don't care to stay with it for much longer than the proverbial 90 days. And so some of our friends stopped pioneering as soon as they acquired a little living space, status, or an instructional materials center. And they became orthodox and fat, and change-resistors. Worse, some of them stopped thinking, worrying and wrestling with the problems of the swiftly developing instructional technology. This is part of the practical sociology of innovation. It hurts more right now because we happen to have hit one of those swift-rising audiovisual cycles.

Another aspect of the sociology of innovation is at the opposite pole. We can have too much novelty, too much change which is neither useful nor ornamental. Hardy Cross, the famous engineering professor at Yale, once gave the engineers some advice that applies as well to us. He said:

The problems of today are in many respects the problems of hundreds of years ago, but these problems deal sometimes with new materials and always with different conditions. When a problem is all solved and the answer is very definitely known in the field of engineering, it is about time to investigate that problem again, because what is known is probably known for certain limited materials. But novelty should not be pursued for itself alone. The novelty often consists in merely doing another thing in about the same way that other things have been done before. (Cross, 1952, pp. 22-23)

We don't need orthodoxy, but we do need the law of conservation of audiovisual energy. We need hard thinking, based upon our general intellectual posture, as to which is baby and which is bath. We also do *not* need any additional self-anointed experts springing from nowhere to lead a charge for a fad which, for a few short months or a year or two, makes them into some kind of local savants. This matter of conservation, of distinguishing dross from gold, is also part of the practical sociology of innovation.

In this context, I should also like to deal with a point which Ray Carpenter recently called to my attention. This point, too, underlines the fact that the way of the educational innovator is hard. If a man discovers a new vaccine, develops a new method of transportation, discovers a new fuel for compact cars or a new base for milady's make-up, even if he should write a new poem—he is, for the most part, highly honored. He receives money, a gold watch, is listened to as he comments on education or the state of relations with Ghana, and is invited to sit at the head table.

Innovation of certain kinds is highly rewarded. Not so with educational innovation; in many cases the educational innovator is punished. I know personally, for example, of a friend with a reputation as an educational innovator who was blocked from a large and important job merely because he had carried out his ideas. Other friends have been castigated by their colleagues and have been denounced in professional circles for the same reason. Until such time as educational innovation is justly rewarded, the practical sociology works against us. The network of groups back of the educational innovator is not necessarily one to which he can turn for approval. This is a great deterrent to us in our business of educational innovation.

Innovation and Cultural Change

Instructional technology, which has become the major basis for educational change now and in the foreseeable future, also presents us with professional problems at a level somewhat higher than that of practical sociology. I refer to the fact that the kind of innovation we are facing in the American culture as a whole and in our specific educational culture is of such a nature that it can change these cultures overnight.

In his new book, *The Two Cultures and the Scientific Revolution*, C.P. Snow states flatly: "I believe the industrial society of electronics, atomic energy, automation, is in cardinal respects different in kind from any that has gone before, and will change the world much more" (Snow, 1959, p. 31).

When we consider the audiovisual segment of education and the new media with which we deal, the words of Marshall McLuhan are as plain:

The children of technological man respond with untaught delight to the poetry of trains, ships, planes, and to the beauty of machine products. In the schoolroom, officialdom suppresses all their natural experience; children are divorced from their culture. They are not permitted to approach the traditional heritage of mankind through the door of technological awareness; this only possible door for them is slammed in their faces. . . .

Photography and cinema have abolished realism as too easy; they substitute themselves for realism.

All the new media, including the press, are art forms that have the power of imposing, like poetry, their own assumptions. The new media are not ways of relating us to the old "real" world; they *are* the real world, and they reshape what remains of the old world at will.

Official culture still strives to force the new media to do the work of the old media. But the horseless carriage did not do the work of the horse; it abolished the horse and did what the horse could never do. Horses are fine. So are books.

Technological art takes the whole earth and its population as its material, not as its form.

It is too late to be frightened or disgusted, to greet the unseen with a sneer. Ordinary life-work demands that we harness and subordinate the media to human ends. (Carpenter & McLuhan, 1960, pp. 181-182)

If, as a profession within education, we are to control the new instructional technology, we must realize and live with this fundamental fact of cultural change. It should be neither a surprise nor the crack of doom to us that the role of teacher will probably change drastically in the future; it should be a challenge to solve the problem so that the nation, the students, and the profession will all benefit.

To use another exmaple, there is no question in my mind that we cannot afford to retain an orthodox position on reading instruction. The culture has changed. Children are beginning school after six years of adult education presented them by television, motion pictures, radio, advertising, and a host of other media. Their vocabularies are simply not those discovered by Thorndike and others years ago. Their stock of meanings has changed. This probably implies a new system of reading instruction involving audiovisual approaches, phonetic analysis, and yet-undiscovered methods; it certainly does not mean the slow addition of 150-200 words a year.

The basis of one of the hard-core audiovisual concepts—the idea of some kind of a linear relationship between the concrete and abstract symbols—is increasingly being called into question by Gerbner, McLuhan and others. This, again, is due to basic cultural changes brought about by the technology of the new media. The sensory-realist position of 1900, still defended by some of our members, is rapidly giving way. This is but another example.

What, then, does all of this mean to the audiovisual profession? It means, first, that we have to understand and live with accelerating technological change in our own business; it means, second, that we must learn to master these changes and help integrate them into the educational process so that their benefits may be useful and rewarding; it means, third, that we cannot let the practical sociology of the situation deter us or we will lose control of the movement; it means, fourth, that we must lead an intellectually demanding professional life in order to deal with the forces of technology with which we daily work.

Finally, it means that we should remember the words of Edgar Dale from a recent article in which he said:

> The disease in all professions is stagnation, a failure to grow in professional wisdom and competence. The curve of growth is not typically a constantly rising line, but one in which plateaus are soon reached. We wonder why children don't want to learn, yet their teachers may exemplify persons who have stopped learning, who have little feeling of the need for disciplining themselves to high standards of professional excellence. (Dale, 1960, p. 400)

DAVI, if it turns out to be nothing else, must be an organization of innovators. In it we need no Canutes; more important, we can have no Ethelreds who are unready. The way of the innovator can be exciting, if hard. I invite you to live it.

References

Carpenter, Edmund and Marshall McLuhan. *Explorations in Communication: An Anthology.* Boston: Beacon Press, 1960.

Cross, Hardy. *Engineers and Ivory Towers.* New York: McGraw-Hill Book Co., 1952.

Dale, Edgar. "What Is the Image of Man Tomorrow?", *Childhood Education,* Vol. 36, No. 9, May 1960, pp. 398-401.

Finn, James D. "Professionalizing the Audiovisual Field," *Audio-Visual Communication Review,* Vol. 1, No. 1, Winter 1953, pp. 6-17.

Finn, James D. "Automation and Education: 3. Technology and the Instructional Process," *Audiovisual Communication Review,* Vol. 8, Winter 1960, 5-26.

Holinshed. *Chronicles of England, Scotland, and Ireland.* (Book VII, Chapter 13). Printed for J. Johnson et al., London, 1807.

McCuen, C. L. *The Challenge of Engineering's Second 100 Years.* Detroit: General Motors Corporation, 1952.

Simpson, Roy E. Address to the Spring Conference of the California Association of County School Superintendents at Asilomar, March 14, 1960, Mimeo.

Snow, C. P. *The Two Cultures and the Scientific Revolution.* New York: Cambridge University Press, 1959.

Teaching Machines:
Auto-Instructional Devices for the Teacher

The American teacher may soon need a new professional dictionary. These days an auto-instructional device is not an Aetna Driver Trainer. A motion picture projector, contrary to what may be thought, is not a teaching machine. Reinforcement is not something you put in concrete or bring up to save the Lost Battalion; and programing does not refer to the way a conductor sets up a concert or an adviser a series of courses for a student.

These terms and some others like them—"repertoire," for example—have been invented or adapted to be used in the new field of teaching machines, self-instructional devices, or auto-instruction (the latest term under consideration by psychologists and audiovisual experts to describe this rapidly growing field).

Because of the national publicity some of the experiments with automated instruction have achieved, teachers are naturally concerned with these developments and are raising many questions. It would help everyone in the teaching profession desiring to assess the teaching machine movement if it were clear exactly what was being talked about, what has been claimed for these devices, what their advantages and limitations are, and what they will do and will not do for teachers and students.

What Is a Teaching Machine?

A teaching machine or auto-instructional device is a piece of apparatus designed to be operated by an individual student. There are many types and varieties of teaching machines, but all of them have the following characteristics in common:

1. The student is presented with a question or problem by some form of display on the machine.
2. The student is required to respond *overtly*—that is, he must do something about the problem such as writing an answer or pushing a button to indicate an answer.
3. The student is informed, one way or another, as to whether his answer is right or wrong and, in some cases, why he is right or wrong. (This is called immediate knowledge of results or *reinforcement.*)

Reprinted from *NEA Journal*, Vol. 49, No. 8, November 1960, pp. 41-44.

4. Often an account is kept of the response to each item—not for testing purposes, but for teaching purposes as, for example, when the machine has a provision to repeat items that have been previously missed.

The device per se is not important except as a vehicle for the *program*, which is the heart of the auto-instructional concept.

The content to be taught is analyzed and developed into a program. The program is the series of items which is presented to the student in the form of questions, problems, blanks to be filled in, pictures and diagrams to respond to, and so forth.

Programs are designed, taking into account a theory of learning, the nature of the student for which the program is being designed, the subject matter to be worked on, whether or not a teaching machine will be used to present the program, and (to a lesser degree) the capabilities of the particular machine, if one is to be used.

The thoughtful teacher should understand something of the different concepts of programing.

Theories of Programing

Perhaps the foremost theory of programing in vogue among research workers today is one which requires that the student *construct* his own response.

This is usually done by exposing to the student a question or problem which contains a blank covering one or more words.

The student is required to write these words, not in the blank itself, but on a piece of paper in another part of the machine (on some machines he can compose a response by pushing numbered or lettered buttons or by writing a complete answer on a typewriter keyboard). Once he has committed himself, he operates a lever or button which exposes the correct answer and, at the same time, moves his answer under a piece of glass or plastic so that he can compare but not change.

A second aspect of this theory of programing is that the material is presented in small discrete steps, each one depending upon the preceding step. Very often information is presented in the item—or *frame* as it is usually called—before the question is asked. This information (called a *prompt*) can be presented in several ways.

Here is a simple form of prompt in a high school physics program written by B. F. Skinner:

A long candlewick produces a flame in which oxygen does not reach all the carbon particles. Without oxygen the particles cannot burn. Particles which do not burn rise above the flame as_____. . (smoke)

Among many others ways this is done, here is a slightly more sophisticated form from another program (one developed by A. A. Lumsdaine):

Gross profit less overhead equals net profit; so to get net profit you subtract_____from_____.

The above examples, because they somewhat resemble test items, may seem to give away the whole business to the student—something which teachers have been trying to prevent for years in the tests they construct. However, the purpose of the item is to *teach*, not *test;* hence, prompts are necessary as the program progresses.

Teachers must remember that we are dealing here with hundreds or even thousands of discrete, interlocking steps to be worked on by the student over a period of time so that he may learn—in the case of constructed response programs—certain kinds of verbal behavior.

Items repeat, come at the student from different directions, and constantly add to his store of information and vocabulary (his repertoire). The program, if it is a good one, starts where the student is and gradually leads him into unknown territory by short enough jumps so that, theoretically, he never makes a mistake. (All program items have been pre-tested on students, until they are satisfactory or until it has been determined that the program is not useful.)

At this point in the constructed response program, several things can happen, depending upon the kind of machine used. In certain devices, the items the student answers correctly are dropped and only those he missed reappear. In others, there are self-scoring provisions where he can punch a paper tape on which he has written his answers so that his errors are recorded. In still others, an intermediate stage is exposed, giving the student additional prompts before revealing the whole answer.

Another form of machine uses a multiple-choice approach to programing. Here the student is presented with a classical multiple-choice question which contains some prompting information as well. He responds by pushing a button beside the proper number, pushing a pencil through a special hole, etc.

The primary difference is that this program requires a different type of response from the student—a reaction to alternatives rather than the construction of an answer. While the difference between these two types of programing may seem superficial, a certain amount of controversy exists in psychological circles over the relative merits of each.

A third form of programing involves what is called "branching." In devices incorporating branching techniques, if the student makes a mistake, the machine may take him off the main track of the program onto a "branch" in order to build up information or background before he returns to the main program.

In a sophisticated machine, branching also permits a bright student to move ahead rapidly after he has demonstrated competence by answering a certain number of questions correctly. Again, because the field is so new, there are proponents and opponents of the branching-type program.

So far, all the programs mentioned have been concerned with verbal behavior—the use of words and abstract concepts. However, there is nothing intrinsic in any of the three types of programing described which confines them to verbal patterns alone.

Construction problems can be based on pictures, maps, charts, and diagrams. The same thing is true of multiple-choice programing. Intrinsic or branching programing in one current machine makes use of short motion picture sequences.

Recent developments in the field indicate that audiovisual approaches will enter more and more into the teaching machine and, hence, the programing picture. Experiments have been made, for example, with motor-skill teaching. Machines now exist using slides and tapes and motion pictures.

While the basic programing techniques used with these audiovisual stimulus materials may be one or combinations of the three programing theories just described, the introduction of these nonverbal materials will probably lead to still other forms of programing.

Types of Machines

In order to reach the student with the program, various kinds of technological arrangements are being experimented with. These are the teaching machines referred to in the first section and the programed textual materials which will be discussed later.

Auto-instructional devices exist in a wide range of technological sophistication. This range includes no machine at all—for example, merely a set of cards in a cardboard or plastic case, or a mimeographed sheet; a write-in (constructed response) machine; a machine using slides and tape; a multiple-choce machine; a film machine; a machine using a combination of microfilm and motion pictures; and a set of machines electronically tied in with a television broadcast.

In addition to the machines, two kinds of textual materials are now available that make use of programing concepts. One is the programed textbook, in which the items are presented much as in a construction machine with the answer usually on the page following the item.

The other is called a scrambled textbook. Here the text makes use of the branching technique. The student begins and, after some instruction, works a problem with a multiple-choice answer.

Each of the choices directs him to a different page in the book. If he is correct, on the page he turns to he finds more instruction and another problem. If he is incorrect, he is directed to a different series of pages depending upon the nature of his error. Both of these textbooks are different from traditional workbooks and should not be confused with them.

What Is the State of the Art?

While teaching machines and the programing concepts they represent are not yet widely used, or even widely available, they have moved from what might be called the laboratory phase into the field-testing phase. There are experiments being carried on throughout the country with relatively large numbers of students in classroom situations.

These experiments cover a wide variety of subject content and extend from the first grade level through college. There have been experiments on the teaching of reading, science concepts, and mathematics concepts to very young children. Arithmetic and other subjects have been experimented with in the middle grades, physics and other subjects at the high school level, and everything from logic to philosophy at the college level.

Teaching Machines

In general, the early results are encouraging. With certain subject matters under certain conditions, total mastery has been exhibited by most of the experimental students.

Students have reported enjoying the experience of learning with auto-instructional devices. Problems of individual differences seem to be closer to solution. Although many problems remain to be solved, it is already apparent that programed learning, for certain objectives with certain kinds of subject matter, is a breakthrough whose dimensions are yet to be assessed. No one, by the way, claims that programed learning will ever supplant all other kinds of instruction.

Because of the difficulty and expense of devising programed learning, the greatest current lack is in the programs rather than in machines. However, there are signs that this bottleneck may soon be broken. A large educational film producer is now going all-out in a programing effort (of verbal materials, by the way); several publishers have started or are considering starting; one teaching machine manufacturer, sensing the programing problem, began some time ago to develop programs and has several on the market for this fall.

Programs and machines represent a chicken-egg situation at the moment, but there are indications that there may be quite a few chickens *and* eggs available within the next two years.

The Technology of Individual Instruction and the Teacher

If programed learning does become available generally in the near future, what does this mean to the student, the teacher, and to education generally?

First, auto-instructional techniques are a promising effort to find a technological solution to the problem of individual differences. Whether a teacher has 35 children in an elementary school class all day long, or 150 pupils in five high school classes, he cannot really handle individual differences even though he uses all known human techniques for doing so.

Under present conditions, no teacher can be a tutor to each individual student. Teaching machines, properly programed and wisely used, can, for certain subject matter and for portions of a school day, provide tutorial experience for individual students. For auto-instruction is tutorial instruction.

From the point of view of the student, this tutorial relationship has other advantages. The student may proceed at his own rate. If he is quick, he is not held up. If he is slower, he is dealing not with a teacher, but with a program that has as much time as the student requires and that never tires.

In all cases, the teacher can be relatively sure exactly where the student is, which will help in planning other work. And programing of the type described here can institute controls on such things as homework and individual study, which will alleviate problems in connection with other important types of classwork.

Second, any wide-scale introduction of auto-instructional techniques will change the role of the teacher somewhat. It is almost inevitable that the availability of such powerful tools will make the teacher into much more of a professional than he now is.

Any introduction of technology, however, with its accompanying higher professional status, will demand more of the teacher in terms of education, experience and professional growth. Teachers will need to understand much more than they do about learning theory and communication; and they will have to exercise sound judgments—in terms, for example, of selecting programs and in determining which educational goals can be best reached through programed instruction and which can be best achieved by other methods.

Further, new types of instructional systems will evolve as the technology of individual instruction is gradually combined with existing conventional techniques of the classroom and with the mass presentation techniques involving film, television, etc.

This is the direction of the future. The machines, the technology, the systems—crude as they are today, improved as they will be tomorrow—will help man become more human *if* the teachers who will manage them understand instructional technology and make use of it to build teaching into the most human of all professions.

New Techniques of Teaching
for the Sixties

Without doubt, the most significant educational development of the next decade will be the increasing application of technology to all parts of the educational system, including teacher education. When examining this statement, however, there is a tendency for many educators to think exclusively in terms of machines and to react, therefore, to a machine versus human concept.

In thinking through the problem of teacher education and the new technology, the first step that needs to be taken is to understand the term "technology." Technology, as has often been emphasized by students of the field, is much more than machines; technology involves systems, control mechanisms, patterns of organization, and a way of approaching problems. The naive educator who reacts to "machines" when he should be considering technology is a little like a squirrel who chooses to worry about the shell of the nut and pays no attention to the meat. Squirrels are smart enough not to do this.

It might be helpful in this connection to refer to a little-known definition of technology which was stated by Charles Beard many years ago. Beard said:

What then is this technology which constitutes the supreme instrument of modern progress? Although the term is freely employed in current writings, its meaning as actuality and potentiality has never been explored and defined. Indeed, so wide-reaching are its ramifications that the task is difficult and hazardous. Narrowly viewed, technology consists of the totality of existing laboratories, machines, and processes already developed, mastered, and in operation. But it is far more than mere objective realities.

Intimately linked in its origin and operation with pure science, even its most remote mathematical speculations, technology has a philosophy of nature and a method—an attitude toward materials and work—and hence is a subjective force of high tension. It embraces within its scope great constellations of ideas, some explored to apparent limits and others in the form of posed problems and emergent issues dimly understood. (Beard, 1955, pp. xxii-xxiii)

Reprinted from *Teacher Education: Direction for the Sixties*, American Association of Colleges of Teacher Education, National Education Association, 1961, 31-42.

New Techniques of Teaching

We have been very gradually developing, within education, a technology of instruction. Since I do not have time today to consider all the ramifications of this development, I refer you to a major paper on the subject (Finn, 1960). Some of the devices developed and put into use during the last 10 years include language laboratories, self-contained audiovisual teaching kits, television, teaching machines of several types, and the Corrigan Teletest System. I would like to emphasize once again that these are the *symbols* of instructional technology—machines—and must be thought of in connection with systems, organizational patterns, utilization practices, and so forth, to present a true technological picture.

Let us look more closely at the trends of the last 20-25 years seen as a whole.

Considering the entire educational system as a whole (I will return to the higher education segment a little later), the introduction of audiovisual materials and devices for the purpose of improving instruction was a slow and painful process until the middle 30s. If we would begin to draw a curve at that point, it would look something like Figure 1. The infusion of money, principally from the Rockefeller Foundation for research in this field, the attention given to instructional technology by a group of very talented men, and the availability of better materials and machines all contributed to a rise at that point which did not stop until the beginning of World War II.

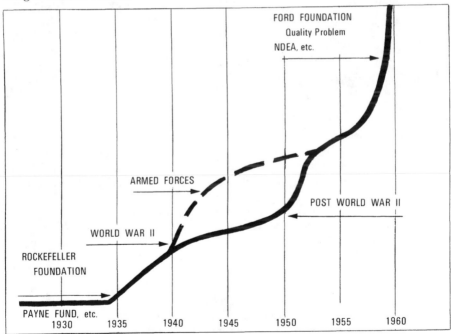

Fig. 1 GROWTH OF INSTRUCTIONAL TECHNOLOGY

As with all educational activities, a slowdown occurred during the war, and the lack of equipment and materials set back the movement somewhat. However, the technology of instruction, as everyone knows, moved over into the areas of industrial and military training during the war. This move relied principally on the previous findings of the educational research and development activities of the 30s and succeeded brilliantly by supplying the necessary money and talent for successful implementation. Incidentally, it was during this period that self-instructional devices came to the forefront even though, again as is well known, S. L. Pressey and others had been working on this problem long before.[1]

Following World War II, a great public interest developed in the use of audiovisual materials and, during the decade 1945-1955, another upsurge occurred as this technology was introduced into education with some force. Here, we must stop to say that of all the levels of education, higher education was the least affected. Oriented to print technology and the lecture system, higher education more or less successfully resisted this movement.

Since 1955, due to a variety of causes—the application of television, the influence of Ford Foundation projects, the attempts to find solutions to the problems of quality and quantity in education, the National Defense Education Act, and similar efforts—our curve of instructional technological development has again started up sharply with no leveling-off point in sight in the immediate future. We can now turn to the third perspective—a look at present trends.

The teaching machines movement is merely the latest development in instructional technology that is forcing itself into the educational arena. Actually, we now have a double-pronged technological growth (Figure 2). The first prong is in the field of mass instructional technology and is concerned with those systems, materials and machines suitable for instruction with large groups of people, for the most part *as groups*. Mass instructional technology includes several types of television, both closed-circuit and broadcast; massed film systems; several types of automatic, programmed projection systems such as the Harwald and Teleprompter units; overhead projection of all types, including the polaroid animated materials; and other equipment.

The other prong consists of a technology designed for individual instruction teaching the individual *as an individual*. Here we come into the area called "teaching machines." Actually, a more general term is necessary. Some prefer to use the term "self-instructional devices"; another quite widely used expression is "automated teaching"; psychologists are now suggesting "auto-instructional devices."

Skipping, for now, over the problem of terminology and accepting the concept of the technology of individual instruction, which I find very useful, we can list classes of these devices in an ascending order of sophistication from the psycho-technological point of view. They are: (a) individual reading pacers and similar devices; (b) individual viewing and listening equipment for slides, filmstrips, motion pictures, tapes, etc.; (c)

---

[1]Ben Perez of the Viewlex Corporation, for example, built two very sophisticated machines for the Navy that contained all of the principles of feedback of knowledge of results, timed responses, and so forth.

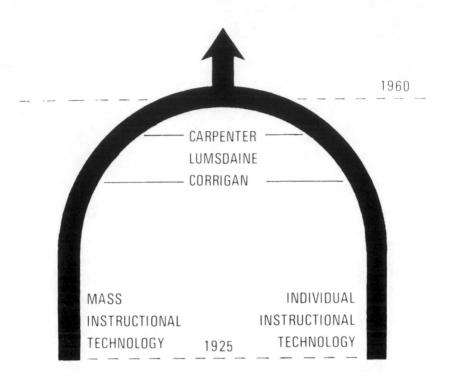

Fig. 2 DOUBLE-PRONGED
TECHNOLOGICAL GROWTH

language laboratories of all types; (d) specifically programed printed verbal materials such as scrambled textbooks; and (e) teaching machines of the Skinner, Pressey or Crowder types containing carefully worked out verbal or pictorial programs with various ingenious mechanical or electronic arrangements to test student reaction and to inform him of his progress and errors.

Reference to Figure 2 will indicate that I am suggesting that the possibilities of combining these two technologies of mass and individual instruction are under development. The work of A. A. Lumsdaine at American Institute for Research, Pittsburgh, Pennsylvania, is one example. In an experiment, Lumsdaine combined instruction with a programed textbook—a form of teaching machine—with the mass use of the Harvey White physics films on television. Over 10 years ago, C. R. Carpenter did extensive experimentation at Pennsylvania State University with a "classroom communicator"—a device enabling him to get individual feedback from students while viewing films, listening to lectures, and so forth. Dr. Carpenter is at present contemplating new work in this area. The invention and development of the so-called Teletest System by Robert Corrigan and Dean Luxton in California is the most sophisticated approach to this problem. This device enables any class receiving television broadcasts to be equipped with individual "teaching machines" which will take silent signals from the TV receiver as to the proper answers to questions and feed them into machines at students' desks so that a student, in responding by pushing a button, will receive immediate knowledge of results while, at the same time, a permanent record is made of his performance.

The next step in this procedure—one that is only now being suggested by a few activities—is illustrated in Figure 3. Here I would like to introduce the concept of instructional systems. Given a technology of mass instruction and a technology of individual instruction, the next step is to

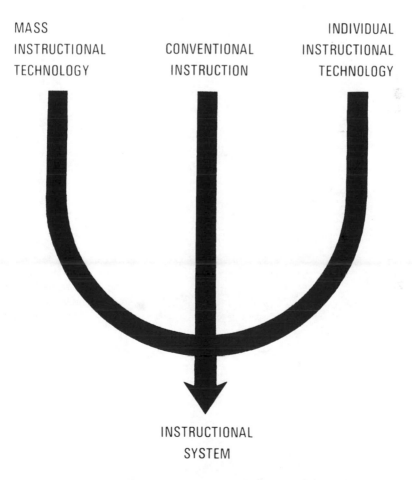

MASS INSTRUCTIONAL TECHNOLOGY   CONVENTIONAL INSTRUCTION   INDIVIDUAL INSTRUCTIONAL TECHNOLOGY

INSTRUCTIONAL SYSTEM

Fig. 3 INSTRUCTIONAL SYSTEMS CONCEPT

combine these two elements on a planned or programmed basis with conventional instructional methods into an instructional system.

A prototype model of such a system exists in the physics program for high schools developed at Massachusetts Institute of Technology by the Physical Science Study Committee. After revising the content of the physics course with a new textbook the group has created or is planning to create: (a) paperback supplementary books, such as the one on soap bubbles; (b) a laboratory manual with experiments for students; (c) a manual on how to build certain apparatus not available from conventional sources; (d) some apparatus; (e) approximately 80 motion pictures; (f) other audiovisual materials such as filmstrips and tapes; and (g) a teacher's manual which is, in effect, a program for the whole system. Such a system is obviously very costly to create and, at least in the eyes of its creators, must be used according to the program laid out for the system.

As I am discussing trends here, the systems concept can be further elaborated. In Figure 4 can be seen the possibilities of such an idea—an idea which, I think, will inevitably come into being in some parts of our educational program, providing that the emphasis on the quality and survival values of education in our society continues at its present level or increases. If we consider, for purposes of this discussion, that the instructional process can be broken down into the elements of (a) mass pre-

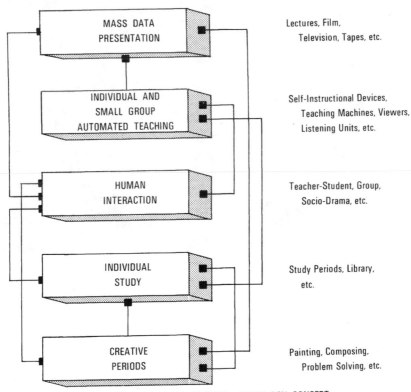

Fig. 4 INSTRUCTIONAL SYSTEMS—BLACK BOX CONCEPT

sentation techniques; (b) individual automated teaching; (c) human inter-action; (d) individual study; and (e) creative periods, then, in a true in-structional technology, these elements would be treated as black boxes in a system. For every instructional problem, assuming the mastery of the technology of each of these elements, the professional teacher and/or curriculum director would create the proper system designed to achieve the agreed-upon objectives.

We have spent some time on the current burst of instructional technol-ogy in the form of language laboratories, teaching machines, massed films, and similar developments. And we have examined the trends which take us into the possibilities, at least for some kinds of teaching and learn-ing, of the development of instructional systems. Our problem now is to relate this to the specific business of the education of teachers.

It might help in making this connection if we examine the recent past and present status of teacher education in the audiovisual area. Obvious-ly, what is known as audiovisual education is merely another term for instructional technology. Although there has been some effort at teacher education in audiovisual techniques since the late 20s, the major effort did not begin until after World War II and is by no means completely ex-tended through the teacher education system to this day.

The first general textbook for teachers, by Anna Verona Dorris of San Francisco State College, is dated 1928. In 1935, Pennsylvania adopted a re-quired course for teachers in audiovisual (then *visual*) education. In 1946, California became the second state to require such a course of all teach-ers. No other states have, at this time, "followed suit."

In the meantime, however, the number of courses and audiovisual cen-ters at teacher education institutions has grown to modest proportions. Robert de Kieffer made two studies of audiovisual programs and audio-visual courses in colleges and universities, in 1947 and again in 1957 (de Kieffer, 1959). In 1947, less than half of the institutions studied had any kind of an audiovisual program; in 1957, almost two-thirds of the respond-ing colleges had programs. The difficulty here is that a "program" is just as likely to be a service operation as a program of instruction. In fact, de Kieffer shows that, in 1947, 28 percent of the June graduates of a rela-tively small sample (126 institutions) had one course in audiovisual edu-cation as part of their preparation for teaching; in 1957, with a larger sample (436 institutions), 38 percent were so prepared. Eleven years after World War II, and with all of the audiovisual developments which grew out of that period, only about one-third of the graduates of leading teacher education institutions had taken one course in the field of audio-visual education.

Several patterns have been developed to accomplish the pre-service education of teachers in audiovisual techniques. I commend to your study the January 1959 issue of *Audio-Visual Instruction*, one of the offi-cial journals of the Department of Audio-Visual Instruction of the Na-tional Education Association. The entire issue is devoted to teacher edu-cation and includes excellent statements, not only of various types of pro-grams, but also of expected competencies that should come out of such programs.

New Techniques of Teaching

These patterns, for the most part, hinge upon the concept of "integration"—not in its Mason-Dixon Line sense, but in the sense that, theoretically, unless audiovisual materials are integrated throughout a student's experience, he will not be inclined to use them once he becomes a practicing teacher. There are, however, several kinds of integration and also there is a very good argument that integration is *not* the answer.

One pattern, which might be called *complete* integration, assumes that what is needed is the effective and extensive use of audiovisual materials in the entire course experience of the teacher-to-be. This, of course, is based on the old concept which contains a good deal of truth, namely, that teachers tend to teach as they are taught. This pattern assumes that the psychology of audiovisual materials usage will be taught in educational psychology, the methods in the methods courses, and so on. Such an organization usually ends up with only the equipment problem relatively untouched and this is usually handled by some sort of compulsory, noncredit laboratory as a pre- or co-requisite to student teaching. Obviously, there are many places where the ball may be dropped in such a pattern, and the only such program I, personally, ever saw which was successful required that the responsible audiovisual professor teach all of these separate units in all of these separate courses. All I can say for that program is that it gave the professor "a lot of mileage" and that he met many interesting people.

A second integration pattern assumes that all that is necessary to know about audiovisual education may be covered in existing methods courses. Here, the crucial factor is the professor. If, as many secondary methods teachers do, he spends a good deal of the course time teaching content, then audiovisual and other things have to go by the board. If he is "hepped" on group dynamics, the same thing can happen. Many elementary methods courses, for example, spend a good deal of time on the problem of motivation and repeat material which should have been dealt with in educational psychology. The weakness in the methods integration approach is symbolized, I think, by an experience I had some years ago in a major university (not my own). A committee of students from a methods course came one day to one of the staff members of the audiovisual center to learn everything about educational television because they had to make a report on it the next day in class.

At a few institutions which have rather extensive operating audiovisual centers serving instruction, a third pattern is followed. Here, there is no organized program, but everyone, including professors and students, is involved in projects at the center. Students prepare materials for class reports; professors look up films and preview them for their own courses; everything is one happy audiovisual family among the staff and students, with the only control, when any, being the requirement for an operator's card for participation in student teaching.

The fourth pattern is the separate course at the pre-service level. Here, following one or more very well-organized textbooks and laboratory manuals on the market, an effort is made to develop the desired competencies on an organized basis. The weakness here, as is probably true in most college courses, can lie in the quality of the instructor and the presence or absence of proper facilities. Some such courses are merely training for equipment jockeys and do not deserve the name of college courses. Others are so abstract and impractical as to be without effect. Practice varies. In California, for example, the time devoted to the laboratory phase of such a course runs, in different institutions, between six and 30 hours a semester. In the first instance, the laboratory includes only equipment operation, while in the second it covers such items as the preparation of materials and evaluation.

My own preference is for the separate course, but I recognize that institutional differences may govern such decisions. I would put one caveat on the separate course idea. It needs to be combined with a well-planned program for the use of these materials in student teaching and a general atmosphere of wide use of the audiovisual materials in all the courses in the college.

All of these patterns and all of these institutions we have been discussing are, however, conducting teacher education in this field within the pattern of conventional audiovisual devices and materials. And our major consideration here today has been that the impact of the new technology will make the audiovisual program of the 60s a different animal than the conventional "cat" now being regularly "belled" in many institutions. Teaching machines, instructional systems, and massed films present nonconventional problems to the teacher-to-be.

The first generalization we can derive from this fact is, that if the need existed right after World War II to develop all these programs of teacher education in the audiovisual field, that need is now presented to you as administrators in a form magnified tenfold—perhaps even 100-fold. Remember, gentlemen, the 22-year-old girl you are graduating next June will not be 65 until the year 2004! Some of you in this audience have no conventional audiovisual programs to brag about; others have been content to see their programs drift with the status quo. I suggest that all administrators and faculty members of teacher education institutions ought to start worrying seriously now about their obligations to the school children for the year 2004.

While I make no claims that my crystal ball is even 50 percent accurate, I am sure that some of the events I suggested at the beginning will come to pass. What, then, are some of the implications of these predictions in terms of the demands that instructional technology may make upon the education of teachers which is designed to prepare for the future?

First, the role of the teacher is obviously going to change. If systems even remotely like the Trump Plan, for example, come into being; if we go into part individual and part mass instruction; if we do many of these things that are now on the horizon, we face the problem, not of educating teachers, but of educating different kinds of teachers. This is a problem you should start working on this year.

Second, as we move into instructional systems, the first professional question is: "How do you perform in a system?" At present, our teachers are oriented to performance almost of a random nature. We face a whole new concept here. What does it mean to teacher education?

Third, the concept of programming is already premeating education from textbook publishing to filmmaking. I even heard the other day of a "programmed teacher"—by which the speaker meant a teacher who had worked on a subject matter hard and long to program it for a teaching machine. As a result, when the teacher got back into the classroom, he not only performed differently, but was measurably more effective than a "nonprogrammed" teacher. What are you doing about programming and the reviving psychological concepts of re-enforcement in your teacher education program?

Back of all these ideas of systems, programming and reorganization lies the concept of a higher level professional—providing always that the control of education remains in the hands of the profession. A true professional teacher of the future should be able to program, should be able to put parts of a system together, should be able to create systems. All of this involves making sophisticated, professional judgments; it involves a high-level knowledge of psychological theory in its applied sense and a knowledge of communication theory and social psychology; it also involves an important value system. Consider, for example, the judgments a television teacher has to make in the preparation of one lesson. While these judgments ought to be of the same order that every teacher does or should make, the fact is that they are not. That is why one-half hour of television teaching is equivalent to one day's standard teaching.

You are probably more familiar with the suggested changes that may take place in school administration and in related fields. Reorganization of the school day, additional services, as well as expanding and redefined roles for administrators are part of the pattern. So also are the uses of data processing equipment for student programming as well as for more conventional business functions. These developments will also affect the teacher and require a different education.

Finally, the last of these implications is related to the fact that we have been going through, and will continue to go through at an increasing pace, a period of innovation in education. The causes for this are complex, but the fact of change is not. One of our great problems today is that teachers and administrators are, in many instances, blindly hostile to innovation. Now, I am not maintaining that all that is new is good. However, neither is it all bad. This means that, if this phenomena of change is to become a continuing factor in educational life, our programs for the education of teachers and administrators must develop a new attitude toward innovation—an attitude that combines careful judgment with a lack of fear.

I would like to suggest in conclusion that, while all of these developments I have discussed are somewhat strange and overpowering, the final result in the teacher education field is simple and familiar. Teacher education used to deal with methodology. The critics of education have, in the last 15 years, continually accused us of emphasizing method to the exclusion of subject matter and other important things. Such an accusation merely shows their ignorance. The fact of the matter is that we have neglected the methodological problem in favor of curriculum magic, unstructured group dynamics, mechanical administrative problems, and a host of other items. A technology of instruction forces us back to our basic mission—methodology. There is, however, a difference. This time we have the tools and we are rapidly gaining precise know-how. I urge you to return to your own tradition in these new dimensions.

References

Beard, Charles A. "Introduction," in J. B. Bury *The Idea of Progress*. New York: Dover Publications, 1955 (First published as the American edition by Macmillan in 1932.)

de Kieffer, Robert E. "AV Activities of Colleges and Universities in Teacher Education," *Audio-Visual Communication Review*, Vol. 7, No. 2, Spring 1959, pp. 122-137.

Finn, James D. "Automation and Education: 3. Technology and the Instructional Process," *AV Communication Review*, Vol. 8, No. 1, Winter 1960, pp. 5-26.

Instructional Technology

... The most significant factor of the coming technological age may well be the new man-machine partnership in intellectual activities. The next decade will see the advent of the period in which electronic machines become highly active in the intellectual activities of the world, in which the capacity for gathering and using information, the overall brainpower of the world, is increased by many fold. ...

—Simon Ramo (1961)

The next decade of which Simon Ramo speaks is now but five years away. These years will be, as we grow older (individually, and collectively as DAVI), five years that will pass with an acceleration which is incredible. Already the sites for the 1970 and 1971 national conventions are being considered. And in the meantime, of course, the physicists will have to deal with twice as much information as they have today, the biologists and biochemists may have cracked the riddle of cancer, and the supersonic aircraft will be an approaching reality, if not an actuality.

The speed and, as Robert Oppenheimer once said, the newness of change itself are characteristic of our times. In facing this time of accelerating change in which nothing will remain totally static, the membership of the Department of Audiovisual Instruction has an option. It can elect to state its self-image in a name which is static or backward-looking, or it can opt for the future and say to the world in general and the educational community in particular that its members (in terms both of their self-image and their consequent function in education) bear the standard of the educational future. *That is all the choice there is.*

My proposal, pointed only to the future, therefore, is that the membership vote to adopt either one of two names: the Department of Instructional Technology or the Department of Educational Technology. I should like now to present the argument in detail for this point of view—under the constraints stated by the chairman of this symposium, Donald Ely. These constraints require that other name proposals not be criticized, but that positive arguments be offered for the point of view of the advocate. These are commendable and virtuous restraints, but they would have made the

Reprinted from *Audiovisual Instruction*, Vol. 10, No. 3, March 1965, pp. 192-194. Prepared for a discussion of the DAVI name change at the 1965 Milwaukee Convention.

writing of the *Declaration of Independence* rather difficult. For when something new if offered, it is almost necessary to criticize the old in order to show why the new fits the situation better.

However, I shall attempt to abide by the restraints as far as possible and argue that the DAVI that has been my professional life for 25 years *must now* connect with the future and announce to the world that its members are the technologists of the educational profession. I feel, in a word, that we have no choice whatsoever; we must become the Department of Instructional or Educational Technology or end up as little better than "audiovisual technicians" as Maxwell Rafferty would have us (Stoops & Rafferty, 1961, p. 312). And there is a world of difference between the technologist and the technician—roughly the distance between Simon Ramo and one of his draftsmen.

I suspect that of all the suggested name changes for the Department of Audiovisual Instruction presented in this symposium, the proposal that the Department be called either the Department of Instructional Technology or the Department of Educational Technology tends to make the membership the most uncomfortable. Such a feeling is regrettable, hardly conclusive, and probably commendable. For, as Anna Hyer has said, we are (or, at least, we once were) adventurers. There is no such thing as a comfortable adventurer; and if comfort is to be the criterion, what is the matter with borrowing a good name (with apologies to Stan MacIntosh) like Teaching Film Custodians? Aye, there's comfort for you! In fact, with a little imagination, we can all siesta it out as the Department of Audiovisual Custodians and in New York it might be possible to make more money than the District Superintendent.

I do not believe that the DAVI membership has lost the spirit of adventure. The concept of an educational or instructional technology is not comforting; on the other hand, a thorough examination of the concept will remove a whole series of artificial objections and make way for the adventure of the future.

Consider "technology" itself. Apparently some people object to the word because it means, they say, (shh) *machines.* If these people are audiovisual types, I wonder what they do; do they have special bushel baskets in the rear of the room under which they hide the projectors? As I shall show later, the exclusive machine definition of technology is completely wrong and the product of literary-oriented scribblers, but let us follow it for a while anyway. If our business has nothing to do with machines, we are the greatest collection of witch doctors this side of the Congo. We can cause images to appear in the air and sounds to emerge from walls by burning the tail hair of a Siberian Mammoth.

Ah, say the anti-machinesters, you are kidding. We know we have machines, but let's keep it quiet. It's the materials that count—you know, films, tapes, slides, that sort of stuff. The machines are just *there.* Now, that is an interesting argument when you think about it. It presupposes that other machines work on nothing, but that our machines are something special in that they work on materials—and it's the materials that count. Do drill presses punch holes in air? Do telephones carry silence? Does a tractor plow ozone? Functionally, does a film exist without a projector, a TV program without a tube, or a still picture without a camera? And you can turn it around, you know. The projector *is* a nothing without the film (and without light control and a lot of other things, we might add). As a postscript, the projector isn't much without a *good* film; on the other hand, the walls of this philosophic trap close in when we can't project a good film because of a burned-out bulb.

The argument is turned on the point that machines and materials together must be part of the system. Ah, say our mythical opponents (I think I'm successfully skirting the Ely restrictions), but you are forgetting the most important item of all—people. Without them, your materials and your machines are meaningless—and, since technology, after all, only deals with machines and materials, we must find another name such as communication or instruction or something that says only people count. This argument is a little more complex.

It hinges on such ideas as, "It's what you do with (insert the film, the TV program, the filmstrip, the programed book) that counts." Such a statement might come close to winning a prize for the cliché of the century if it didn't appear in professional educational writing, where it continues to be treated as if it were emblazoned on the shield of Henry IV freshly discovered in some new diggings under Westminster Abbey.

This point is freely admitted. But it is so obvious as almost be be nonsense. Even the simple technician in the automobile factory does not use his drill press to punch holes in Swiss cheese, and very few farmers plow sea water. Any teacher, trainer or Harvard professor who uses the wrong machines, materials and techniques for an instructional job is a poor workman; he is as much a part of the technology as the machine or the material and should be put to right, one way or another. The point is that any technology includes humans and the human element. That is what is meant by a *man-machine system.* And I challenge any member of this symposium or any member of DAVI to show me that a teacher using a film in a classroom is not a man-machine system in miniature.

I shall skip over matters of students, classroom environment, and so on and come to one final point along this line. That is the question of organization and administration. There are those who will say that the job of most members of DAVI is, somehow, administrative and that, therefore, technology has little to do with the job description. Again, I must demur. Organization is, as the French say, *technique.* What is being organized (audiovisual, instructional communications, or what have you) is also *technique.* Any concept of instructional technology must, in addition to men, machines, materials, and the like, include organization patterns.

We are now ready for a technical definition of technology. In other discussions of this problem, I have cited Beard (Finn, 1960, p. 10) and Hoban (Finn, 1963, pp. 100-101) among others and no one seems to pay a great deal of attention. Let me try again—this time with Henry B. duPont, who said:

> Technology is a term which is frequently applied loosely and sometimes erroneously. It means actually more than scientific development, more than engineering achievement, more than mechanical force or the harnessing of energy. Technology is, in fact, the sum total of all the

tools and techniques through which men have added leverage to the human effort since the beginning of time . . . [it] is a force which brings together invention, skills and techniques, machinery and equipment, men, money and methods. It is motivated by the question, "How can we accomplish a desired result at the lowest cost in terms of money, time and, above all, of physical output to the end that our resources, our time and our energies can be made available for further advances?" Technology has succeeded spectacularly in America because we have rejected the thesis of man as a beast of burden and valued his ingenuity above his physical strength. (duPont, 1963, p. 2-3)

The pursuit of a definition of technology and then the application of that definition to education can be carried great distances. For example, the French sociologist Ellul takes somewhat over 50 pages on definition and then keeps throwing in additional material throughout the first part of his book (Ellul, 1964). Machines, materials, methods of use, systems are all part of the pattern of rational mechanisms operating as means to educational ends. And, as Hoban has said, machines are central to this concept even if alone they are not technology. This is my answer to Lumsdaine (Lumsdaine, 1964, p. 372).

If, as an organization of professionals, we understand what technology is and understand that our jobs, whatever they may be—producer, designer, administrator, research worker, teacher, educator—are the management and operation of the technological function within the social institution of education at any level, we have a future, we have a great adventure, we have a job to do. If we do not understand this fundamental, we will be assailed by each new device and process; we will continue to splinter; we will be left behind either managing a warehouse full of materials, stuck in some audiovisual corner watching others pass us by, or chasing the will-o'-the-wisp of something called communications.

Properly constructed, the concept of instructional or educational technology is totally integrative. It provides a common ground for all professionals, no matter in what aspect of the field they are working; it permits the rational development and integration of new devices, materials and methods as they come along. The concept is so completely viable that it will not only provide new status for our group, but will, for the first time, threaten the status of others. We can, I repeat, make no other choice but to call ourselves the Department of Instructional (or Educational) Technology.

Such a suggestion is, of course, a big step—a break with tradition. The idea does not satisfy our nostalgic sensibilities. For those who feel this way, I suggest they ponder the words of Ellul: "Nostalgia has no survival value in the modern world and can only be considered a flight into dreamland" (Ellul, 1963, p. 19).

References

duPont, Henry B. *Technology: Everybody's Business*. A speech given at the Engineering Convocation and Dedication Ceremonies, College of Engineering, University of Rhode Island, Kingston, R.I., April 20, 1963, pp. 2-3.

Ellul, Jacques. "Ideas of Technology: The Technological Order," in Carl F.

Instructional Technology

Stover (Ed.), *The Technological Order.* Proceedings of the Encyclopaedia Britannica Conference, 1962. Detroit: Wayne State University, 1963, pp. 10-17.

Ellul, Jacques. *The Technological Society.* Translated by John Wilkinson. New York: Alfred A. Knopf, 1964.

Finn, James D. "Automation and Education: 3. Technology and the Instructional Process," *Audiovisual Communication Review*, Vol. 8, No. 1, Winter 1960, pp. 5-26.

Finn, James D. "Instructional Technology," *The Bulletin of the National Association of Secondary-School Principals*, Vol. 47, No. 283, May 1963, pp. 99-103.

Lumsdaine, A. A. "Educational Technology, Programed Learning, and Instructional Science," in *Theories of Learning and Instruction*, The Sixty-third Yearbook of the National Society for the Study of Education Part I. Chicago: The University of Chicago Press, 1964, pp. 371-401.

Ramo, Simon. *Peacetime Uses of Outer Space.* New York: McGraw Hill, 1961.

Stoops, Emery and M.L. Rafferty Jr. *Practices and Trends in School Administration.* Boston: Ginn and Company, 1961.

The Marginal Media Man
Part I: The Great Paradox

In spite of the clichés currently in vogue about the progress and challenges of professionalism in the educational media field today, a hard, honest look at the problem reveals that our business, field, profession—whatever you choose to call it—is actually in serious trouble.

It is the thesis of this article that traditional concepts about the professionalism of the educational media field[1] will probably have to be changed and that the organization of DAVI ought to be changed in order to accommodate the demands of the times on the media man. For the media man *must* be, and must always remain, marginal to many fields, organizations, events, and institutions, and any move toward greater professionalism must take this stubborn fact into account.

The statement that we are in trouble can be countered by overwhelming evidence that the educational communications field as a professional specialization is not only *not* in trouble, but is enjoying its best days. And here is the beginning of what might be called the great professional paradox. On one hand, it is true that the audiovisual communication field has been growing; its practitioners are becoming more professional; and its many professional contributions to American education over the last 30 years finally are being recognized. On the other hand, increasingly it seems, our destiny is being taken out of our hands; we are gradually being demeaned; and the educational media field, once our domain, is now either too important to be left to educational media specialists or too insignificant to admit the importance—or even the existence—of the need for an educational media specialist, except as he temporarily may be useful as a technician in repairing a projector, hooking up a TV set, or, in extreme cases, locating a film that the librarian could not find.

Let us first make the case for progress toward professionalization within DAVI. In January 1952 there was no *Audiovisual Communication Re-*

---

[1] In the course of this discussion I shall use all of the names now under consideration by DAVI (see *AVI*, March 1965 and previous chapter in this book) more or less interchangeably for purposes of style. Incidentally, the name itself is of almost crucial importance in this problem of professionalism.

Reprinted from *Audiovisual Instruction*, Vol. 10, No. 10, December 1965, pp. 762-765.

view. As of this year, *AVCR* appears in Volume 13, embracing almost 60 issues, including six special supplements. In January 1952 *Educational Screen* was the official publication of DAVI and, as such, carried little professional content directly related to the concerns of the organization. The year 1965 marks the tenth year of publication of DAVI's own professional magazine, *Audiovisual Instruction.*

In the span of about 13 years, the publication program of DAVI has had some remarkable successes, topped by the monumental Lumsdaine and Glaser work, *Teaching Machines and Programmed Learning: A Source Book* (1960), and by the subsequent publication of *Teaching Machines and Programed Learning, II:Data and Directions* (1965). There have been yearbooks, monographs and other substantial contributions, such as the Williams' book, *Learning From Pictures* (1963). By any standard, the publications program argues that tremendous professional growth and success have been achieved.

A list of other outstanding successes can readily be put together: the growth and increasing influence of the national convention; the contributions of the Okoboji conferences; the effect of the influential joint media conferences held with ASCD in 28 states a few years ago under a Title VII contract; the invitations to representatives of DAVI to participate in meetings overseas; and the great influence of DAVI thinking in the Educational Media Council (at one time there were five DAVI past presidents sitting on the Council[2]).

The list could go on to embrace legislative activities and many other important contributions. A word should be said concerning DAVI's relationship with the NEA hierarchy and with the complex of NEA affiliated organizations. Over the last 13 years or so, because the Division of Audiovisual Instructional Service of the NEA and the Department of Audiovisual Instruction were, in effect, one and the same, the counsel of the educational media specialist has been heard increasingly in the general educational community; in some cases, as with the current copyright situation, leadership for the whole educational community has fallen to us.

So far the discussion has centered principally upon the professional growth of the field through activities and developments within DAVI and its symbiotic partner, the NEA Division. There are, of course, many elements within the overall field of instructional technology itself that project change, growth in importance, and the move toward professionalism. In that same January of 1952 used as a benchmark before, educational television did not exist except on paper; teaching machines and programed instruction had not been exhumed from their premature burial at Ohio State University 20 years earlier; the language laboratory was a gleam in its developers' eyes; computers were discussed in popular science articles as "giant electronic brains" with no conceivable relation to education; and even such simple devices as the overhead transparency projector (over a decade old at the time) were almost unknown to the majority of American educators.

---

[2]Cochran, DeKieffer, Schuller, Witt, and Finn.

The Marginal Media Man

Money, which is equivalent to energy, was inserted into American education during the years following 1952 by the great foundations and the federal government for the development and use of audiovisual communication devices and materials. Title III of the National Defense Education Act, for example, which provided the school systems of America with a great deal of equipment and materials, forced a certain amount of professional direction upon both state departments of education and school districts. Title VII, the research and dissemination title, had been expanding the intellectual horizons of the media field since its first implementation in 1958 or 1959. Other legislative programs associated with the Great Society concept of the Johnson administration are all having these same effects.

In response, professional programs for training and development have shown some increase. At this writing, approximately 20 NDEA fellows are working on doctorates in the educational communications field in four institutions of higher education. There were over 35 institutes for educational media specialists in the summer of 1965, funded under Title XI of the NDEA. Other graduate programs within a small cluster of institutions have shown increases. Placement requests for instructional technologists with the doctorate are impossible to fill at the present time.

Finally, the professional organizations have grown as well. DAVI had approximately 1,200 members in 1952. It now enrolls 6,000. The National Association of Educational Broadcasters was reorganized and refinanced during this period and has now become very powerful. The National Society for Programmed Instruction was founded. Both NAEB and NSPI have engaged in joint projects with DAVI. The Educational Media Council was created about 1961 and now includes representatives of 14 national organizations and maintains headquarters in Washington, D.C. Professional organizations outside of the field, such as the American Association of Colleges for Teacher Education and the Association for Higher Education, have become interested in media.

Space prevents a further catalog of events, organizations and achievements which point to great professional development on the part of the educational communications field. Enough has been said, however, to make the case. And the case sounds great. The only trouble with the case is that it is true only if you look at it from the optimistic side. *For the paradox in which we live our professional lives is that while things professional in the audiovisual field seem very good, they are also very, very bad.*

Let us now make the case for retrogression of professional development in audiovisual communication. When I use the term "retrogression," in a way I mean it to be thought of in a "what-might-have-been" sense. The positive development of the field described above *did* happen. However, while it was happening, somebody else picked up and carried off most of the balls in the park. We had a few short years—from about 1952 to about 1958—to achieve true professional status and, at least, give some effective professional direction to the field. The truth of the matter is that we didn't make it. At this writing it remains in the balance whether we ever will.

The case for retrogression may be seen most clearly by examining two factors: the downgrading of the educational media field by the U.S. Office of Education and the complete lack of attention given to educational media at the White House Conference.[3]

The Downgrading of the Educational Media Field by the U.S. Office of Education

The Commissioner of Education, Mr. Francis Keppel, recently reorganized the U.S. Office of Education. In the process, he wiped out the Educational Media Branch, separating the functions of Title VII-A (Research) and Title VII-B (Development-Dissemination) within the larger bureaucracy. At least for Title VII-B projects, the monitoring of the projects will be scattered throughout the Office; for example, the monitor of the new contract of the Educational Media Council is to be a nonmedia person.

Since Dr. Walter Stone left the U.S. Office a few years ago, where he was the head of the Media Branch, the Commissioner has never replaced him. Seth Spaulding, Thomas Clemens, and John Gough all occupied the position on an acting basis. Several other excellent professionals have paraded swiftly through the Media Branch, either back to where they came from or on to better things, notably Gerald Torkelson and Hugh Mc-Keegan. Obviously, if Commissioner Keppel had any concern at all for the media field, he would have obtained and appointed a permanent head long ago and made conditions such that the parade of experts might have stopped short of a general exodus. Now the Commissioner doesn't need to concern himself—the Branch is gone.

The Commissioner's recent decision not to allow the proposed revision of the *Educational Media Index* to continue as the unique partnership it was between the government, private industry (McGraw-Hill), and the Educational Media Council, but rather to force it into the public domain, was a direct blow at the field of instructional technology and a direct measure of the significance the field has within USOE thinking.

Finally, the support and conduct of the New Media Demonstration Center within the U.S. Office is almost a scandal. The Center was grandly conceived as a place where school people from the United States and abroad could see all types of new equipment and materials in a demonstration situation. To that end, many manufacturers and producers, with high hopes, I am sure, placed their equipment and materials in the Center. The Center has gradually declined into a sort of unofficial audiovisual service for USOE agencies, and the room is considered an adjunct meeting room for USOE staff functions. In my last two visits there, much of the demonstration equipment was piled into the corners gathering dust. I hasten to add that it was not the fault of the staff assigned to manage the operation. The staff was too small, overworked and lacked the proper logistical support to do a good job. It can be concluded that the Commissioner does not even care about the public image the media field has withint the U.S. Office of Education.

---

[3]The White House Conference on Education, 1965.

The Marginal Media Man

The White House Conference and the Media That Weren't There

The much publicized White House Conference on Education held in July of this year erased any doubts as to the group now running American education for any person at all knowledgeable concerning the national educational scene. The New Educational Establishment which has been put together in the last 15 years was revealed in all its glory from Grandfather James Bryant Conant on down the line.

Although there were political, geographical and other dimensions associated with the invitations in addition to educational eligibility, the fact remains that only two members of the old-line audiovisual field were invited as participants—and they were there by virtue of the fact that they had worked on the Educators for Johnson and Humphrey National Committee during the last campaign, not for their connections with the educational media field. As far as I know, the only other people associated one way or another with educational communications were a couple of programed learning specialists from industry, several big-time publishers, and, of course, Senator Benton and his associates from Encyclopaedia Britannica.[4]

The program was also revealing. Many of the panel topics might or might not have had representatives from the media field, as, for example, "Pre-School Education." However, it would be expected that the two panels on educational innovation—one for elementary and secondary schools and one for higher institutions ought to have had such representation since much of what is referred to as educational innovation either is or is related to educational technology. The closest one might come to this requirement was the fact that Alvin C. Eurich, formerly of the Fund for the Advancement of Education of the Ford Foundation, was on the panel for innovations in higher education. While Dr. Eurich has had more influence on the audiovisual field than most of us put together, it is stretching a point to consider him as a professional representative.

Further, the vice-chairman of the Conference in charge of these two panels—Dr. Ralph Tyler, director of the Center for Advanced Study in the Behavioral Sciences—could hardly plead ignorance. After all, his brother Keith has had his entire professional career in the educational broadcasting field, and one of Ralph Tyler's oldest friends is Edgar Dale. The conclusion is inescapable: From the point of view of the planners and directors of the White House Conference, the educational media field deserved no attention whatever.

One of the major addresses of the Conference was given by Dr. Tyler under the title, "Innovations in Our Schools and Colleges." In this address he did mention educational technology as one of five promising areas of innovation. It could be argued, therefore, that there was some consideration. However, from the professional point of view of people who have pledged their careers to educational media, his words are hardly comforting. Read this quotation carefully:

---

[4]William Harley, president of the National Association of Educational Broadcasters, was allowed in as an "observer"—one grade lower than a "participant."

At the present time, however, the yield from the innovative efforts [in educational technology] has been small. Too many of the projects undertaken have been guided by those whose training and competence are in the technology and they have not been wholly familiar with the educational tasks, the aims sought, the conditions of learning to be maintained, and the like. However, today some experiments have been started by persons who have the educational competence as well as having knowledge of the technology being used. We need many more efforts of this sort. . . .(Tyler, 1965, p. 188)

Dr. Tyler is plainly saying that, in his opinion, the audiovisual or educational media professional isn't good enough and that other educators must take over his function.

There were also two consultants' papers (published prior to the Conference) which had bearing on the media field. These papers paralleled the two sections on innovation. One, by Professor Lewis B. Mayhew (1965) of the Stanford School of Education, was concerned with innovation in higher education and devoted a great deal of space to the newer educational media. Professor Mayhew quoted extensively from the Stephens College report of 1959, a well-known document in our field. He also referred to some newer developments, such as Florida Atlantic University. Significantly, there is no evidence that Professor Mayhew was acquainted with the DAVI-AHE book on newer media in higher education. That, and his reliance on a 1959 statement in 1965 during a period of accelerating change, suggests that his acquaintence with our field and with the central literature is woefully inadequate. Of course, a scholar would not go looking for literature generated within a profession that he didn't know existed. . . and that is hardly his fault.

Professor Dwight W. Allen, also of Stanford, prepared the paper on innovations in elementary and secondary education. He devoted about two paragraphs to media, but since he was considering innovation broadly, this seems fair enough. In general, he took the Trump position with respect to audiovisual communication devices and materials. *Nothing in either of these papers suggests the need for professional direction of media programs, the existence of a professional media literature, or even of media professionals.* The conclusion one is forced to come to is that, in the eyes of the Establishment, such media programs, after all, are pretty simple and can be managed by any amateur (with the possible help of a "technician").

Finally, the discussion in the two meetings on innovation merely reinforced the general trend of downgrading instructional technology. In effect, all of the discussion could be summarized as stating somewhere fairly early in the proceedings that there were such things as films, television and teaching machines; that such things had their small place; and that it was now time to get on with the business of talking about innovation. If I may editorialize even more directly for the moment, I thought both innovation meetings were very poor, particularly when compared to the excellent coverage given the subject at the last DAVI national convention. The cavalier shunting of the audiovisual field to one side early in the meetings and the general low quality of the meetings themselves were especially horrifying under the circumstances, as when Dr. Tyler announced that 35 of the 50 state superintendents of public instruction were in that particular audience!

Quite a lot of wordage has been devoted to the White House Conference because it is significant as a bellwether for American education and will be for some time to come; and the media field, as a field or as a profession, came off badly. There were, it should be added, two positive factors—one minor and one major. The National Audio-Visual Association (NAVA) supplied the briefcases for the participants, and Vice-President Humphrey (still perhaps influenced by the Minnesota DAVI Convention) gave educational media a plug in his speech.

Basically, the White House Conference must be seen as an expression of the new power structure in American education—the New Educational Establishment. If the National Education Association, the loose coalition of state education associations, the chief state school officers, professors of education generally, and the American Association of School Administrators continue to think of themselves as *the* Educational Establishment (so often criticized in the decade beginning in 1950), they had better forget it. There is a New Educational Establishment, and what this power group thinks about educational media or audiovisual communication is absolutely crucial to the professional development of instructional technology.

Therefore, the New Educational Establishment and its general attitude toward and performance with educational media needs further exploration as a major part of the negative side of the professional paradox. And there are other elements, such as the absolutely decisive role of Jerrold Zacharias in media development, the disappearance of some of our old friends, the low condition of state-level audiovisual administration, the fundamental lack of understanding and support of our field among teacher and administrator organizations, etc.

Once these somewhat dismal factors have been explored in the next article, attention can be directed to the mixed and troublesome situation within the media profession itself. While we may not yet be in the position of appreciating our enemies more than our friends, we will probably get to that stage fairly soon. Such incidents as the recent anti-intellectual memorandum submitted by the Audio-Visual Education Association of California to the DAVI Delegate Assembly need to be scrutinized, as well as the state and nature of our training and certification programs, the national organization power struggle, the inability of the profession to develop standards, and the general commitment to footdragging.

After the exasperating state of affairs within the profession of instructional technology is located with the paradox of success and failure, the problem will have been defined. It then may be possible, in the concluding article,[5] to suggest new dimensions for the marginal media man and how these dimensions may be drawn. All in all, it is an uncomfortable pattern of thinking that we face, but, if there is to be salvation in

---

[5]Because of other commitments, Dr. Finn was unable to complete Part II of this article as planned.

media, it will be because we persisted with, as the sportswriters say, a hard nose.

References

Allen, Dwight W. "Innovations in Elementary and Secondary Education," *White House Conference on Education: A Milestone for Educational Progress* (prepared for the Subcommittee on Education of the Committee on Labor and Public Welfare, United States Senate). Washington, D.C.: Government Printing Office, 1965, pp. 136-142.

Audiovisual Instruction. Vol. 10, No. 3, March 1965. Theme: "The Media Specialist—Agent and Object of Change."

Balanoff, Neal. Study on Telelecture. A Stephens College report prepared under NDEA Title VII.

Educational Media Index. Industrial and Agricultural Education. New York: McGraw-Hill Book Company, 1964.

Glaser, Robert. *Teaching Machines and Programed Learning, II: Data and Directions.* Washington, D.C.: Department of Audiovisual Instruction, National Education Association, 1965.

Lumsdaine, A. A. and Robert Glaser. *Teaching Machines and Programmed Learning: A Source Book.* Washington, D.C.: Department of Audiovisual Instruction, National Education Association, 1960.

Mayhew, Lewis B. "Innovations in Education," *White House Conference on Education: A Milestone for Educational Progress* (prepared for the Subcommittee on Education of the Committee on Labor and Public Welfare, United States Senate). Washington, D.C.: Government Printing Office, 1965, pp. 126-136.

Tyler, Ralph W. "Innovations in Our Schools and Colleges," *White House Conference on Education: A Milestone for Educational Progress* (prepared for the Subcommittee on Education of the Committee on Labor and Public Welfare, United States Senate). Washington, D.C.: Government Printing Office, 1965, pp. 185-190.

Williams, Catherine M. *Learning from Pictures.* Washington, D.C.: Department of Audiovisual Instruction, National Education Association, 1963.

A Possible Model for Considering the Use of Media in Higher Education

A rather recent and somewhat extensive experience on the national level working with professors of several disciplines on the problem of using the newer educational media in their respective fields has developed what we believe to be a useful way of thinking about the application of media to college level instruction. Professors of various subjects approach these new instructional processes in several different ways, depending in general upon their degree of naivete and, to borrow a concept from Professor Bruner, the structure of their disciplines. Thus, geographers as a whole are very knowledgeable about a wide range of media because of the nature of geographic investigation, information storage, and communication; however, within the universe of geographers, there is a wide range of knowledgeability and performance with media extending from unsophisticated to sophisticated. We have found historians, on the other hand, much less sophisticated in general in the media field, but with a great desire to learn about it, beginning at a performance level somewhat below that of geographers. This is due, we believe, to the nature of historical investigation, information storage, and communication.

With that concept as background, we can turn to a classification of media relationships in college instruction.

In one sense, educational media used at the college and university level can be thought of as existing in several categories or levels. These are:

1. The tool level
2. The data level
3. The behavior control level
4. The meaning level
5. The research level
6. The systems level.

The Tool Level

The tool level may be defined as providing the instructor with certain devices and materials with which he may do better what he is already accustomed to doing. The prime example of a tool-level device (and the

Reprinted from *AV Communication Review*, Vol. 15, No. 2, Summer 1967, pp. 153-157. An edited extract from *Educational Media and Harvard University*, a memorandum prepared at the request of David Purpel, Graduate School of Education, by James D. Finn with the assistance of Robert O. Hall and Bruce Humphrey.

material to go with it) is the overhead transparency projector. In its simplest form (projecting writing or drawing made by marking pencils on sheets or rolls of transparent plastic done in front of the class), it is a much more efficient chalkboard; in a more sophisticated form using prepared transparencies, colored overlays, etc. it provides a lecturer with a powerful device. Other examples of tool-level devices are automatic slide projectors; semiautomatic photocopying devices; and individual single-room, closed-circuit television systems in which the system is operated by the instructor in lecture-demonstration situations. Characteristically, tool-level instructional technology is, at all times, under the direct control of the instructor, and the materials used are generally created by or for him.

The Data Level

This concept refers to the fact that information of all kinds is no longer stored exclusively in conventional print form. Data may be originally recorded on or transferred to various forms of photographic, magnetic or even thermoplastic storage. Microfiche cards, computer tapes, videotapes, high-speed motion pictures, audiotapes, and colored still pictures are among the many examples of media used in this manner. Obviously, such sources of information set new requirements for student assignments, "library"-type research, and the use of such information in the classroom situation. Devices of many kinds and different approaches to materials are required.

The Behavior Control Level

Here we refer to the area known as programed instruction—considered broadly (without commitment to a particular programing theory). It is characteristic of all such programing that great energy is spent on a *prior* basis encoding the content into some form of program designed either to shape the behavior (mainly verbal) of the student on a controlled basis or to lead the student through carefully planned sequences of subject matter, branched or unbranched. The program material may appear in many forms—print, colored slides, 8mm film loops, audiotapes, etc. The language laboratory is a special case of this category. From the point of view of the instructor, the work (instruction) is all done before contact with the student is made. The instructor becomes, in the words of Robert Heinich, a "mediated teacher" (Heinich, 1965).

The Meaning Level

The term "meaning level" is used to describe the applicability of a wide range of educational media to the problem of building meaning into abstractions. In a simple case, a motion picture might be used to convey a certain amount of concrete background so that a student may build meaning into an abstraction. Some college instructors argue that after a bright student has passed the age of five, you no longer "have to show him pictures." This theory has two fundamental weaknesses. First, it assumes humans are equally at home in all fields and hence can be communicated with at a high level of abstraction in all fields simply because they can manage verbal and mathematical symbols. This is simply not true. Secondly, it assumes that all humans learn in the same way through the same sequences of symbolic manipulation. In contrast, there is plenty of evidence, for example, that students vary even within a profile of listening skills in a similar manner to the way they vary in reading skills, and that these two patterns do not match in the same individual.

Meaning, however, goes beyond grounded understanding of abstract symbols. There are affective, emotional characteristics. There is the world of the concrete, the flesh, per se. Here, again, the newer media permit the instructor to deal with affective meaning and the other dimensions of life on a controlled basis.

The Research Level

The research level is the complement of the data level. The only difference, really, is purpose and, perhaps, time. Today, many forms of instrumentation are used in research to record, to measure, to observe. Much of this instrumentation is, in fact, communication technology applied to research instead of instruction. A phenomenon may be photographed at thousands of frames per second; dialects may be recorded; documents may be x-rayed. The list could go on, but the point is that the university faculty member doing research needs a range of media support merely to do his research. The ultimate results of such recording and measurement may eventually end up as "library" materials at the data level. Their generation, however, is a different function.

The Systems Level

An interesting tendency of the last decade has been a move toward the so-called systems approach to instruction. Generally, this has meant the creation of systems of instructional materials—books, manuals, recordings, films, programed materials, tests, etc.—covering an area of subject matter in a systematic way designed to achieve rather precise objectives. Computer-assisted instruction, when designed to cover whole courses or large segments of courses, is also referred to as an instructional system. In either case, the highly systematic design required presents a new level and use of media. At this point in time, the experimentation with instructional systems at the university level is not extensive, but there are signs that with the advent of computer-assisted instruction, large-scale use of television, dial-access systems such as that now in operation at The Ohio State University, and similar approaches, a much more systematic use of media—perhaps at the whole course level—is in the offing. Again, the implications for the instructional staff are the same as those appearing under the behavioral control-level concept; much of the activity of the instruction is prior to the instruction and, in the case of systems, also involved in the feedback-evaluation loop.

A Comment on Distribution Efficiency

There are, as we all know, many ways to describe the elephant, and the category system just used cannot accommodate another media function relating particularly to higher education. Institutions of higher learning have, for the most part, extensive campuses; many of them house students as well as instruct them, so that the entire plant is spread out, and in it are performed many functions. Certain of the media can be viewed as transmission systems as well as processors of instruction. Television, audio lines, low-grade signal lines, and computer links are examples. All of these can be used to connect the separate units of the physical plant and transmit mediated instruction throughout.

It follows that any concern with media on a university campus should take into account this possibility. Recently, for example, in the midwest, dormitories are being connected with the central closed-circuit television system, and instruction is reaching students in the housing areas. This is also true of dial-access audio instruction. Computer terminals will follow.

In summary, for university functions, the newer educational media can be applied at several levels—as individual instructional tools, as data storage, for behavior control-type instruction, to build meaning, as research tools, as the core of instructional systems, and as means to increase the distribution efficiency of learning experiences of all kinds. If available, the individual university faculty member can select one or more of these levels at any time.

References

Heinich, Robert. *Systems Engineering of Education II: Application of Systems Thinking to Instruction.* Los Angeles, Calif.: Department of Instructional Technology, University of Southern California, 1965.

Educational Innovation and Title III

It is within this general context of the development and present state of the art of educational media that the remainder of this paper will consider Title III. Basic to this consideration was the detailed examination of approximately 40 proposals of various types which had been funded under the initial Title III grants and a site visit to a California school district which had begun operation under a Title III innovative grant.

The proposals studied evidenced some confusion as to their objectives. This I believe to be not so much the fault of the proposal writers as the inherent confusion in the law, the guidelines, and the proposal forms themselves. The writers had to struggle with a form which was obviously put together by the same committee that created the camel. The most fundamental problem, however, is the fact that Title III has two parts—almost equal and opposite. Even more confusing, when one of the parts (the so-called supplementary service centers provision) is analyzed, it too has, potentially, two parts.

The first part of Title III, as everyone must know by this time, provides for the formation of supplementary educational service centers on some sort of regional basis. Within this provision, however, are traditional services which are often not available or not sufficiently available to schools—guidance, health, media, etc.—plus central services embracing art, music, dance, etc. While to a legislator or to an abstract organization administrator such a conglomeration seems reasonable, from the point of view of the realities of the educational system as it exists, these are two kettles and they both don't even contain fish.

In the stated objectives of the proposals which I analyzed, objectives relating to cultural services were dragged in by the heels in order that the proposal might be kosher and obtain other needed services for the district. Nothing in what I have said should be construed that I am against cultural services; far from it, but when a district wishes to establish needed health and/or media services, the cultural requirement muddies the water and makes the objectives somewhat less than true.

This paper is part of a paper called *Educational Technology and Innovation*, 1966. The part omitted here is an introduction section on a basic history of Instructional Technology. The paper was prepared for PACE (Projects to Advance Creativity in Education).

Educational Innovation and Title III

The second part of Title III is variously referred to as innovative, exemplary, demonstrative. It is to provide funds for programs which develop innovations and exemplary programs. Obviously, such programs would have different objectives from those for supplementary service centers. They did not, in many cases. Again, the cultural elements plus unreal service objectives showed up in the proposals. I understand there was some difference in the proposal forms between the two parts of the law, but this difference was rarely reflected in the objectives. Whether the proposers could not read or whether they were touching all bases just to be safe come to mind as alternatives causing this objective mishmash. I am inclined to lean toward the touching of all bases theory which, of course, confused the objectives.

I am not saying that the proposers did not know what they wanted to do. There were a few cases of this, at least in part, but mainly the real objectives were buried among a pile of irrelevant baggage. Hence, in many cases, the program was not related to a portion of the objectives as stated in the proposals. Programs did seem, however, to be well related to the real objectives—stated or unstated.

So much for objectives. Since the main thrust of Title III for the present is to be in the direction of educational innovation with its concomitant functions of example and dissemination of results, it would be useful to continue this discussion of the proposals with some analysis of the problem of educational innovation as it relates to Title III. Some years ago, Hoban pointed out that the agricultural model for innovation probably did not apply to education, since education was a bureaucracy and that no one really knew very much about innovation in a bureaucracy. Actually, the way Title III operates with the proposal generating within a local bureaucracy and with one copy going to the State Educational Agency—another bureaucracy—and the final award being given by the U.S. Office of Education, after due cognizance of the state reaction, we have here an attempt at educational innovation within bureaucracy cubed.

Now there may be nothing wrong with this procedure providing certain realistic views are taken of the whole process. The question has been continually raised, "Were the proposals reviewed really innovative?" This problem has been well discussed among members of the task force, and, of course, the answer has to be, "Innovative for whom?" In the field of educational technology, almost every possibility discussed above has been experimented with somewhere, and, in that sense, unless we are considering a proposal to use broadcasts from Telstar, or lasers as information transmitters, or ESP, nothing is likely to be, strictly speaking, innovative in the Kuhn sense of creating a new paradigm never before set forth.

At a level slightly below this, however, there is a field for innovation that goes beyond something merely new locally. For example, the library learning centers being installed at Temple City, California, are, in fact, quite a new concept (actually, they have a couple of elements completely new) even though a portion of the setup is in the classification of audiovisual access systems for individual students—forms of which are being experimented with at several places in the United States. Educational innovation, then, in my view, cannot be classified either as something completely new or something that is only new within the locality or region. Rather, it should be thought of as existing on a continuum.

If the concept of an innovative continuum is acceptable, obviously the judgments required of the Title III staff must be more discriminating if the avowed effort is to encourage innovation somewhere in the abstract as opposed to innovation within the local environment. On this basis, I would have to say that, of the proposals that I have read that were stated to be innovative, they were in fact on the side of the generally innovative as opposed to locally innovative, for the most part. These proposals tended to be oriented to individualized learning, computer utilization (both for data processing and for instruction), and various forms of dial-access systems with some attention to television. This merely reflects the "in" elements of instructional technology as of the fall of 1966.

On the other hand, one proposal dealt with filming certain teaching practices which could then be used as media for in-service and pre-service education of teachers. It was claimed in the proposal that the practices to be filmed themselves were innovative, but the major effort of the proposal was for the filming—hardly a new idea, yet bound to produce useful and new material if done properly. This project would have to be rated as noninnovative; however, if the film proved useful and the practices were new, the films could be used all over the country and would be presenting an innovation. (I am inclined to doubt that either of these conditions will prevail in this particular project, but that is a technical matter that should have been picked up by the initial readers.)

It is when we come to the supplementary service centers that received operational grants (as opposed to planning grants) that the question of innovation gets in the way. As noted in the introductory section above, many school systems lack new media and equipment and the logistics support to apply them to instruction. Help—a great deal of help—is needed. Two of the proposals examined were from one or more county units literally thousands of miles apart. Both had the same problem—they had nothing in the way of new media and needed tooling up even to reach a minimum. Both felt they had to justify their proposals on the grounds of innovation and went through some twisting and turning in the writing in order to prove this. Both were innovating locally, but, more than that, they were improving instruction and relating their programs to the modern world. It seems to me that no more justification is needed. In one case involving two counties with 23 school districts in an isolated area of a northwestern state, the annual expenditure for audiovisual materials to service about 12,000 students amounted, on a three-year average, to about $5,500 per year. Innovation might be nice, but a much better technological base is a necessity for these two counties.

This brings up two related aspects of this problem that should be considered. Both are concerned with instructional technology and innovation. These two aspects of the problem are:

1. The claim that proposing districts or other entities are requesting too much money for equipment and materials.
2. The claim that curriculum development (whatever that may mean) is much more important than any effort to do anything about instructional technology.

In the case of the first claim (which has been widely publicized, supposedly as current policy in the U.S. Office of Education), one of the major functions of a technological society is innovation through technology—so why not through instructional technology? The implication of this claim is that, somehow, tooling up is a second-rate approach to innovation, probably more influenced by projector salesmen or TV peddlers than educational objectives. The assumptions behind this claim are based on the old-fashioned view, discussed above, that technology is a "thing," not a process function.

It is highly likely that an infusion of technology (this includes organization patterns, objectives, management controls, etc., remember) will institute more change per dollar spent than all the curriculum meetings, in-service education bashes, and professional library book purchases a district could generate during twice the period covered. There is a further overtone of that same educational puritanism here which Anna Hyer once described as the philosophy of new uses for old coat hangers. Such puritanism is neither efficient nor in gear with any kind of modern economic thinking.

In the site visit made, the superintendent was questioned closely with regard to the genesis of his project. His point was that he had spotted a very great need and could see no other solution to meet it than to come up with a new technological system—which he did. The need was for a certain kind of individualized instruction and the solution was to develop a highly technologized learning center adjacent to the library. He is now finding that other innovations are occurring as a result of this capability and he and his staff are busy generating proposals to both public and private sources for further development.

The second claim that curriculum development is more important than technological development is, first of all, the direct result of the "addendum" viewpoint referred to in the introductory section. More important, it is completely out of touch with reality. One only needs to look at the great curriculum revision projects that now occupy the national scene. With the exception of one or two, such as Project English, the entire curriculum reform movement is accomplishing its objectives through instructional technology.

Many years ago, Dale and Corey reminded the audiovisual field that what children do in school is usually a consequence of the instructional materials used by the teacher. This is Professor Zacharias' main credo. Once the content has been set to the satisfaction of the experts in the discipline, it is converted to educational media—films, paperbacks, programed instruction, etc.—and moved into the schools via these vehicles. This is true curriculum reform because what the students do is now totally changed. It would be well for those making Title III policy to study this lesson very carefully.

All of this argument suggests that educational technology is, perhaps, the best possible machine for instituting the changes thought desirable. Even if this idea were accepted, certain roadblocks remain that show up in the proposals from certain states with strong educational bureaucracies. Although the law provides that the proposing entity applies directly to the federal government, a copy must go the state agency and be reacted to. In strong states, it is possible for these agencies to, in fact, control this process through state regulations, pressures, etc. In California, for example, there is sort of an unwritten, oral requirement that all proposals must contain statements of behavioral objectives. The assumption seems to be that the success of all innovation hinges on the acceptance of Skinnerian theory. This would be a hard case to make. I say this, believing rather strongly in the efficacy of behavioral objectives.

In New York, there are rumblings about the controls existing through regional committees controlled by the state. The problem here is that innovation is difficult at best; under vigorous bureaucratic control it may be well nigh impossible. The state bureaucracies, however, through controlling the proposals, have an efficient means of controlling the direction of new developments in their states while, at the same time, piously claiming local autonomy. As one superintendent put it, "...the trick is to nurture creativity at the local level." Nurture and control are two different things.

Both at the federal level and at the state levels, we are attempting through Title III to institutionalize educational innovation through bureaucracies; further, when dealing these days with cities of any size, we have to add a third bureaucratic level with much greater controls than both other levels put together. It is still a very moot point whether or not bureaucracy is the exact opposite of innovation. Apparently it needn't be; all of the proposals examined had some elements of innovation, existing somewhere upon the continuum. Time will tell as to the extent, lasting power, and future generating power of these innovative efforts sitting, as they do, within bureaucracy cubed.

There are some other notes that ought to be made in this somewhat rambling discussion of the proposals. Questions were raised with the task force as to the scholarly elements the proposals contained. "Did they," for example, "indicate a reasonable familiarity with the literature?" Frankly, they did not. The best proposal from the point of view of sheer innovative power, also contained the best literature backup. Many of the supplementary center proposals indicated no literature work-up whatsoever. In the case of a supplementary center, however, I would have to ask whether or not that drill was really necessary. The condition of the proposals does suggest that, for the planning grants, the subsequent operational proposals should be scrutinized more carefully for literature familiarity due to the fact that time has been available for an adequate search.

With the exception of one or two units that were located near or were cooperating with higher institutions, very little was said about evaluation in the proposals examined. This is due, I believe, to the fact that evaluation procedures were not required in the initial proposals.

We can turn, then, to the question as to the provisions for dissemination of the results of the innovative efforts as they come to pass. There were some modest efforts to provide for dissemination in some of the proposals; by far the best was a plan submitted by a county in southern Illinois to use television as an instrument for dissemination. In general, however, dissemination plans were weak or nonexistent. Many of the weak ones proposed the same old thing of teachers' meeting, attendance at courses, etc. with, perhaps, a mention of a newspaper story or two.

The entire American educational community would hardly take first prize for its dissemination activities—and this particularly includes the U.S. Office of Education and most state departments. No doubt legislative restrictions play a part in this, but that cannot be the only reason. The U.S. Office of Education has funds and a unit designed to disseminate the results of research and other information; its greatest achievement to date is to come up with the beginning of what will probably be a very good information storage and retrieval system for research information. However, an information storage and retrieval system is not a dissemination machine; it is the fuel for such a machine. The machine does not exist and, if the innovative reports from Title III go into ERIC under the theory that they will then be disseminated, the most we can say is that we have given them a more decent burial than is usual for such documents. There was nothing particulary useful, let alone spectacular, in the proposals concerning dissemination; I see nothing at either the state or federal levels which will compensate for this. Dissemination remains a major problem.

Finally, we were asked to comment on weaknesses or gaps which showed up in the proposals other than those already discussed. With respect to those proposals dealing with educational media, I found four:
1. Manpower
2. Lack of adequate consultants
3. A relatively naive attitude toward production problems and an unrealistic treatment of production problems, including budgeting
4. Requests for advanced technology without an adequate technological base, including the all-important logistical support.

Of these weaknesses or gaps, the manpower problem is the most serious. There is today, as noted above, a very grave shortage of qualified manpower in the educational media field. Not only the proposals I examined, but many others propose to make use, one way or another, of educational technology.

When technology is applied on any reasonable scale and at some depth, specialized manpower is required. Further, if modern views of the technological function in education are accepted, the key types of manpower needed are in the design-engineering (educational)-consultant area. Other types, of course, are also needed, principally technical administrators and technical specialists ranging from television producers through graphic artists to electronic technicians. Perhaps the greatest problem faced by those charged with making all Title III proposals operate successfully is to recruit the specialized manpower required from a minimum, poverty-stricken market. If the slots cannot be filled with the properly qualified people, they will be (and are now, in some cases, I believe) filled with unqualified or minimally qualified personnel—and the whole program will be in jeopardy.

The second gap noticed was, in many cases, the lack of adequate consultants available to the proposers during the process of developing the proposal. Several proposals embraced rather sophisticated technological approaches, but had, apparently, no outside expert consultants, and insufficient attention was paid to providing for them in the proposal itself. Others used, or planned to use, people whose qualifications I, personally, would question.

Some of the better proposals, on the other hand, made use of a locally developed expertise and definitely reflected that expertise. For example, a group of school districts in Texas associated themselves with a sophisticated local university computer center, and this excellent proposal reflected this thinking. Experts do not have to come from thousands of miles away, but they ought to be experts.

Relating to the consultant problem, perhaps a note should be added on the anti-commercialism syndrome that exists, to some degree, sub rosa in the U.S. Office of Education. This view is certainly legitimate in many cases; it is expressed in the general worry that substantial parts of many proposals (extending far beyond Title III) are written, not by qualified educational experts, but by salesmen working on a commission for some audiovisual dealer or book peddler—with all that that implies in terms of unnecessary equipment, etc., etc. Some of the time this is no doubt true. In the proposals examined, however, where the fine hand of commercialism was detected—as it was in perhaps 10 percent of the lot—the proposals were still good. (One possible exception to this was noted, but the actual facts in the case would have to be determined by a site visit.)

The issue here has to be, not where did the advice come from, but was it competent? There are many well-qualified commercial representatives calling on schools. In some cases that I know about, I would be fully prepared to take their advice over some professional university consultants that I also know, every time. In other cases, the salesman-consultant represents every bad stereotype that haunts the somewhat puritanical educational frame of reference—including elevator shoes.

In the third place, most proposals and committees generating proposals that I have contacted take a very naive view of the problems inherent in generating new materials of any kind. For example, one proposal intends to use teachers to prepare a certain kind of "program." Even though some provision was made for time off, etc., and it was stated that the teachers concerned were somewhat experienced in preparing such materials, I believe this part of the proposal to be deficient. It takes much more time, effort, energy, skill, and know how to produce an extensive range of educational materials than was provided. This gap—or deficiency—was present in many proposals. A dial-access system for audiotapes is of little use if all of the tapes are to be prepared locally or copied surreptitiously from phonograph records or other tapes using extra teacher time.

The point is not so much the old hardware-software problem so often expounded on in recent pronouncements. (Hardware is bad or easy, as the case may be, and software is superior and a somewhat mystical problem.) Rather it is that adequate provisions in terms of manpower, facilities, and time *must* be made available in generous amounts in order to do the job of producing new software. It takes much longer and requires much more effort than the proposers seem to think.

There is another important sidebar here. Based on the outmoded theories referred to above, many school administrators feel that it is somehow immoral to provide time off for teachers to do these jobs or to pay for substitutes or to provide extra compensation or to set the teacher up on a separate summer salary in order to create new materials. It is true that such provisions have been and are being made in many places, but it is my feeling that the majority of administrators still feel that such extra work may be extracted out of the "professional contribution" hide of the dedicated teacher. Proposals need to be carefully scrutinized from this point of view. We are skirting the edge of a teacher revolt in this country anyway, and the effort to innovate should not contribute to it.

The last gap or deficiency that needs to be explored is revealed in the fact that some of the proposals examined—and others, no doubt, furnished other members of the task force—are designed to develop a sophisticated technology upon an almost nonexistent technological base. For example, I suspect that one school district had a proposal for a complex dial-access information storage and retrieval system that probably did not have an adequate conventional audiovisual service. This means that the necessary minimum technological backup is not present in the district. The proposal did not include provisions for an adequate backup. Such a lack of foresight is undoubtedly due to this lack of a conventional technological base.

A final suggestion might be made about a gap that is not so much a gap, but a possible illusion. Dr. Thomas James, dean of the School of Education at Stanford, recently commented in a speech in Los Angeles that much of the alleged innovation that was going on—particularly that relating to educational technology—was more or less a fake. James' idea seemed to be (as reported in the press) that districts were engaging in so-called innovative practices because federal or state funds were available; that once these funds disappeared, the machines would go on the closet shelf, dust would gather on the materials, and the districts would go back to older, more comfortable, and certainly less expensive ways. While I do not go along with Dr. James all the way (assuming, of course, the reporters were accurate), he has a point. There is, I think, some of this feeling abroad in school administration circles today. This superficial frosting of innovative practices on the old school cake, if really true, represents not so much a possible gap but a potential reversal of the whole innovative business. This may be a serious matter and should be treated as such by the Title III staff.

Recommendations

1. Instructional technology be viewed as defined in this paper—a complex pattern of man-machine systems and organizations based on concepts of feasibility and that proposals centering on hardware and materials be required to show some understanding of this concept on an operational basis. This is to say that the proposals cannot treat hardware and/or materials in isolation from the more complex technological process.

2. In connection with Number 1, a complete and workable logistical pattern be required in every proposal centering on a technological approach to instruction.

3. The proposing institution be required to describe in detail the technological base upon which it is intended to install new technology.

4. The proposing institution be required to show the consultants used in preparing the proposal and the consultants it is intended to use in carrying out the project. This description of the consultants should be so designed so as to establish their competency and no consultant should be considered incompetent solely on the basis of the position he holds whether that position be commerical, educational or religious.

5. The proposing institution be required to show precisely where it intends to obtain its technological manpower and to establish some competency for this manpower.

6. Where production of any kind of audiovisual or programed material is contemplated, the readers be cautioned to examine carefully the provisions of the proposal for this production. They are to specifically look at budgetary, manpower and time arrangements along the lines suggested in the main body of the paper. Generally, they are to avoid underfinanced and unrealistic approaches to production. If a proposal has merit and is not properly set up for this function, care should be exercised by the Title III staff that the proper support is provided in the negotiations.

7. Detailed equipment and material lists be eliminated from the proposals. (**NOTE:** While these provisions may make auditors and contracting officers happy, they are much more suited to the production of missiles than of educational materials and programs. I would suggest that such contracting officers and auditors be transferred to the Department of Defense, and that budgeting, particularly for materials, be confined to so much per student.)

8. A panel of media experts be convened to set up evaluative standards for the states to use in reacting to media proposals. If the states desire to add additional reactions, they should be required to state very clearly, in operational terms, what additional criteria they are using.

9. The Title III administration, since it is investing in the tooling up of American education, begin immediately to press for stan-

dards, first for equipment. It is suggested that this might be done through a university or similar institution working with concerned groups, as, for example, the Educational Media Council, the Department of Audio-Visual Instruction of the NEA, the National Association of Educational Broadcasters, and the related commercial, industrial and professional groups such as the Society of Motion Picture and Television Engineers. Every effort ought to be made to begin this program at once and attempt to schedule its completion in two or three years. As standards become available, they should be included in the guidelines of proposal readers.

10. Some Title III money be made available, either through the Title XI (NDEA) program, the Title VI-B (higher education) program, the experienced teacher fellowship program, and/or Title III itself to improve the manpower situation in the media field. Programs of some length designed to retread general school administrators, update conventional media personnel, and, particularly, to recruit new people are badly needed. It may be that money could be put into awards for projects which would provide for extensive training of school personnel already involved in the projects. This training level ought, due to the general situation in education, be at the advanced masters (specialist) degree level.

11. The Title III staff immediately initiate studies of dissemination and the relationship of information to behavior change. Further these studies should be oriented to using communication technology to carry out this dissemination. (**NOTE:** I have a very strong feeling about this. In our work we have recently had occasion to examine what is known about the flow of information among medical practitioners and scientists. We have located approximately 200 studies. Nothing like this exists for educational personnel. A real crash program is needed here combining basic research with invention and field testing of existing ideas. What little work has been done in education is oriented to the agricultural model which I believe does not hold at all—any more than the medical model holds for school administrators.)

What Is the Business of Educational Technology?

Perhaps it might be useful to summarize Mr. Muller's (1968) arguments, although any reader can certain understand them from his clear and well-written exposition. The summary will, however, assist in delivering the reaction. Essentially, Mr. Muller argues that educational technology has been oversold at every point along the line and has had little if any real effect on the educational enterprise; that people, not technology, can provide better education; that technology can never solve our basic problems; and that the computer can make three contributions to instruction: (a) through computer-managed instruction,[1] (b) through providing the computer to the student so that he may manipulate data and thereby learn thought processes, logic, etc., and (c) through providing direct instruction in the use of data-processing and computer equipment both as vocational training and (by implication) for general education.

There certainly is little to quarrel with concerning these three potential uses of the computer. Personally, I am in agreement with a certain segment of the field of instructional technology that is slightly disenchanted with the state of computer-assisted instruction (CAI) *at present*. Further, I would agree that CAI has been oversold in certain quarters, up to and including the Bureau of Research of the U.S. Office of Education. This is not to say, however, that the potential is not there, probably 10 to 20 years in the future. The trick, of course, is to figure *what* potential. It would have been helpful if Mr. Muller had engaged in some speculation on this point.[2]

---

[1] This means, essentially, to provide continuous data to teachers and others on each student's progress so that decisions as to instructional procedures can continuously be remade.

[2] Mr. Muller's reluctance to discuss CAI may be related to an ambivalent reaction to the subject at Harvard. Professor Anthony G. Oettinger—a man who really knows computers—recently worked over computer-assisted instruction and the systems approach to educational problems in an article in the *Saturday Review* (Oettinger, 1968). Professor Oettinger's eye was jaundiced, to say the least. On the other hand, Dr. Lawrence Stolurow, an authority on CAI, has joined the Harvard faculty.

Reprinted from *Planning for Effective Utilization of Technology in Education*, Reports prepared for a national conference "Designing Education for the Future: An Eight-State Project," Denver, Colorado, August 1968, pp. 37-48.

What Is the Business of Educational Technology?

However, one cannot discuss education and the new technology and concentrate on three relatively peripheral uses of the computer—that is, one cannot do it when, in the introductory pages of the paper, the whole philosophical problem of technology and education is raised and examined. Such a stance rather requires showing a more thorough understanding of the vast range of technological devices, materials, patterns of organization, and systems relating to education than Mr. Muller presented.

Muller presented, for example, a quote from Dean Sizer of Harvard that appeared in his report to the president to the effect that the large companies that had recently entered the field of education had backed off because the school system is both broke and conservative. This is a gross oversimplification, and I do not see how it could enlighten the president of Harvard.

Many things have been involved in these abortive adventures including, for example, some cases of incredibly bad management, demands to turn a profit in 12 to 18 months, the whole theory of conglomerate management which is very reluctant to integrate its own divisions in new projects when they sit making profits the way they are (hence publishing and electronics, for example, are never really integrated), the unbelievable conservatism of the publishing companies in this country—some of them may not yet be using movable type, the lack of a viable marketing organization, and so on.

I am sure that Dean Sizer is right in that General Learning, Inc., for example, has yet to turn a profit; nonetheless, I am acquainted with two men who formed a company in the educational technology field two and one-half years ago with $6,000 that will gross $1 million this year. While, as Mr. Muller suggests, there may be some cold eyes turned from the educational community to educational technology, the situation is much more mixed and infinitely more complicated than his paper implies.

I might suggest here that we need, perhaps, a little less "high-flying rhetoric" concerning *both* the promises and accomplishments *and* the failures and dangers of technology in education. We can no longer afford these Ciceronian luxuries; unbounded and unquestioned enthusiasm for this or that bit of instructional technology on the one hand or a failure of nerve in the face of the problems of technological development in education on the other are both conditions under which the cement of the educational status quo will harden until it cracks into bits that will have to be swept up by the societal broom. The medicine man promises of technological panaceas for education will only, as Mr. Muller indicates, result in regression and retreat on the part of the educational community when the hopes that have been aroused meet the hard-nosed reality of the confounding educational system. The failure of nerve will bring about the same result (perhaps a little quicker) in that it will amplify the inertia of both the professionals and public bodies that comprise the system in the face of the exacting and difficult task of technological development in education. As the inertia increases, the system will fall apart. Such is the penalty of failure of nerve.

I am concerned that the Muller paper reflects a line of thinking that I believe to be both pessimistic and one-sided. The genesis of this line of thinking, as well as its opposite—the "2mm filmstrip will save American education for the Indians" type—lies in the fact that many commentators since about 1958 who have chosen to speak about educational technology do not know the business. They do not know the business in the sense that they did not sweat through the scrambling visual-audiovisual movement into television, programed instruction, and multi-screen, multi-media techniques and face the sunset looking at computers, simulators and probably holographic lasers useful in conveying patterns of ESP.

Perhaps there's a little dramatic license here, but not much. The pronouncements on educational technology come and go—exceedingly optimistic to gloomily pessimistic—often unrelated to the infinite realities of the educational system and the possibilities of instructional technology, and, at best, concentrating on experimental results or under- (or over-) funded developments. And the relationship of many of the pronouncers to the business at hand—educational technology in the sense that I used it in the paragraph above—is tenuous, to say the least.

The overcautious pessimistic view of technology in general and instructional technology in particular tends to come, in my opinion, from two postures. They are (1) the literary side of the Snowian dichotomy of the two cultures—science and literature,[3] and (2) an anachronistic puritanism toward business, etc. which continues to pervade the ranks of public education. I would not put either Mr. Muller or Dean Sizer in the second category, but I believe them to be excellent representatives of the first.

To clinch this point, I should like to quote Sir Robert Watson-Watt, the father of radar and one of the originators of operations research. In a conference sponsored by the Center for the Study of Democratic Institutions at which he was definitely in the minority, he had this to say:

Because I suspect some here of sharing Ellul's pessimism, I am moved to comment very personally—with appropriate apologies—on the imbalance of judgment which is almost inevitable between those who have been brought up mainly on a diet of printed paper, and those nurtured on wood shavings, iron and brass filings, propulsive gases and other base mechanic offal. *It is quite clear to me that pessimism is nurtured by paper, optimism by offal of the kind I have just mentioned.* [italics mine][4] (Watson-Watt, 1963, p. 3)

The business of knowing or not knowing the business of educational technology is illustrated by Muller's rapid generalizations concerning the history of the claims for a series of technological devices when they were introduced into education and his conclusion that the influence of these devices has been marginal—without assessing the milieu in which the exaggerated claims were stated. Thus, he says that similar claims were made for radio in the 20s, film in the 30s, television in the 50s, and computers in the 60s. Actually, we have old Thomas Edison to blame for exaggeration in the case of films. He said much the same thing several times,

---

[3] A slight oversimplification since, in a little-noted section of Snow's book, he points out that most scientists don't understand technology, either.

[4] There is also the temptation to point out that Marshall McLuhan is saying the same thing in a different way and is, perhaps, closer to the mark since we are talking about the specialized technologies of media.

but here is an authenticated statement which appeared (note the date) in the New York *Dramatic Mirror's* issue of July 9, 1913:

> Books will soon be obsolete in the schools. Scholars will soon be instructed through the eye. It is possible to teach every branch of human knowledge with the motion picture. Our school system will be completely changed in ten years. (Saettler, 1968, p. 98)

We can turn now to the 30s, the period Muller states was the time in which the claim was generally made that films would revolutionize education. The best book on the subject covering the 30s came out in 1942. Consider these two excerpts from this report covering four years of experimentation by the Motion Picture Project of the American Council on Education:

> The fact that motion pictures are a pictorial language with a universal appeal, that they arouse a wide variety of student interests, that students of varying mental maturity respond to them in characteristically different ways, and that faulty concepts and superficial understandings sometimes develop from films has important implications for the curriculum. . . . (Hoban, 1942, p. 79)

> It is the function of motion pictures to provide pictorial representation of the "sensible world," and it is the function of the teacher, with the aid of the book and the lecture, the laboratory and the discussion group, the library and the study period, to direct and guide the student in "penetrating within" in order to find "substantial reality," "meaning," and "truth." (Hoban, 1942, p. 27)

No doubt there were salesmen that were making extravagant statements during the 30s about films in education (echoed by a chorus of faddist educators), but responsible comment on films and education during that crucial decade followed the Hoban line. The audiovisualist, then—as the instructional technologist, now—occupied the gray land between the white and black mountains of irresponsible rhetoric and stygian inertia—admittedly a little closer to optimism than pessimism.

Mr. Muller introduced the concept of irresponsible rhetoric making high-blown, unacceptable claims as to the value and usefulness of this or that technology in education only to be followed by disappointment in the educational community. Admittedly, we have had some of these claims and the effects on the educational community Mr. Muller suggests since the scanty beginnings of an educational technology. A man by the name of Josiah F. Bumstead in a book on the blackboard in 1841 stated:

> The inventor or introducer of the blackboard system deserves to be ranked among the best contributors to learning and science, if not among the greatest benefactors of mankind. (Anderson, 1961, p. 23)

It would be more useful to consider educational claims of all kinds as a general phenomena, occurring at a given time in the technological development process, set within a given technological milieu and complex of attitudes, training levels, and political-social forces existing within the educational system, and uttered by a particular class of educator or promoter. Care is needed to assess the source of the claims and the effective response of the educational system, if any, particularly by comparison with what was said and done by that small band of instructional technologists who were *actually* in the business at the time. Such a study would, I think, be of great assistance in helping design educational technology for the future. A few tentative suggestions as to answers may be made in the hope that this phenomena might be investigated thoroughly at a later time.

For example, the tentative idea could be advanced that people close to the initial invention or development of a device or system such as the motion picture would have a sudden shock of recognition and project claims far beyond the ability of the times to carry them out, the educational system to absorb them, or even the (then) primitive state of the technology to deliver. This would account for Edison's claims. And here is another example. In January 1933, Professor E. B. Kurtz, Head of the Department of Electrical Engineering of the State University of Iowa, and his colleagues successfully began broadcasting educational television programs. (These experiments were conducted through 1939.) On the occasion of the inauguration of the second series of educational broadcasts (apparently still in 1933), Kurtz said:

> ... this new instrumentality [television], which bids fair to become the most potent agency for universal education ever conceived. For in due time, every home will have its own classroom, with professor, blackboard, diagrams, pictures, and students. (Kurtz, 1959, pp. 70-72)

Note that this is about as far back in time from the claims for television in the 50s as Edison's remarks were from the claims for films in the 30s (1933—1954; 1913—1935).[5]

Accepting for the moment the shock of recognition idea on the part of the inventor or pioneer developer, our tentative exploration can now enter another area—that of the educational leader and his role in prophecy or claim. Casual observation suggests that somewhere along the line certain far-sighted educational leaders pick up the ball and make public pronouncements far in advance of educational thinking that receive a certain amount of currency—even if not acceptance in the form of action. In this connection, I find four quotations fascinating:

> (1940) ... George F. Zook [at the time President of the American Council on Education and former U.S. Commissioner of Education], in his report to the American Council on Education in 1940, described the motion picture as "the most revolutionary instrument introduced in education since the printing press." (Hoban, 1942, p. 16)

> (1957) Dr. Thomas Clark Pollock of New York University said more recently, "It now seems clear, however, that television offers the greatest opportunity for the advancement of education since the introduction of printing by movable type." (Stoddard, 1957, p. 27)

> (1962) "It [programed learning] is the first major technological innovation in education since the development of printing." (Woelfle, 1962,

---

[5]This has, to some extent, to be qualified. The *silent* film had been fairly thoroughly investigated as an educational instrument in the 20s (Wood and Freeman, 1929; and Knowlton and Tilton, 1929). When Muller used the decade of the 30s, he was referring, I assume, to the sound motion picture; 16mm sound became available for educational use at that time. Note, however, that Edison thought in terms of the sound film, regardless of the state of the art in 1913.

p. 503)

(1967) "The impact of the computer on society, and hence on the curriculum, has been compared to that of movable type and the printing press since Gutenberg." (Caffrey & Mosmann, 1967, p. 12)

Mr. Muller may feel that I have uncovered enough to document his entire argument; I believe otherwise—that such claims at the leadership level are a part of a leading process and, in the time stage in which they are made, merely get the idea into the broader system, giving it currency. Effects on practice at that point are inconsequential, and expectations at that same point are only slightly stimulated. In fact, as is well known in the case of television, the efforts of the Ford Foundation produced more of a "Ho-hum, what else is new?" attitude than anything else.

Somewhere between initial recognition and the prophetic public statements of certain educational leaders, work by the technologists has begun. Ideas have been picked up and laboratory or field development started. Here, again, the description of this process would be well worth the effort to dig it out. This will not be easy because, for example, there is the phenomena of false starts which are revived later and which must be taken into account (Pressey and then Skinner, for example). The nitty-gritty of technological development may remain more or less unnoticed by the educational community, or experimental centers may become more public relations-demonstration arrangements than true developmental efforts.

Since educational technology usually involves some kind of product—hardware or software (or even, these days, services)—the possibilities of commercial exploitation are always present. This results, sometime after development begins, in various enterprises that may or may not succeed, but which contact the educational community and spread the word, unfortunately often in the extravagant sense noted by Mr. Muller.[6] However, in all fairness, it should be pointed out that industrial representatives have, also, in many cases done an excellent, professional job of dissemination of information on technological developments and have helped teachers and others at the grass roots level with problems of using the new technology.

Before proceeding with the end of the tale of reaching the educational community and discussing the community's response, it is necessary to note that changes in this process have occurred in the last few years. Today, for example, with the larger corporations, at least, market research is conducted to determine whether or not the company should enter the "knowledge business." Research and development on new devices, materials, processes, and systems is much more programmatic, being funded principally from federal sources; large industry, too, is investing slightly more in research and development in educational technology. These processes (market research and control and programmatic research and development) will, no doubt, change the fundamental flow of discovery, in-

---

[6]The early teaching machine industry was a wonderful example of this. See Finn and Perrin (1962) for some idea as to the volume of companies that entered the business. Very few are left today.

vention, development, and adoption. Whether or not it will change the process of claiming is open to question.

It should also be noted that—as part of a general, though weak, programmatic attack on the problems of education which moves the whole process away from the older, laissez faire approach—educational innovation is (a) being studied *itself* in order to develop an applied science, and (b) the process of innovation is being formalized and encouraged by federal law and state action in the form of, for example, the Title III innovative centers set up under the provisions of the Elementary and Secondary Education Act. The second development, of course, is designed deliberately to encourage dissemination of information about new developments, many of which fall into the classification of educational technology. These deliberate processes, though, run exactly contrary to the extravagant claim situation—insofar as educators and research workers play a part in it—as the system emphasizes evaluation and caution.

Returning now to the response of the educational community, eventually adoption, assimilation and vulgarization (in the sociological sense) of the processes, materials, devices, and even organization patterns occur. Although the subject of much study in recent years by Rogers and others, the innovative-adoptive process remains essentially a mystery in which a crystal ball is as useful a predictor as statements of the characteristics of "early" and "late" adopters. No one has satisfactorily explained, as far as I know, why the language laboratory and the systems of language teaching it requires was adopted so rapidly throughout the United States, while the overhead projector languished for a decade or more and then took off in a startling fashion.

The response of the educational system to technological development is another crucial point in the Muller paper. To repeat, he indicates that instructional technology, to this point in time, has had only a marginal effect upon the system. I believe that both the facts and experience show that instructional technology has, in reality, considerably exceeded the stage of marginal influence without, by any means, reaching the stage of saturation.

Some data on these matters do exist. In addition to the base-line data generated by the Finn, Perrin and Campion study cited above, the studies of Eleanor P. Godfrey (1967) and the Bureau of Social Science Research on the state of audiovisual technology in the first half of this decade are especially illuminating when the question of "marginal effect" is raised. Were, for example, the estimated 7,000,000 filmstrips on hand in the public schools in 1961 marginal or something more?

The key item in the Godfrey studies, however, was the comparison made on her sample for 1961 and 1964. In this connection, she says:

Some of the individual gains were quite spectacular: increases of 12, 10, 9, and 7 units per school were reported for tape recorders, record players, television sets, and overhead projectors.

As a result of the general increase in amount of equipment per school, the median inventory *level* shifted upward for all items except radio, where the level remained constant at one unit per school over the three-year period. In 1964, assuming that he had all nine equipment items, the superintendent of the hypothetical median district had one overhead projector for each two schools, seven television sets for each ten schools, two language laboratories for three secondary schools, two opaque projectors for every three schools, one radio in each building, and enough tape recorders, 16mm projectors, slide-filmstrip projectors, and record players to put one unit in every school and multiple units in some, presumably the larger ones. (Godfrey, 1967, p. 81)

Another source of comparative data is presented by two studies covering the status and growth of closed-circuit television in 1962 and 1967 (Campion & Kelley, 1963). As Jorgenson indicated in the preface to the 1967 study:

> In 1962, the NEA's Technological Development Project's study of the status of closed-circuit television concluded that CCTV would "move forward at an accelerating pace during the next decade." It had identified 462 CCTV installations in 403 institutions. It is apparent at this half-way point in the decade that the prediction was accurate. Four years later, the number of institutions reporting CCTV installations in the present survey has nearly doubled. Seven hundred and seventeen institutions now report 812 systems. An additional 178 institutions indicate they operate CCTV and/or ITFS systems or would do so by January 1968 although they did not respond to follow up inquiries. Taking these two figures in combination, it is fair to estimate that approximately 1,000 CCTV or ITFS systems will be in operation by 1968. (*A Survey of Instructional Closed-Circuit Television*, 1967, p. 1)

In the field of computers in higher education, Caffrey and Mosmann, using an earlier study by the National Science Foundation and information from an annual administrative data-processing survey among colleges and universities, state:

> ... More than 30 percent of American colleges now use computers of some kind. In recent years, 80 to 100 institutions (or about 4 percent) acquired their first computer each year. ... The number of computers installed at colleges increases at the rate of 45 percent each year. (Caffrey and Mosmann, 1967, p. 13)

These computers, of course, are used for a variety of purposes, but they must be having some educational effect.

A recent report by the NEA Research Division using a carefully controlled sample of 1,609 teachers throughout the country shows that from approximately 80 to 95 percent report the availability of what we call conventional audiovisual devices and materials—still and motion projectors, audio devices, etc.[7] (*NEA Research Bulletin*, 1967, pp. 75-77). The more exotic devices, or "newer media"—CCTV, CAI and multi-media devices, as well as programed instruction—fall in much lower ranges (from about 10 percent to 30 percent). What is seen is a profile of availability. Even recognizing this, influence cannot really be measured. Industrial estimates which I have obtained privately indicate, for example, that in the specialized field of reading devices, there are about 25,000 units of the Craig

---

[7]Percentages vary with each device.

What Is the Business of Educational Technology?

Reader in the field and an estimated 100,000 to 150,000 Controlled Readers manufactured by Educational Development Laboratories. And there are other devices such as the hardware and program in reading (recently put on the market by Hoffman Information Systems) which is being well accepted. Is all of this merely marginal and having little or no effect? I doubt it very much.

Current technological development in education is obviously not yet equal to the task and demands that are being made upon it. Nonetheless, it is, I think, much more than marginal. In addition, it bears an importance that is sometimes missed. I refer to the development of the technological base in the educational system necessary for the further development and application of technology—for, if you will, the adoption of the more exotic devices and systems such as computers.

This argument, together with the data to support it, was developed some years ago when, as part of the Technological Development Project of the NEA, we did a study of the growth patterns in the acquisition of audio-visual equipment by the public schools from 1930 to 1960. Analysis of the data suggested that Rostow's concepts of the five stages of economic growth could be applied to the American school system and that the system could be thought of as a primitive or underdeveloped culture from a technological point of view. It then followed that Rostow's conditions for "take-off" into a modern technological culture would apply. At that time, I wrote:

Prior to 1930, there was little in the form of communication equipment available to education. The growth patterns of acquisition then take on additional significance, as they reflect the general communication revolution as it entered education. The main point, however, is that *we view the build-up of audiovisual equipment and materials in education as one of the principal preconditions for a technological revolution in education—for take-off in the Rostow sense.*

The level this growth has attained and the speed and pattern of attainment are, we think, the best indicators of the future. Judgment has to be applied to the data in order to estimate whether or not the take-off point has been reached. It is clear, however, that, for example, the psychological preparation—the expectation of and aspiration for technological solutions to educational problems—cannot exist until a technological *milieu* exists to create this preparation. The extent to which audio-visual equipment is available to the working school teacher is a valid test of the strength of the *milieu*. If, for example, teaching machines or the electronic learning laboratory take the educational system by storm, it may be due more to the widespread availability of the 16mm sound motion picture projector than all of the theories and research of experimental psychologists, the efforts of industry or the thousands of words and pictures appearing in the mass media. (Finn, Perrin and Campion, 1962, pp. 9-10)

Since that time, the technological capital of the American school system has increased markedly due to the infusion of federal funds, changed attitudes on the part of educators, and the pressures of problems such as those generated in the inner city. It should be emphasized that the educational system is still highly underdeveloped technologically, but I would maintain that the present technological build-up, which is somewhat more than marginal in its effect, has provided a sufficient base for take-off. In other words, we are farther along the line than Muller envisions without yet reaching a dollar volume which would encourage General Electric to bid on computer-assisted instruction for the state of Iowa rather than trying to catch a Navy contract for a flying submarine.

As sort of a corollary to his analysis of the claims-effects relationship of educational technology, Mr. Muller shifts the discussion temporarily to the effects of the mass media on children outside of school. This, he admits—with a passing reference to McLuhan—is considerable, and then he proceeds to the somewhat hoary issue of whether or not the machine will replace the teacher.

The machine-teacher question is handled by Muller with the sentence, "Our traditional education system will endure, and our teachers will teach—and our computers will never be a substitute." This is followed by a comment that only people and not machines can provide better education.

If one is sensitive to all of the recent accounts of ghetto and poverty-rural education and the traditional educational system operating therein, all one can do upon reading Mr. Muller's sentence is to invoke the Deity for the country and take to the hills. It was no accident, for example, that a Black Power group in Los Angeles recently approached the principal of Manual Arts High School with a demand that several teachers be dismissed —on the grounds that they were totally insensitive. In any large city in the United States today, a neat question can be raised as to how many human teaching machines are in service.

This concept, of course, is only a small chunk of the teacher-machine-child relationship. The general issue has been widely discussed, and, no doubt, most readers are familiar with the arguments. The literature on differential staffing and discussions of proper roles of several levels of educational personnel within an organization structure that permits the use of a technology of instruction wherever it is applicable. Further, it was no accident that the great curriculum reform projects of this decade elected to use films, film loops, television, etc. to "preserve the integrity of the content." It appears that, as Professor Zacharias has said many times, children are also entitled to the truth.

Basically, I hold that, with the population explosion, knowledge explosions and the great, inchoate revolutionary drive by all suppressed peoples for a place in the sun, we can no longer afford the luxury of the traditional system; that the system needs a vast overhaul and, in order to solve some of the problems presented by the three revolutions, we must develop a technology of education that will carry a greater share of the load than the simple idea that the teacher may be helped out a little now and then by some kind of an aid (or aide, for that matter). Those designing education for the future must keep this in mind.

Two other reactions to this general problem should be mentioned. First, if technology applied to instruction does assist in making differential staffing possible, it is not too clear as yet what master teachers or leading teachers of one kind or another are supposed to do. These actions are still

clothed in broad generalizations such as "guidance," "individual attention," etc. The key, here, I believe, lies in the cognitive growth, inquiry
teaching field where supporting technology is an absolute necessity, and in
the existential-therapy psychology area where, today, teachers are un-
trained and probably, for the most part, temperamentally unfit.[8]

Secondly, everyone seems to be forgetting the teachers themselves within
this whole picture of the teacher-machine relationship. This seems peculiar in a year that has been filled with teachers' strikes, militancy and
hard-nosed negotiation. One of the great issues of the future as far as educational technology is concerned is the role that teachers' organizations
will play in its adoption through negotiation. The day of the arbitrary decision on the part of boards and administrators in many fields of education
is fast coming to an end. The open question is: Will or will not teachers
take a Luddite position?

Leaving the teacher-machine-child question, we can briefly look at the
other key question raised by the Muller paper. This question is concerned
with the whole issue of the technological society and its effect on human
beings. I am very much in sympathy with Mr. Muller at this point: He
properly notes the generalized fear of standardization and conformity
which a culture organized around a high technology generates with its
punched cards, data banks, bureaucratic and organizational systems, and
alienation of the person. He claims that the key questions of the social
order—tolerance or bigotry, for example—have no technological solutions.
The alternatives, of course, are ethical solutions. The situation is probably
very mixed, but such a general hypothesis must, of necessity, be drastically
revised if it can be shown that some problems of this nature have been
resolved or partially resolved by technological means as opposed to ethical
means.

Gerard Piel calls our attention to one such case—the case of slavery. Here
is what he said:

Slavery was the technological underpinning of high civilization and of
history until very recent times. Slavery was abolished only when the
biological energy of man was displaced by mechanical energy in the
industrial revolution.

What about the role of ethics in this story of man's inhumanity to
man? Slavery is condemned in the Old Testament only when the
chroniclers are concerned with the Egyptian captivity. There is no
condemnation of slavery in the centers of Jewish culture in Palestine;
there is no suggestion of the immorality of this peculiar institution.
In the New Testament, Christ assures us that the poor will always be
with us. In Plato's view, some men were natural slaves. His opinion
was echoed by no less a figure in our own history than Thomas Jefferson. *Slavery became immoral when it became technologically ob-*

---

[8]A very sensitive account of what can be done by teachers using the existential
approach can be found in Dennison, 1968. It is obvious from the account that the
time, effort, skill, etc. required, as well as the very small groups involved, would
require enormous technological support if the humanistic objectives of the school
were to be realized on a vast scale.

solete. [italics mine] (Piel, 1963, p. 81)

While I certainly would not claim that technological solutions can be found to all human, national and international problems, I believe that the categorical statement that such solutions are absolutely impossible has in it a fair-sized element of error.

Finally, I agree with Mr. Muller that educational technology must never be used to condition young people to live in a world of standardization, conformity and alienation. I fail to see why technology cannot be used as a powerful force in the educational system to move in the opposite direction—toward freedom, creativity and a worthwhile sense of self. Education for the future should carry this design. Robert Redfield, the great social scientist from the University of Chicago, once indicated the full-scale bias he carried into his work. He said:

> . . . I have placed myself squarely on the side of mankind, and have not shamed to wish mankind well. (Redfield, 1953, p. 141)

References

Anderson, Charnel. *History of Instructional Technology, I: Technology in American Education: 1650-1900.* Occasional Paper No. 1. Washington, D.C.: NEA, Technology Development Project, 1961.

Caffrey, John and Charles J. Mosmann. *Computers on Campus: A Report to the President on Their Use and Management.* Washington, D.C.: American Council on Education, 1967.

Campion, Lee E. and Clarice Y. Kelley. *Studies in the Growth of Instructional Technology, II: A Directory of Closed-Circuit Television Installations in American Education with a Pattern of Growth.* Occasional Paper No. 10. Washington, D.C.: DAVI, NEA, 1963.

Dennison, George. "The First Street School," *New American Review, Number 3.* New York: New American Library, April 1968, pp. 150-171.

Finn, James D. and Donald G. Perrin. *Teaching Machines and Programed Learning, 1962: A Survey of the Industry,* Washington, D.C.: NEA, Technological Development Project, 1962, Occasional Paper No. 3.

Finn, James D., Donald G. Perrin, and Lee E. Campion. *Studies in the Growth of Instructional Technology, I: Audiovisual Instrumentation for Instruction in the Public Schools, 1930-1960—A Basis for Take-Off.* Occasional Paper No. 6, Washington, D.C.: DAVI, NEA, 1962.

Godfrey, Eleanor P. *The State of Audiovisual Technology: 1961-1966.* Washington, D.C.: DAVI, NEA, 1967. Monograph #3.

Hoban Jr., Charles F. *Focus on Learning: Motion Pictures in the School.* Washington, D.C.: American Council on Education, 1942.

Knowlton, Daniel C. and J. Warren Tilton. *Motion Pictures in History Teaching.* New Haven, Conn.: Yale University Press, 1929.

Kurtz, E. B. *Pioneering in Educational Television: 1932-1939.* Iowa City: State University of Iowa, 1959.

Muller, L. A. *"Education and the New Technology,"* from *Planning for Effective Utilization of Technology in Education,* Reports prepared for a national conference, "Designing Education for the Future: An Eight-State Project," Denver, Colorado, August 1968, pp. 30-37.

NEA Research Bulletin. Vol. 45, No. 3. October 1967, pp. 73-77.

Oettinger, Anthony G. "Myths of Educational Technology," *Saturday Review,* Vol. 51, May 18, 1968, pp. 76-77+.

Piel, Gerard. "Ideas of Technology: Commentary," in Carl F. Stover (Ed.) *The Technological Order.* Detroit: Wayne State University Press, 1963. pp. 79-81.

Redfield, Robert. *The Primitive World and Its Transformations.* Ithaca, N.Y.: Cornell University Press, 1953.

Saettler, Paul. *A History of Instructional Technology.* New York: McGraw-Hill Book Company, 1968.

Stoddard, Alexander J. *Schools for Tomorrow: An Educator's Blueprint.* New York: The Fund for the Advancement of Education, 1957.

A Survey of Instructional Closed-Circuit Television, 1967. Washington, D.C.: DAVI, NEA, 1967.

Watson-Watt, Sir Robert. "Technology in the Modern World," in Carl F. Stover (Ed.) *The Technological Order.* Detroit: Wayne State University Press, 1963. pp. 1-9.

Woelfle, Dael. "Prophecy Fulfilled," *Science,* Vol. 135, No. 3503, February 16, 1962, p. 503.

Wood, Ben D. and Frank N. Freeman. *Motion Pictures in the Classroom; an experiment to measure the value of motion pictures as supplementary aids in regular classroom instruction.* New York: Houghton, 1929.

Section IV

Commitment to the Future

Commitment to the Future

"We have tended to emphasize order and system. I believe we must now move in the direction of freedom—freedom for everyone in the system."

In this section, Finn confirms his commitment to the future as he carefully examines the relationship between education and technology. With a mighty reach, he draws together and analyzes a conglomerate situation, forecasts trends, states his concerns, and offers a direction.

The emphasis Finn suggests is toward freedom and openness in a world in which technology and bureaucracy can and must be mastered for the benefit of mankind. The need for leadership in finding solutions to technological problems through technology is re-emphasized.

"The Emerging Technology of Education" (1966) is a major article which serves to analyze and summarize the current situation. In forecasting trends for the next decade, Finn sees the period as one for consolidation of gains in instructional technology but with new developments in the political-social-economic sphere. The era of groundwork and foundation building for educational technology has passed, due principally to the infusion of large amounts of capital through foundation, state and federal sources.

The extensive review covers past events and present developments in media, educational theory, research and development, industry, and government as phases of the scientific-technological revolution. The period from 1955 to 1965 is given particular emphasis as a crucial time in the emergence of educational technology. Using this as a base, Finn concludes with a statement that prediction is possible and without claiming the precision of probability, he forecasts 11 different categories of future trends.

Maintaining his earlier position that ". . . it is impossible to be a professional teacher without technology," Finn states that the increasing commitments of government and industry to education will revolutionize the educational enterprise and its administration. He sees the traditional resistance to change becoming less intense as the time span from idea to practice shortens from the customary 35 years down to less than 10 years. He forecasts a rediscovery of educational methodology brought about through a systems approach using a combination of empirical bridges and operations research.

"Henceforth," Finn predicts, "the new establishment will orient American education more in the direction of science and technology as associated with its own processes and will absorb only that part of the older establishment which will fit this overall scientific technological pattern."

"Dialog in Search of Relevance" (1967) summarizes a keynote address in which Finn outlines, for national leaders in the field of educational technology, four criteria for relevance and focuses on the nature of the dialog needed with industry. "If no direction is given, technology will take off on its own."

Finn suggests a change of emphasis in the media field with a commitment to human purposes. He explores new margins within which the "marginal media man" should operate. The five suggestions he makes emphasize the need to alter the field's theoretical framework and demonstrate his concern for freedom and creativity.

In "Institutionalization of Evaluation" (1968), Finn takes the opportunity to provide a broader perspective for the field of educational technology by examining in depth the central component, the subtechnology of evaluation. "The whole concept needs to be explored and assessed within the technical-social-political context of the entire educational enterprise."

The importance Finn attaches to this concept is considerable and obvious. "In a fundamental sense the entire Title III[1] effort at educational innovation is a move toward bringing the educational enterprise into the modern industrial state. Title III *is* educational technology . . . I submit that the very selection of evaluation as a field to analyze in connection with the Title III program is evidence of this move toward technology."

Questioning the institutionalization of evaluation, Finn explores the problem and the various forces at play. He sees evaluation as inevitably relating to the control of human behavior in one direction or another. The need for quality control is examined in relation to the potential negative effect of another bureaucracy.

Five reasons for evaluation are analyzed and related to the diversity of projects in Title III. The need for a broader based evaluation theory is discussed and the possible incongruency between behavioral and operational systems of objectives is explored.

The evaluative process is next considered at the philosophical level and related to the imperatives and characteristics of technology. The emerging, and often heated, dialog attributed to a "lack of human fit of many technological developments" is examined, and several observations are made about the recurring theme of impersonalism and bureaucratic control versus the human condition.

Finn focuses on the relevance of this dialog and cautions educators: "I believe unless we listen to what some of these bright young people are saying, to what the New Left is trying to expound, to what some artists are expressing, we, as educators, may fail this country and all of the young people in it."

Finn concludes with a series of recommendations to create a system of evaluation which will allow for and encourage both human variation and the release of creative energy.

---

[1] Title III programs—a part of the federally funded National Defense Education Act.

The Emerging Technology of Education

Introduction and Definition

Education in the United States can be viewed as an institution—social, political, economic. It can also be viewed as a process whereby humans learn and are taught. Subject matter content or skills are also sometimes loose referents to the word "education," in the sense of such phrases as scientific education, physical education, or teacher education. Finally, there is always a definition which represents a description of the effect of personal experiences, as in the classic *Education of Henry Adams*, or in such a phrase as "that experience taught me a lesson in (or gave me an education in) human relations."

These separate but somewhat overlapping concepts (and they are by no means exhaustive) suggest that care need be taken when addressing the problem of the relationships of technology to education. It would appear that the Commission should be interested in the bearings that technology has on education as an institution, a process, and its content.

There is also some disagreement within the educational profession as to what is meant by the expression *educational technology* (or *instructional technology*). A group of experimental psychologists would like to confine the term to an empirical approach to the process of teaching and learning.[1] That is, they refer to the work in experimental psychology as a developing "science of learning," and applications of this science—for example, applying the principles of operant conditioning to teaching through programed instruction—as the "technology of instruction."

While these psychologists admit that there is, by no means, a one-to-one relationship between their version of the science of learning and the technology of instruction, they view this relationship as the ultimate goal. There are some real weaknesses to this point of view and it will not be the one from which this paper is written.[2] However, since this relationship

---

[1] See, for example, Lumsdaine (1964) and Finn (1963).

[2] Among other problems associated with this point of view is that the "science of learning" must be a particular psychological position; e.g., Skinner or Bruner. More important is the fact that psychology is not the only science of learning. Social and economic factors introduce other variables.

Reprinted from *Educational Implications of Technological Change*, Appendix Volume IV; Technology and the American Economy. Studies prepared for the National Commission on Technology, Automation, and Economic Progress, February 1966.

does exist to some extent, it will be considered within a broader context of educational technology.

Educational or instructional technology is sometimes too narrowly thought of as confined to hardware. Thus, for example, closed-circuit television or language laboratories represent to many educators the sum total of instructional technology—making it very easy to discuss or dismiss. Actually, much more is involved in the concept. If the point of view is taken that the institution of education in the United States is now the subject of and is moving into the general scientific-technological revolution, then hardware, materials, systems of organization, and new roles for teachers and administrators are all a part of educational technology. Within this broad umbrella, there are subtechnologies; for example, psychological testing, the development of new hardware such as the cartridge-loading 8mm film viewers, and the creation of whole systems of materials such as those designed for the Physical Science Study Committee by Professor Jerrold Zacharias to teach high school physics.

Further, when education is viewed as an institution, technology of a sophisticated type can (and is, to a small extent, beginning to) play a part in the management and operation of the entire enterprise. For example, data processing machinery and techniques are being used for scheduling, pupil accounting, etc.; modular construction of school buildings is being tried; information storage and retrieval processes are being considered as applicable to age-old library problems.

There are, in addition, economic and other aspects to this general concept of educational technology. These will, in this paper, be interwoven as the discussion crisscrosses between education as an institution and as a process. The general emphasis, however, will be on the subtechnologies most closely related to instruction, as it is felt that this is the current interest of the Commission. These include hardware, materials, systems, organization, and psychological applications.

Sources of Present Concern

Since World War II the national concern with education has been increasing, a concern that took a quantum jump with the unveiling of the first Sputnik. Since that time, the national effort to improve the education system has redoubled several times. However, from the point of view of instructional technology, all concerns cannot be said to originate in the space race. The efforts of the Ford Foundation in the development and promotion of educational television, for example, antedated Sputnik I by several years.

The concerns for education represent a sort of catalog of problems and issues relating to instructional technology. These can be briefly listed:

1. The population explosion with all it implies in terms of more students and fewer teachers and the necessity for new educational arrangements.[3]

---

[3]A recent report from the TEMPO Division of General Electric in Santa Barbara, California, projects the total U.S. population under 20 years of age in the year 2000 at 143 million.

2. The information explosion, which presents great problems in curriculum construction (what to teach) and requires greater efforts in increasing the efficiency of teaching and learning (more learning in less time).[4]

3. The general rising throughout the world, including the United States, of the depressed sectors of mankind. Education is seen as the main weapon in the war on poverty, for example, and the requirements of such an educational war include new methods and techniques to reach these people—here and overseas.

4. The urge to raise the quality of life for all, which, in turn, requires raising the quality of education even though the institution is pressed on all sides by population, knowledge and various kinds of special demands. This situation also sets a requirement, although in a different way, for a more efficient educational process.

5. Research and development on all aspects of education have generated more knowledge about the process and the institution as part of the general knowledge explosion. From this situation rises a drive to introduce all kinds of innovations—most of them relating to instructional technology—into the educational system. Practice in education has, until the last few years, lagged about 50 years behind research and theory. Today the process of educational innovation itself is under study in an effort to speed up the rate of change.

6. The need for education and re-education and for training and retraining of manpower to meet individual personal needs and national needs increases as society becomes more automated and technologically sophisticated. If, as is often said these days, an individual may have to learn several jobs in a lifetime, the same problems of educational efficiency involving possible uses of instructional technology arise—both on a formal and an informal basis.

7. The ill-defined drive toward a new value system, symbolized by the student revolts at Berkeley and elsewhere and by very personal concerns on the part of a segment of the intellectual community, suggests, among other things, dissatisfaction with the abstract, status-oriented, large-scale organization of the college and university characteristic of a technological culture. Since a return to the colonial college is impossible, solutions to these problems, too, must be found in technology, although the student spokesmen would not understand this. The effort to find technological answers, however, is a function of a technological society as noted by such students of technology as Ellul (1964).

8. There is a cluster of economic pressures pushing for more technology in education. Several foundations, particularly the Ford Foundation, have poured money into the system for developments in this area. The drive to diversify on the part of the aerospace and defense industries

---

[4] A paper on the explosion of knowledge will soon be released by the Instructional Technology and Media Project at the University of Southern California. In the meantime, a good general reference on a portion of the knowledge explosion may be found in Price, 1963.

(possible sources of educational hardware, materials and services) as part of the general move into the public sector is another little understood aspect of this problem.

No claim is made that the catalog above is complete or that the application of technology to the educational system or to instruction will solve all the problems listed. It is true, however, that a component of educational technology will have to be present in any attempt at a broad solution. It would, perhaps, be well to examine some of the dimensions of instructional technology as it has developed in order better to understand its possibilities.

A Short Historical Background

For the problem at hand, it is crucial to understand the history of the development of instructional technology for the last 50 years. However, the interested scholar may chase this development back much further than the overworked Greeks, if he desires. The great Sumerologist Samuel Noah Kramer puts the invention of writing on clay tablets at Sumer at about 5,000 years ago and then notes that "The Sumerian school was the direct outgrowth of the invention and development of the cuneiform system of writing" (Kramer, 1959, p. 1). To some extent, the development of devices and techniques of communication—the development of a technology of instruction—has affected education ever since.

Skipping over thousands of years, the history of 19th century American education is replete with effective technological developments—including the development of textbooks (to which there was some objection), the blackboard, and even improvements in pens and ink (Anderson, 1961). However, from the point of view of the Commission, the crucial period has been roughly from the turn of the century.

The last 60 or 70 years have been marked by social historians as the period of the so-called "communications revolution"—steam-driven rotary presses, photography, motion pictures, radio, etc., with accompanying developments in psychology, mass communication techniques, school organization and finance, etc. The following discussion will break down some of these developments to aid in obtaining perspective.

Conventional Audiovisual Devices and Materials

The best date to mark the beginning of the modern trend of using audiovisual devices and materials in education is about 1920. Before that time, the glass slide, the phonograph record, and to a very limited extent, the 35mm motion picture had been in some use. After 1920 developments speeded up, partly due to the use of films during World War I. By 1926, 16mm safety film and equipment for education were available, and a start had been made by the Eastman Kodak Co. in producing educational films; radio had reached its commercial stage and a beginning had been made with educational radio; other devices and materials were either developed or improved. Another big change—perhaps *the* big change—came in the middle 30s with the development of the 16mm sound motion picture, the workhorse of the audiovisual movement.

Accompanying these hardware and material developments had been parallel developments in the study of learning from audiovisual materials, beginning with the first studies made right after World War I by K.S. Lashley and John B. Watson, the great experimental psychologists of their day. This connection between some very able research psychologists, students of mass media, etc., and the audiovisual field has continued to the present and has resulted in a respectable literature, much of which is unknown to general educationists, other psychologists, and to many vocal subject matter experts outside the field of education. It is also safe to say that while these developments did occur—partly through the generosity of the Carnegie and Rockefeller Foundations—the application to American educational practice was very limited. There were, for example, only slightly over 600 sound motion picture projectors in the schools of this country in 1936 (Finn, Perrin, & Campion, 1962).

By the time the United States entered World War II, an instructor could have had an adequate instrumentation at his disposal. It included still and motion projectors of several varieties, recording devices and players, various forms of printing, chart production, etc., and sound and communication equipment. In addition, a fair amount of material to use with this equipment (educational film, for example) was available, and quite a lot of know-how derived from research and practice was at hand. In fact, however, very few American school teachers had even seen much of this technology, let alone having it available. (In some of the larger cities, of course, it was available, but not too extensively.)

The wide—almost saturated—use of audiovisual materials in the military and industrial training programs during World War II has been well told in several places (Miles & Spain, 1947). This was the first successful mass application of the technology of instruction to the training of large groups of men and women.

Following the war, a renewed interest arose in the American public schools in the use of audiovisual materials. Some states, such as California, made a great effort and supplied funds in quantity for this development. Across the United States, however, the reaction was spotty, and the additional know-how derived from the war effort had little effect in many places.

In the meantime, hardware developments added to the impetus and the confusion. The development of wire and then magnetic tape recording, the long-playing record, improvements in the optics of projection equipment, and the adoption of the overhead transparency projector which had been developed during the war all contributed. Producers rushed in to turn out motion pictures, recordings, filmstrips, slide sets. Research—some of it financed by the armed services—in the problems of learning from these materials was carried on. However, as we approached the middle of the decade of 1950-59, educational interest in what has come to be called conventional audiovisual materials tailed off somewhat. Present developments date from that time and will be discussed below.

Psychological, Sociological and Educational Developments
Even a brief account of the history subsumed under the title above could fill one or more volumes. The intention here will be merely to outline some of these related developments. The work of Thorndike at Teachers College,

Columbia, early in this century set the stage for the development of a scientific psychology of education. Thorndike even anticipated teaching machines and programed learning (Finn and Perrin, 1962). S. L. Pressey of Ohio State University, as is well known, discovered certain principles of machine instruction while attempting to produce a test scoring machine in the 20s.

In the early days of modern educational psychology, there was a tight connection between laboratory experimental psychology and educational practice. Gradually this connection eroded. The experimental psychologists shut themselves up in their laboratories. Some educational psychologists turned their attention to psychological, achievement and other forms of testing and developed a formidable subtechnology that has been the object of some recent criticism (e.g., Hoffmann, 1962). Others turned to the field of guidance. The effect on practice, however, was highly variable. By 1955, this trend toward separation of experimental psychology and practice reversed somewhat and added a forceful element to the developing educational technology. This, too, will be discussed below under the present state of the art.

In the meantime, other research workers, principally in the field of social psychology (and, to some extent, related fields), had turned their attention to the mass media of communication. The work of Paul Lazarsfeld both before and after World War II is an example. This, together with the work of communication and perception theorists and industrial psychologists interested in group dynamics, had certain fallout value for educational practice. An educational technology began to develop in the group field which had quite a vogue until the middle 50s. The work of the mass communication specialists and perception psychologists has had some effect on the theory back of the educational use of audiovisual materials.

With respect to the process of education itself—learning and teaching taking place in an institutional setting—theorists in the 20s had great hope of developing something called "educational engineering." The great leader of this movement was W. W. Charters of Ohio State University.[5] Charters' ideas were attacked and pretty well demolished during the 30s by both the social reform and the child-centered philosophers of education. From the middle 50s onward, the Charters concepts were revived, although in a new form and were seldom attributed to him.[6] Interestingly enough, Charters also had a great interest in audiovisual communication and founded the Institute for Education by Radio and Television at Ohio State in the 30s.

Industrial and Governmental Developments

Industrial developments relating to a technology of education can be briefly discussed with reference to (a) the design, development and marketing of hardware and environments, (b) the production and distribution

---

[5]One of the writers has on loan from Ohio State University a manuscript for a book being written by Charters when he died. The title is "Educational Engineering"; the date approximately 1952.

[6]An exception is Silvern who has related Charters' ideas to the present situation and credited him for his contributions. See, for example, Silvern, 1963.

of materials of instruction, and (c) intra-industry use of instructional technology.

Generally speaking, hardware, at least until very recently, has not been designed for use in schools. A manufacturer would produce, for example, a motion picture projector designed for the industrial, home or military market. Small changes might be made for the school market, but the design and reliability needs of educators went unheeded. There was many a tragic joke about the 110-pound school teacher who had to carry a 45-pound projector up three flights of stairs. The whole question of the design of instructional equipment is discussed in a paper of the Technological Development Project of the NEA (Leverenz & Townsley, 1962). Badly designed equipment and equipment mismatched to educational needs were among the causes for the failure of schools to rush to adopt innovations during the post-war period.

School buildings provide spaces for learning—learning environments. As such, they are part of the technology that should aid in learning and obviously should be matched with other technologies that have the same end. Since before World War II, this had not been true. Recently, some signs of change are evident.

School architects have done a pretty good job in providing space for certain special educational functions—gymnasiums, auditoriums, shops, etc. They have, over the last 30 years, immensely improved the aesthetic quality of the school environment—color, light, etc. They have made the environment more healthful with such things as better lighting and ventilation. They have done almost nothing to help the general instructional process by providing for modern instructional technology. Until recently, the concentration on using natural lighting through acres of glass windows has prevented full use of projection equipment, and many schools today still do not have adequate light control, acoustic treatment, or wiring. Beginning about 1955, this situation, too, has changed for the better.

Problems of the design of instructional equipment over the last 50 years and, to a certain extent, problems of the design of buildings are inextricably related to the economics of education. The manufacturer could not invest the necessary capital to create special educational equipment because his chances of recovery in a reasonable time were almost nil. The manufacturer, therefore, adapted existing commercial equipment and made very few changes in the fundamental models marketed for educational use once they were set. One manufacturer used the same stampings for the body of a projector for about 20 years, ignoring both the needs of the teacher and the technological developments that had occurred during that time.

The economic problems involved in providing an educational technology during this period (1920-55) were complex. The educational establishment had no precedent and no great inclination for investing in a machine technology at all. The local and state system of finance left little money in educational coffers for investment in technological capital. The local nature of the educational market (Robert Hayes, with insight, once called the educational market "clumped," as opposed to homogenous or monolithic) together with local bidding procedures prevented large-scale selling. The distribution organizations of the manufacturers were primitive, ranging from one-man camera and music stores to a few large audiovisual dealers with service departments.

The materials side of the developing educational technology over the same period also suffered from a lack of market on the part of the schools and a lack of capital, vigor and imagination on the part of the producers. The first three large-scale efforts to produce educational films were subsidized for many years by industry—partly in the hope of eventually developing a profitable market and partly for what have to be philanthropic reasons. Eastman Kodak subsidized Eastman Teaching Films in the 20s; Western Electric subsidized ERPI Classroom Films (now Encyclopaedia Britannica Films) beginning in the 30s and running for about 20 years; David Smart began producing Coronet Films on his own estate, using the profits of *Esquire*.

The textbook market, while no doubt profitable over the same period of time, never realized its potential. School systems sometimes kept the same textbooks in use for periods as long as 25 years—with all that implies in terms of outdated learning. It is only very recently that the development of libraries for elementary schools has taken hold, and many junior high school and high school libraries still are pretty bad.

In 1948, the textbook industry had a study made by a consulting firm to see if the industry should expand generally into the audiovisual field. The conclusion was that it should not (Harcourt, Brace and other publishers, 1948). McGraw-Hill, a firm that had already made a beginning in this direction, ignored the recommendations and now occupies a commanding position as an overall producer of all kinds of teaching materials. In general, the publishing industry is still not facing up to the general technological developments that are occurring (Redding, 1963).

As more of the newer media—long-playing records, tapes, etc.—came in, the same marketing procedures applied. The supply a school had of new and older materials could only be said to be fortuitous; it often depended upon who called on the superintendent. Maps had to be obtained by mail order or from a salesman who worked a large territory out of his house; the supply of textbooks depended upon the vagaries of state law—in Kansas students had to buy them, but in California they were supplied free (and this had certain political implications in selling); the demand for films, slides, filmstrips, etc. was conditioned upon equipment available and the extent of library service; this is a classic chicken-egg situation and is, incidentally, one faced with the introduction of any new media or technology of instruction (teaching machines—programs, language laboratories—practice tapes).

Beginning with the National Defense Education Act in 1958, the general marketing picture changed radically and with it improvements in the overall technological capital of the educational system. Figures available to 1960, however, indicate that the technological capital of the American education system is still in a primitive state; I have suggested that the Rostow concept of five stages of technological growth of a primitive society can be applied to the educational system and that the American educational system may be about ready for technological "takeoff" (Finn, Perrin and Campion, 1962).

The Emerging Technology of Education

Some time and space have been spent on this very sketchy account of the development of some aspects of educational technology in American education in this century because many educated laymen (and for that matter, educators and subject matter experts now attending to the educational problem) are not aware of these developments. Many seem to assume that the concept of educational technology sprang full blown from the forehead of B. F. Skinner about 1954, or began when the Ford Foundation blessed instructional television with money the same year. To some degree, judgments as to what will and should happen with educational technology have to be conditioned on its history—even if every decision made up to this point in time has to be reversed.

These developments, set arbitrarily from 1955, have led us to the present time. All the evidence suggests that the middle of the 50s marked a shift, a quantum jump, in technological development for American education. This shift brings us to the present state of the art, and it will be described in the following section.

Present State of the Art and a View of the Future

As was indicated above, the year 1955 has been arbitrarily set as the beginning of modern developments in instructional technology that have given rise to such expressions as "the technological revolution in education." This period was examined in some detail by me in 1960. A brief overview of developments in the 10-year period after 1955 should provide a reasonable description of the state of the art.

Hardware

More or less all technological developments during this time—machines, materials, techniques, organization patterns, etc.—may be conveniently classified into two large categories:

1. A technology of mass or large-group instruction.
2. A technology of individual instruction.

Developments in instructional hardware fall neatly into these categories.

Television is the instrument par excellence of mass instruction, even if the receiver is watching alone. The promoters of educational television in the beginning viewed television as an instrument designed to help solve the teacher shortage by providing quality instruction with fewer teachers. Television may be either broadcast, closed-circuit or broadcast on the 2500mg band (very short-range broadcasting).

There are now about 100 educational television stations in the United States. The tendency in recent years is for these stations to transmit more cultural programs and more programs of a general educational nature for both adults and preschool children. School broadcasting (instructional television) still occupies some time on these channels. For reception, there were about 50,000 television receivers in the schools in 1962 (Finn, Perrin and Campion, 1962).

Closed-circuit television and the adoption of the 2500mg band provide a more direct means for using television in school instruction. The adoption of closed-circuit television has not been spectacular. As of 1963, there were approximately 600 such installations of all kinds in schools and colleges in the United States (Campion and Kelley, 1963).[7]

Related to television development has been the development of the video-tape recorder. In the beginning, the price was prohibitive for educational channels, but the Ford Foundation eventually supplied the nation's educational television stations with professional videotape recorders. In the last five years, efforts, including competition from the Japanese, have tended to reduce the size, complexity and price of this equipment. There are still problems of broadcast compatibility (the smaller, less expensive units produce tapes that cannot be broadcast over standard broadcast equipment without some difficulty; they can, however, be used on closed-circuit systems). Videotape recording, of course, permits the freezing, storage and free scheduling of television instruction.

The other large development has been the language laboratory, a sophisticated combination of sound equipment centering on the tape recorder, now often called the learning laboratory. The language laboratory is a device making it possible to drill many students in speaking and listening at the same time but as individuals. Further, since there are oral (and aural) aspects to other subject and skill areas (spoken literature and stenographic dictation, for example), the possibilities of expanding the use of such a facility are being explored in many places. There are about 10,000 units in the United States, making this development the most spectacular from the point of view of adoption and expenditure of funds (Finn, Perrin and Campion, 1962). These laboratories range all the way from small portable units that can be set on a table or wheeled around on a cart to very large and complex installations.[8]

A third technological approach to large-group instruction is called multimedia, or multi-media/multi-screen. Here the hardware innovations are in the form of control equipment which can operate several different types of projectors, projecting on two or more screens, singly or together. Projection may be from the rear, as in the case of the Multi-Media Laboratory of the University of Wisconsin, or from the front, as in the case of a system developed at the University of Southern California. This technique permits the projection of slides, motion pictures, and overhead transparencies as well as the playing of tapes and recordings, very often with teachers or lecturers playing an integral part in the presentation.

There have been small improvements in other forms of mass instructional technology as the conventional audiovisual devices have undergone some development, as, for example, the semiautomatic threading 16mm motion picture projector and the semiautomatic or automatic slide projector. In some ways, principally due to improvements in the equipment

---

[7]This figure is now somewhat low due to expenditures in the last three years provided under the National Defense Education Act.

[8]About two years ago the American Association of School Administrators launched an attack on the language laboratory using some research done at the Institute of Administrative Research of Teachers College. The Association is opposed to categorical federal aid to education, from which funds (NDEA) were for the most part, used to provide these laboratories to the schools. The NDEA was up for renewal at the time.

which made it much easier to use and to a superb approach to marketing by one large company, the Minnesota Mining and Manufacturing Co., the overhead transparency projector has now become a great contender for the educational technological innovation of the decade, after languishing since its development from the bowling alley projector by the Navy in 1941. Coupled with the tremendous increase in the equipment has been a tremendous increase in eggs or chickens, as the case may be. Transparency materials of all kinds, some prepared transparencies, some materials making it easy for teachers to prepare, and some sort of in-between arrangements, like cake mix, have become available. By far the most widely used process recently has been the heat-transfer office copy machine since transparent materials have been developed to use with it.

Principally due to publicity, the technology of individual instruction is thought by many to center on the teaching machine. This is not the case. When B. F. Skinner published his famous article in *Science* in 1958, a great interest developed in teaching machines and programed learning. This interest not only engaged psychologists, but also segments of industry from small inventors to large firms. From late 1959 to 1962, a rash of companies went into the teaching machine business (Finn and Perrin, 1962). Most of the small ones have disappeared and most of the large ones have dropped the projects.

This situation had several causes. The educational community was hardly ready for such a startling innovation; the whole business suffered from the same old chicken-egg situation. There were not enough programs for the machines; the machines were not standardized so that one program could fit another; most machines were poorly designed; most programs violated all principles of programing. Research showed no difference between machine and book-type programs. Programed instruction survived in the form of books, notebooks and other ways of using printed material without machines and is, at present, the captive of publishers who are not particularly interested in a machine approach.

The teaching machine, however, is not dead. Some experimentation continues across the country. In a few places, third and fourth generation machines have been produced. The basic problem in the early days was that teaching machines attempted to present verbal information only. This was all that most programers knew how to prepare. The capability of a machine, however, is in the direction, on the one hand, of a complex stimulus—picture, sound, color, as well as words—and, on the other, of a complex response mechanism relating the student's response instantaneously to the task at hand. This leads to audiovisual stimulus machines and computer control—items that will be discussed in the next section.

In addition to teaching machines per se, a range of devices is in use or development that constitutes a solid technology of individual instruction. Using phonograph records and, later, magnetic tape, some teachers began experimenting with "listening corners" in the elementary school shortly after World War II. Headphones on gang jacks were developed so that several children could listen at once without disturbing the class. Later, illustrative materials in the form of filmstrips and slides and various kinds of workbooks and directions were added. A short of home-grown individual instructional technology was developed in a few places.

Following on the heels of this, several manufacturers brought out well-designed listening and viewing devices for filmstrips, tapes, sound film-strips, etc. (Various types of reading pacers fall into this category.) The materials producers have not kept up with this development too well, although they have produced some materials. A lot of work remains to be done to produce the proper kind of material for individual as opposed to group use.

However, the most significant development on the individual instructional technology front has been the growth of 8mm film. About 1960, sound was added successfully to 8mm film; this event triggered a large number of developments in 8mm, many of them, interestingly enough, in the silent field. Although 8mm may be projected for fairly large groups and, with the new format just announced, will probably generate a projected image about as good as 16mm was 10 to 15 years ago, it is in the field of individual instruction that 8mm seems to be destined to make its mark.

The small, inexpensive technicolor cartridge-loading projector for individual viewing is already finding much use and has started a new materials movement known as "the single concept film." The single concept film is a two- or three-minute film on a self-winding cartridge that may be inserted in the projector and be ready to project. It covers a single idea; for example, nuclear fission. The Air Force recently ordered a quantity of these projectors with stop-frame devices which make it possible to program the films and use the projector somewhat as a teaching machine. Viewing devices for 8mm sound film are also coming on the market. What this will do to television is still not clear, especially since there are other developments in the videotape recording field that bear on the problem of individual viewing and listening.

Materials—Systems

In the last five years, a great deal has been heard about the "systems approach" in education. This interest has increased recently; there are projects at Syracuse University, Michigan State University, and the University of Southern California that are working on certain aspects of the problem. The systems concept in education is related both to materials and to hardware as well as to certain psychological and philosophical concepts.

Technically, if you consider such expressions as the educational system, the school system, and the state system of higher education, then an instructional system is a subsystem. It is this subsystem that is generally the object of the systems approach in education.

One way to understand the developing instruction system is to approach it through the materials of instruction. In this sense, the idea is not too new; for example, with the introduction of graded readers accompanied by workbooks, etc. in the 20s, a sort of system of teaching reading was introduced into the schools. However, the decade we are examining has seen a speeding up of trends toward systematic organization of the materials of instruction. This trend was identified by the Technological Development Project of the NEA as "from kits to systems."

The Emerging Technology of Education

There are now available all sorts of kits for teachers to use which consist of a variety of materials organized around a topic, such as a country (Japan), a process (wheat—bread), or a concept (energy). These kits often include paperback books, filmstrips, realia, phonograph records or tapes, etc. The kit is accompanied by a manual which suggests how the resources may be deployed.

Another systematic approach to instruction began several years ago with the production of a whole series of film lessons (162 half-hour films) in the field of physics. This was followed by other film series in other areas. These were not simply "aids." They were systematic presentations of content accompanied by suggestions for other activities and related, to some degree, to texts or other materials.

The concept of the Physical Science Study Committee project under the direction of Professor Zacharias developed an even more systematic approach. Here, beginning with a thorough textbook revision, laboratory exercises, an apparatus, a whole film series, and some other materials were prepared. This development was accompanied by tryouts in the schools and a program in the summer for training physics teachers to use the materials. While not developed exactly according to accepted principles of systems design, it could be said that the Zacharias group developed a system for teaching physics.

The foreign language field has recently contributed some very systematic approaches to instruction. There is one system for teaching elementary French which consists of books, workbooks, tapes, records, and films. These materials can be supplied in a case with the necessary hardware. The films supplied, while in color, come in both 8mm and 16mm, and can be used on television.

Throughout the materials field, organized and semisystematic collections of instructional materials continue to make their appearance. Eight millimeter single concept films are being marketed in sets and are related to textual materials; even the textbook industry is publishing sets of books, workbooks, tests, and materials from which overhead transparencies can be copied. Some of the newer work, as, for example, by the AIBS (biology) group is even more systematic. Materials now being tested by this group lead the teacher almost step by step through the teaching process (See, for example, Biological Sciences Curriculum Study, 1964).

It is, I believe, safe to say that this move toward organization and system in the instructional materials field is one of the most solid trends of the technological revolution in education. Its implication for the role of the teacher in the future, is, of course, tremendous.

The systems approach to instruction also has a hardware aspect. Up to the present time this has shown up principally in an attempt to bridge the gap between the technology of mass instruction and the technology of individual instruction. In a sense the language laboratory has always done this, as the individual student in the laboratory deals with what is, in effect, a mass communication system.

However, there are several devices now being used that bridge this gap more directly. These devices provide for mass stimulus (television, projection or multi-media presentation) but allow each individual student to react to what he sees by pressing a button, turning a dial, or otherwise inform-ing the system as to his response. These devices go under the general name of classroom communicators. One such device is now being used in several of the Job Corps camps. The data derived from the students' response can be processed in several ways; for example, providing the teacher with an immediate readout of how well the students are performing. The materials for such a system have to be very carefully programed and are, of course, highly systematic. This requirement is the biggest problem in connection with the use of this equipment.[9]

There are several other ways to consider the systems approach to instruction. For example, the small group interested in computer applica-tions to education seems to take the position that the only systems ap-proach to education is through the computer. This will probably be the ultimate systems approach.[10]

Organizational Concepts

As was indicated by the introductory material, organization is a major aspect of technology. As might be expected, organization concepts better fitted to present educational needs are being suggested and applied. We are considering here only organization for instruction—not other proposals having to do with the governing and financing of the educational system as a whole.

In essence, there are two such organizational patterns on the current scene which have many common aspects. The first, usually associated with the name of John Goodlad, reorganizes the elementary (and junior high) school into an ungraded pattern, permitting the individual student to pro-gress much more easily at his own rate. The system of grade classes, im-ported from Germany in the 19th century, is unrealistic in the sense of human variability. It was, in its day, a technological solution to a chaotic problem. Whether the ungraded elementary school is the answer for 1965 yet remains to be seen. The requirements in terms of materials, individual-ized instruction, and the like are frightening.

The second new system of organization is applied to the secondary school and is usually referred to as the Trump Plan after its originator, Lloyd Trump. Trump has attempted to break up the "egg crate," 30-students-to-a-classroom situation in the high school and provide for large-group instruction (40 percent), small seminar instruction (20 percent), and individualized instruction (40 percent). This concept is formally referred to as the Staff Utilization Plan. Inherent in this system is the idea of team teaching where teams of teachers, sometimes under a master teacher, manage these several responsibilities. Various aspects of this plan have been tried out all over the United States, and the idea seems to be moving into practice. It should be emphasized that the Trump Plan is also

---

[9]The best such program the writer has seen is one developed for the retraining of bus drivers by Western Greyhound using films and slides on a system known as the EDEX.

[10]See the paper prepared for the Commission by Donald Bushnell for a complete discussion of this aspect of educational technology.

postulated on a much wider use of instructional instrumentation than now exists, particularly for large-group and individualized instruction. (Many existing team teaching experiments do not have this component.)

Curriculum Reform

Another aspect of the technological revolution in education is curriculum reform. Reference was made above to the Physical Science Study Committee. The PSSC set the pattern for curriculum reform, backed by money from private foundations and the National Science Foundation. Since that time, following the PSSC pattern to some extent, groups in biology, chemistry, mathematics, and many other fields have been organized. As of two years ago, the NEA identified 37 such projects. Almost all of these have an orientation to instructional technology in some or all of its aspects—television, programed instruction, films. In other words, the approach to the radical revision of the educational process is through the materials and devices of instruction.

The Current Situation in Educational Psychology

If the approach to change in education is viewed, at least in some quarters, through the materials and devices of instruction, it follows that the psychological orientation of these devices and materials within the teaching-learning process is of crucial importance. This is the area of applied educational psychology, or, as was indicated in a previous section, what some theorists, such as Lumsdaine and Glaser, consider to be the technology of instruction.

The current situation, however, within educational psychology is, to say the least, mixed. Very few theorists or practitioners would be willing to say that one given point of view has achieved all of the answers. In the general practice of schoolkeeping today, an eclectic point of view as to theory coupled with a large portion of experience-passed-on-down constitutes the operational base.

Within the general literature of learning theory and related matters, there have been many attempts to distinguish, describe, even to reconcile various theories of learning. Hilgard, for example, in a classic work treats nine theories of learning (Hilgard, 1956). Such considerations are important to the scientist exploring this area, but it is highly likely that most practitioners today—teachers, materials producers, etc.—could not make the sometimes subtle distinctions between these points of view, and certainly could not consciously apply these distinctions to real life problems of teaching and learning.

The uninitiated person reading only a portion of the literature might get the impression that the matters of teaching and learning were pretty well settled by the psychologists. This is not the case. However, since the educational process is ongoing and since children and adults come to school daily, some kind of commitment to a point of view is often made, particularly at the growing edge of educational technology. The following brief description of the situation must be read with this fact in mind and also remembering that most practice does not have a coherent theoretical base.

The most dominant point of view within the area of instructional technology today is that of Prof. Jerome Bruner of Harvard. Professor Bruner's point of view has been adopted by the Zacharias group which began the national movement of curriculum revision through the materials and devices of instruction with the physics course of the Physical Science Study Committee. This group has extended its activities to such areas as elementary school science (A Review of Current Programs, 1965), and its point of view and methodology are, it is believed, dominant throughout almost all national efforts at curriculum revision, even including music and the fine arts.

At the great risk of oversimplication, Bruner and his colleagues, because they have carefully and brilliantly studied what is known as cognitive structure (an individual's organization, stability and clarity of knowledge in a particular subject matter field at a given time), have emphasized two general concepts:

1. Knowledge in a given field has a structure that can be taught and on which all individual facts and events can be hung.
2. Such a structure is best learned by the method of "discovery" or "inquiry," which in turn will help the learner be more creative and able to learn all his life.

Bruner's ideas have received great acceptance among scientists, mathematicians and, interestingly enough, from many old-line progressive educators. A popular source for Bruner's ideas is his little book The Process of Education (Bruner, 1965). His concern with other aspects of learning may be found in another small volume, On Knowing (Bruner, 1962).

Again, recognizing the difficulties of oversimplification in a brief treatment such as this, B. F. Skinner, also of Harvard, can be thought of as holding a point of view almost directly opposite to that of Bruner. Skinner has recently criticized the Bruner position in a popular article (Skinner, 1965b) and in a technical essay (Skinner, 1965a). Skinner, recognized as the father of teaching machines and programed instruction in the modern sense, has a theory of teaching and learning known as operant conditioning in which the individual learner's behavior is "shaped" by positive reinforcements administered under certain contingencies and following certain schedules. Theoretically, programed instruction of the so-called linear type is designed to do this. In practice, very few programers write programs according to pure Skinnerian principles. Skinner has had an enormous influence on the thinking of psychologists, and it is safe to say that most psychologists today are behaviorists who would be willing to accept a great deal, if not all, of Skinner's point of view. The fact remains, however, that his influence on educational practice, particularly on educational technology as it is being applied, is small compared with that of Bruner.

A third point of view, for the moment somewhat outmoded, may be described as anti-technological and existential. It is stubbornly held, however, by a portion of the educational community, notably those most concerned with child development and the person, and some of those concerned with creativity. These latter often accept much of the Bruner position in addition. In some ways this position combines a little of the old gestalt concepts with the so-called assumptive view of perception and a great deal of emphasis upon self, self-actualization and self-perception.

The Emerging Technology of Education

These concepts, combined with certain others derived from psychotherapy are often referred to today as "existential" psychology.

As was mentioned, there are, of course, other theories of teaching and learning; the so-called neobehaviorists, for example, adhere to a range of closely overlapping points of view. We are speaking here primarily of the views of practitioners (considered broadly) concerned with developing educational technology. In this connection, several newer positions which appear to have promise for or are related to instructional technology should be mentioned. Studies in neurophysiology have resulted in a theory of thought in behavior dealing with the electrochemical functions of the central nervous system; memory has been studied with reference to changes in nervous structure, etc. Such a viewpoint is appealing to some theorists in instructional technology, probably because of its science-technology relationship. One of the leaders in this field is D. O. Hebb (1949).

Closely related to the interest in neuropsychology and even more closely related to developments in technology in general is what is often referred to as the cybernetic model of behavior and learning. Essentially, this viewpoint undertakes to explain thought and behavior in terms of models derived from the studies of computers. An excellent statement of this position may be found in Miller, Galanter and Pribham (1960). In many ways, this theory can be considered an information theory of learning and is related to an even more elaborate concept—J. P. Guilford's structure of intellect (Guilford & Merrifield, 1960). David Ausubel has recently published a theory relating only to what he calls "reception learning," or the processes of learning meaningful verbal materials, arguing that this type of teaching is what is principally done in school (Ausubel, 1963). All of this work suggests that it will influence technological applications in education within the next decade. If this is true, some changes in current trends will occur.

Finally, it should be noted that both dominant and emerging theoretical positions in the literature (which have been applied in small areas of technological practice in some instances) merely emphasize the fact that the educational practitioner cannot turn to a science of learning for his answers. This has resulted in a technological or empirical approach to solving practical problems of teaching and learning. This point of view is espoused by Ofiesh and is contained in part II.[11] The empiricists hold that materials, devices and processes must be validated in advance in terms of well-stated objectives of instruction on the students on which they are to be used. In other words, materials, devices and processes are produced on a best-guess basis, are tried out, are refined or changed as necessary, and are tested again. The process is repeated until a system that works is derived. Such a concept can be considered as a sheer technological or engineering approach and is so referred to by those who advocate it.

---

[11]This article is part I of a study prepared for the National Commission on Technology, Automation, and Economic Progress, February 1966. Part II, by Gabriel D. Ofiesh, deals with additional material on research and development.

The Emerging Technology of Education

If present efforts at research, development and theory construction continue to expand; if, for example, the national research and development centers and the regional educational laboratories take root and grow, it can be expected that a much greater alignment of theory and practice of teaching and learning will come about within the next two decades. While it is unlikely that the answer will be achieved in this first new thrust, the move toward the classic relation of science to technology in the field of education will be speeded up.

Research and Development

The history of research and development in the field of instructional technology is a long story in itself. Contributing to the quantum jump of 1958 in this field was the passage of Title VII of the National Defense Education Act. Title VII provided both for research in the new media field and for dissemination of information about new media to the educational community. Although the amounts of money available have been small (too small), the effect has been large. Psychologists and other educational research workers have turned their attention to educational media, and fundamental questions concerning effectiveness and use of the technology of instruction have been investigated.

The Cooperative Research Group in the U.S. Office of Education has also funded some studies in the field of instructional technology, and some research has been backed by other government agencies concerned with manpower retraining, etc. Compared to the need, this research effort is still too small and the dissemination of the results of this research leaves much to be desired.

The new Elementary and Secondary Education Act of 1965 contains a provision for $45 million for what are known as regional educational laboratories. These laboratories (actually, they will probably be more like groups of cooperating institutions and school districts) are supposed to do research and then apply the results of this research to practice. The task force, which worked preparing the background for the bill, had in mind something like Professor Zacharias' Educational Services Inc., a nonprint corporation which produces the materials for PSSC and other curriculum projects. However, it is likely that the organization of the laboratories will be much broader. There is no doubt, however, that there will be a large component of activity relating to instructional technology in these laboratories.

While many important research questions need to be settled, the great problem is the dissemination of the results of this work and the application of these findings to practice. To take just one example, there is very little evidence that the research findings on programed instruction have actually been applied in the construction of many programs that are available; further, the use of programed instruction in the schools remains at a minimum level—the movement has not affected practice a great deal.

The great foundations of this country have contributed both to the research effort surrounding instructional technology in this decade and to the development work associated with it. Actually, foundation support has principally been for development and, as such, it has been more of an economic than an intellectual force. Because of this, it will be treated in the next section.

The Infusion of Capital

If technological capability—machines, materials, etc.—is viewed as capital, the educational system of this country is still poverty-ridden. Using hardware as a measure of this capability, Finn, Perrin and Campion found it to be relatively low (Finn, Perrin and Campion, 1962).

This low capital level is why, during this crucial decade of 1955 to 1964, foundation support has in many cases been decisive and has turned the path of American education into new, technological directions. The role of the Ford Foundation in this picture cannot be underestimated. Especially in the field of television, the support from the Ford Foundation was decisive. This included help for weak educational television stations, the establishment of the National Education Television Network Service, support for a national effort in instructional television, the institution of the airborne television program in the midwest (Midwest Program on Airborne Television Instruction—MPATI), etc. Millions of dollars helped build technological capability and, of course, included other activities besides television; for example, the founding and support of the Educational Facilities Laboratories which are designed to improve educational construction of all kinds.

Other foundations, notably Carnegie, also contributed to this buildup. However, other than television, viewing the needs of the country as a whole, all the foundation money put together could only be described as seed corn. It remained for the U.S. Congress to provide funds for a much larger capital infusion into the American educational system.

From the point of view of educational technology, Title III of the National Defense Education Act made all the difference. In its original form, it provided money on a matching basis for states to furnish local school systems with funds to purchase materials and equipment for teaching science, mathematics and modern foreign languages. This provision has since been expanded to cover other areas of the curriculum. Under the older restrictions, approximately 60 to 70 percent of the funds went into science equipment and the balance into audiovisual materials and equipment. These percentages are no doubt changing and, during the last year, the amount of money expended under the provisions of this title was about $90 million.

Almost all of the acts relating to manpower (manpower redevelopment and training, vocational education, etc.) are designed to provide money for research and for teaching equipment and materials. No estimates were available to the writer as to the exact amount of such funds, but they are considerable and will contribute to the overall increase in technological capital for the educational system.

Acts of the Congress relating to manpower were climaxed with the passage of the Economic Opportunity Act of 1964—the so-called War on Poverty. In the act, provision is made for centers for the basic education and vocational training of jobless youths. In the organization and development of these centers, the assumption has been made that a great deal of use must be made of instructional technology because the system of instruction, it is believed, must be considerably different from that used in schools from which these young people dropped out in the first place. Because the main Job Corps training centers are being operated by private industry in many instances, every effort is being made to apply the best that is known in the instructional program of these camps. Money is available. Camp Parks, for example, installed about $90,000 worth of television equipment. This development will have far-reaching effects, but they have yet to be assessed.

The latest act of Congress affecting instructional technology is the Elementary and Secondary Education Act of 1965. Mention has already been made of the laboratory provision. One title of the act provides for the strengthening of library services, and this will add to the technological capital available to schools, as library services will include materials other than books. Another section provides for supplementary educational centers. These are conceived very broadly and will contain everything from guidance and remedial services to resident musicians. These centers will obviously have a media component and may end up as, among other things, regional centers for more sophisticated technological services. Plans for these programs are just being laid.

There are other governmental influences in the field of instructional technology; for example, the wide use of many of these techniques in the training programs of the armed services. Further research in the field of instructional technology is being conducted by HumRRO for the Army and by other defense agencies.

The picture adds up to the fact that an increase in the technological capital of the educational system of the United States can be expected in the next few years as a result of government programs. The extent of the increase, its general acceptability, and its effect remain to be seen. At present, from the point of view of the existing—let alone the future—technology, the American educational system is undercapitalized.

The Response of Industry—Diversification and Reorganization

Accepting the fact that the American educational system remains primitive from the point of view of technological development, it is no different from other parts of the public sector, such as the field of social work. This concept, of course, can be traced to Galbraith who suggested in *The Affluent Society* that investment must move into the public sector and that we should treat such institutions as education with at least as much generosity as we do Las Vegas or the Strategic Air Command.

In addition, the economics of the defense and space programs cannot forever remain at the same levels. These programs have created great scientific-based industries with tremendous technical and manpower capabilities. Reduction in space and defense programs could cause social problems of no mean magnitude. It is no accident, therefore, that the Galbraithian concept of beefing up the public sector of the economy should be linked to the potentials of the aerospace, defense and science-related industries.

With some anticipation in the earlier manpower acts, this policy became reality when the Economic Opportunity Act of 1964 provided for the participation of industry in the Job Corps and other programs. This put the science-related industries into the education business and, in the perspective of the general scientific-technological revolution, was inevitable anyway. Further, many of these industries or industrial groups had made previous passes at private educational developments (teaching machines, computer scheduling services to schools, etc.) and already had created some special educational capability.

The various acts discussed above provide a considerable amount of money and are an enticement for the science-based industries to enter the educational field. Since most of them have a quick-reacting capability, it was to be expected that the last few months have been occupied with tooling up, preparing proposals, reorganizing in the direction of what can only be called "educational diversification," and otherwise preparing to move into this new enterprise.

Several of the large Job Corps camps, as was indicated above, are being operated by these industries, sometimes in combination with a university. However, the most interesting signs are those related to mergers, acquisitions, joint venture agreements, and the like. Raytheon has purchased EDEX, a classroom communicator manufacturer, and Dage Television. Xerox has created a research laboratory in the basic behavioral sciences and has acquired Basic Systems (a programming group) and American Educational Publishers; Westinghouse has acquired the entire programming capability of Teaching Machines Inc.; Litton Industries and Hughes Aircraft have both built up in-house capabilities in these fields and, particularly in the case of an acquisition-oriented company like Litton, acquisitions of various kinds may be expected. This is a partial and very incomplete list, but it is illustrative.

These industrial groups are aggressive and have access to highly educated manpower both in-house and on a consulting basis. The effect of these developments is not yet clear, but some of them ought to be pretty obvious. There is not too much manpower available in the instructional technology field, and hence, there will be competition for manpower.[12] The very existence of technology-based instructional programs in Job Corps camps, etc. will put pressures on the schools. School board members may not be so reluctant to vote funds for instructional technology once the industrial factor becomes evident, and they may even put pressure on school administrators.

Since this is just the beginning of this phenomenon—essentially a phase of the scientific-technological revolution—much more can be expected to come out of it.

Trends—A Forecast

The preceding section covers the current situation with emphasis upon the decade 1955 to 1964, a crucial period for the development of educa-

---

[12]One of the writers had one doctoral candidate in the spring of 1965 who received four job offers in one week. Two were from universities and two were from industries. He took an industrial job.

tional technology. The changes occurring during this period were so strik-
ing (and some of them, such as the language laboratory, so unpredictable)
that forecasting in this field appears extremely risky. Further, it cannot be
overemphasized that developments involving any appreciable degree of
novelty are still very slow to affect educational practice. In effect, events
occur at two levels: An analogy might be to consider the interest in research
and development now occurring in educational technology as the upper
level of ocean currents which can be seen and measured, and actual practice in the majority of educational institutions and systems as the deep,
slow-swelling, cold currents that move in their own time and are difficult
to detect.

Nevertheless, there are probabilities and trends in the situation, and prediction is possible, although its accuracy cannot be stated even with the
precision of probability statements now used in weather forecasting. What
follows in this section is such a prediction; it is the sole responsibility of
the author. For reasons which should become clear, this forecast is divided
into two parts—a short-range forecast and a long-range forecast.

The Next Five to 10 Years

An analysis of the situation suggests that the next five to 10 years will
be a period of consolidation and spread into educational practice of the
technological developments of the last decade. The educational system has,
in effect, been threatened with novelty; in the coming decade the novelty
will be absorbed to the point where it will be no longer novel. There will be
some new developments; these, however, will tend to be in the political-
social-economic sphere and not, as many of the current thinkers in educational technology suppose, in materials, hardware and psychological
breakthroughs. Breakthroughs and novel developments are always possible these days, but the current trends continue to suggest that their time is,
perhaps, a decade or more away.

Based on the generalization that consolidation of gains will characterize
the next decade, the following forecast is made from current trends:

1. *Innovation—Change.* It has become the official policy of the U.S. Office of Education to encourage educational innovation; further, the concept
of innovation is "in" with the entire educational community at state, regional and local levels. Three forecasts can be made in this area:

a. The principal site of education innovation will change from the lower
levels of the school system to higher education. Colleges and universities
will be forced to innovate, principally due to the flood of students, but with
other factors acting as an influence as well. Universities, while sources of
innovation for the whole culture, have been loath to make drastic changes
in their own procedures, particularly their teaching procedures. The innovations which will be forced on the higher educational system during the
next decade will, therefore, cause a great deal of strain.

b. With respect to the system as a whole—particularly the public school
system—educational innovation will become institutionalized, centering
upon the U.S. Office of Education and, secondarily, the several state
departments of education. Existing legislation and plans and the influence
of a new educational establishment will all combine to push this now visi-

ble trend into actuality.

The principal instruments for this institutionalization are likely to be the regional laboratories now being set up by the U.S. Office of Education together with regional educational centers to be set up under Title III of the Elementary and Secondary Education Act of 1965. The regional centers do not have to have this function, but at this writing, it looks as though Part B of this section of the law, which does provide for exemplary (i.e., demonstration centers designed to spread innovative ideas) projects, will receive precedence over regional educational services. However, even if these instruments are not used for one reason or another, others will be found, and the innovative process will be institutionalized.

c. With the institutionalization of educational innovation, two things will happen. First, the innovations themselves as they are picked up by units of the system (schools, school districts, colleges) will become simplified and vulgarized, sometimes beyond recognition, and will lose a great deal of their power. This is principally due to the fact that the system as a whole is not sophisticated enough to absorb many of the new processes and procedures with all of their subtleties and qualifications. Secondly and more important, as the process of innovation is institutionalized, innovation will gradually become little more than change. This is based on the assumption that the "invention of the method of invention" as applied to educational innovation—and that is the avowed purpose of this institutionalization—will not necessarily work with the same force that it has with industrial technology. It is likely that the true innovators will begin to drop out of such an institutionalized system, and the remaining bureaucracy will not be capable of far-reaching innovation.[13]

2. *The Development of the New Educational Establishment.* American education has always seemed to have more common procedures, goals and even buildings and teaching materials than are warranted by existence of 50 autonomous and presumably different state school systems. There are many reasons for this, but among those often cited is the existence of an educational establishment. In the past, it has been stated that this establishment has consisted of the national educational professional associations, the teachers colleges and schools of education in universities, and the state departments of education. After 1950, this establishment came under heavy fire, and beginning about 1955, a new educational establishment began to emerge.

The significance of this new educational establishment for this paper is that it has a scientific-technological base. Essentially, it consists of four or five of the leading higher institutions in the United States, several foundations, a component of the new scientist-politicians that have emerged in the last 20 years, and some able individuals both within and without government. Its relation to the older establishment is almost nil.

The next decade will see the complete domination of educational think-

---

[13]This is why the author has some doubt as to Ofiesh's proposal in Part II. The proposal is for another form of institutionalization. For an interesting discussion of the difficulties of change in an educational bureaucracy, see the article by James (1965).

ing in this country by this new establishment as it develops and consolidates. Nothing in this statement should be construed as suggesting that there is anything conspiratorial about this emergence. It is doubtful, for example, whether individuals now belonging to the new establishment even know it as such. Rather, the emergence of such a group of intellectual leaders for education was almost foreordained by the development of our advanced technological society. Henceforth, the new establishment will orient American education more in the direction of science and technology as associated with its own processes and will absorb only that part of the older establishment which will fit this overall scientific-technological pattern.[14]

3. *The Systemization of the Materials of Instruction.* The already well-developed trend toward more systematic organization of instructional materials will reach fruition in application in schools and colleges within the next few years. Systems of teaching the structure of subject matter and certain skills such as reading will be applied on an increasing scale. These systems will make use of all of the available instructional technology[15] and will absolutely control the curriculum in the areas (such as physics) where they are applied. To some degree, competing systems will be created, and schools and colleges will be asked to choose among systems; however, since these systems are expensive and take years to develop, the choices will be limited. Further, there will be problems of obsolescence and logistics associated with them for which the schools and colleges are ill prepared.

The materials within these systems (films, programed learning sequences, videotapes, books, etc.) will increasingly be tailored directly to learning tasks and will represent much more of a rifle approach than the historic shotgun approach of the standard textbook or educational film. As such, their overall effect should be more more efficient. Further, research now going on in several places should have begun to supply some answers to the general question as to which medium is the most effective for a given purpose. If these answers do develop, the emerging instructional systems will also reflect this knowledge. Increasingly large amounts of money will be spent on developing these systems.

4. *Developments in Hardware.* Hardware, particularly in sophisticated systems, such as military weapons systems, can change or develop rapidly. On the other hand, in the consumer field, such as refrigerators and automobiles, the changes tend to be slower and are often more apparent than real. In the field of instructional technology both possibilities are present.

a. Optical-photographic versus electronic systems. Because of the long lead time that optical-photographic (conventional audiovisual) systems (projectors, film, etc.) have had on electronic systems (television, videotape, etc.), existing instructional hardware is heavily weighted toward the

---

[14]In this connection it is fascinating to note that recent news stories report that the new Russian educational program was prepared by a commission composed of members from the Academy of Sciences and from the Academy of Pedagogical Sciences.

[15]It is believed that forecasts which claim all instructional systems will be computer controlled in the near future are wrong by at least 20 years.

optical-photographic for pictorial (and audio) storage and transmission. Further, photographic information is still superior to electronic by a factor of, perhaps, 100:1. There are other influences as well, such as accessibility.

The next few years will see a continuing invasion of this field by electronic transmission. Improvements will be made in the information capabilities of electronic systems; the transmission of color will become cheaper and easier; accessibility will be improved through cheaper videotape-type storage; and videotape recorders and players will become smaller, less expensive, and easier to operate with reliability. By the end of the decade, a new balance will have been achieved between these two (partially) competing systems. Neither will disappear, but electronic and optical-photographic systems will claim a much greater share of existing hardware designed for pictorial storage and transmission than is the case now.

b. Television, videotape, etc. As indicated above, videotape players and recorders will become smaller, cheaper and more reliable. Whether the current methods of recording on magnetic tape will still be in use might be in question, as there are several other ways to use electronic impulses to record information on some medium; thermoplastic recording is an example. The precise means is unimportant from the educational point of view. What this does mean is that images and sound will be available in inexpensive, easy-to-use form.

In addition, it is to be expected that television will expand in its educational aspects during the same period of time. This expansion will principally be in the closed-circuit and 2,500mg areas and not in broadcast television for schools and colleges. The expansion will occur first in higher education, and it can be fairly confidently predicted that interinstitutional cooperation in the use of television, such as has been experimented with in Oregon, will be extended as the pressures on higher education increase. A professor in one institution teaching a class in another, or several others, will not be at all unique except in small enrollment, prestige institutions. Even in such cases, lectures from Nobel Prize winners and the like will probably be delivered by television.

Along with these developments in television and, for a while at least, overshadowing them, will be an enormous increase in the use of telephone lines to transmit certain kinds of educational materials. The last few years have seen some growth in the so-called "telelecture" technique where the lecturer at one location can speak to a group via amplified telephone at some other location. Recently, as at the Harvard Business School and in connection with various medical education projects, conference-type seminars between groups have been held using the telephone system. A new invention makes it possible for a teacher to draw while lecturing over the telephone and have the image projected by a special overhead projector at the receiving end; slides and other materials distributed in advance have also been successfully used with telelectures. Since this procedure is relatively inexpensive and very useful, it may be expected to grow spectacularly during the decade.

c. Other hardware developments. While improvements and changes may be expected in all audiovisual equipment, it is likely that the major advances will occur in the field of self-instructional devices. There are, at present, several prototypes under development of multi-media machines (still and motion picture and sound) designed as individual instruction devices. While the history of this type of teaching machine has not been too spectacular up to this point, it seems reasonable to predict that the next decade will find several types of these in use. This, of course, will set a new requirement for programing and production of materials.

The 8mm film, particularly with the new format (40 percent more information per frame), will constitute the most important development in the audiovisual field in the next few years. Following Professor Forsdale of Teachers College, it is believed that 8mm will be used primarily for individual instruction, although with the new format classroom projectors for groups of up to 50 in size may be expected to become quite common. As 8mm comes in, a technological lag problem will become apparent with the huge investment the educational system has in 16mm film and projectors. One way around this problem will be to develop individual instruction devices for use with 8mm film, particularly devices that permit student response. Considerable resistance to this development may be expected.

Multi-media/multi-screen techniques for large groups will continue to expand during the next 10 years. Lecture halls and briefing rooms will be built with such hardware requirements in mind. Automatic projection equipment will be redesigned in order to operate in gangs for this purpose, and control equipment will be developed on a miniaturized, high reliability form.

d. Computers and the interface problem. Since a separate paper has been prepared on computer applications to education, computers for computer-based instruction will not be discussed here. However, it may be important to point out or re-emphasize that there are certain needed hardware developments related to computer-based instruction which will probably occur during the decade under consideration. These developments are referred to as the interface between the computer and the student and have been the subject of a recent study by Glaser, Ramage and Lipson (1964). In the next 10 years various ingenious interface devices will be developed so that students may receive stimulation in various forms from a computer (pictures, words, numbers, sounds, graphs), may manipulate the subject matter so presented with instruments such as light pens, and may be informed on other portions of the interface device as to progress, what to do next, etc. Until such interface devices are developed, computer-based instruction will never achieve its full potential.[16]

Essentially, at least a portion of what should go into such an interface device is the result of developments which, in the computer field, go under the general name of information display. Although the existing literature seems to suggest it, there is nothing in the educational picture which would require all such display techniques to be confined to devices requiring student response and controlling student behavior in detail. It is reasonable, in fact, to predict that information display techniques which are essentially electronic or electronic-optical in character will also be used in connection

---

[16]There are, of course, other requirements for the ideal computer-teacher. See the Case and Roe study abstract (1965).

with television, other wave-propagated transmission and telephone to convey teaching materials from one point to another without elaborate response and measuring devices. As yet, with the exception of a few experiments, the techniques of information display in use with sophisticated systems, such as space and space support systems, have not been tried with education problems.[17] The next decade will see many developments in this area, including simulation.

5. *Information Storage and Retrieval.* An area which may develop as spectacularly in the next 10 years as the language laboratory and associated teaching techniques did in the 1955 to 1964 decade is the area of information storage and retrieval. As such, in its educational applications, it could represent an exception to the general orientation of this section.

The problem of the information explosion is well known; the fact that much information today never reaches the book stage in time but remains in the form of documents, articles, etc. has given rise to a whole new profession known as "documentalists"—experts in information storage and retrieval are calling themselves "information scientists"—and there are signs within the old-line professional library field of a deep schism between conventional librarians and information scientists and documentalists.

Obviously, new technological information storage and retrieval techniques can be (and are now, to a certain extent) applied to the problems of a conventional library. However, except for large university libraries, it does not seem likely that these techniques will make much of an inroad into school and college libraries during the next decade.

What is likely to happen is that sort of an end run will occur, and the new technology of information storage and retrieval will reach the educational system in some strength outside of the main library stream. There are several reasons for this. Again, the position of the U.S. Office of Education may be crucial; however, it is the considered opinion of the writer that even if that office were not a factor, this phenomenon would occur.

The U.S. Office of Education, however, will play a large role in this development providing present plans for its proposed Educational Research Information Center (ERIC) are implemented with sufficient funds and personnel. Operating from a center in Washington, D.C., and from, perhaps, up to 200 satellites or clearinghouses located in higher institutions and research centers of various types, microfiche (small sheets of microfilm) chips containing research documents will be supplied educational users ordering from a system of indexes, bibliographies and abstracts also provided by ERIC. This system is now underway on a small scale.

The availability of this information will create a demand for microfilm readers of various types and for equipment to reproduce hardcopy from microfilm. Such readers are all—with some modification—potential teaching machines; further, the ability to reproduce hardcopy presents the possibility of expanding such services to instructional materials too current to be available in any other form. Such a procedure has been experimented with in San Diego County, California, for some years with great success,

---

[17]One example of such an experiment is that undertaken by Licklider (1962).

where local industries and scientific institutions have been supplying schools with current scientific materials produced in this way.

If this development proceeds as suggested—and the probabilities are high —a full-fledged educational information storage and retrieval system may grow up outside existing channels. Further, because the hardware and services are adaptable, it is possible that a new generation of teaching machines will come into being based on the microfiche reader, thus short-circuiting a whole series of obstacles. The presence of hardcopy-producing equipment might speed up the use and adaptation of newer curriculum materials, both in programmed and in more conventional forms. Such a development would tend to restore a certain amount of curriculum inde-pendence to local school districts, providing staff and facilities were made available to take advantage of it. This latter development is highly unlikely. What is more likely is, as the state departments become stronger, curricu-lum materials will be supplied by the state departments in this easier-to-use form and curricular autonomy will be lost, not gained, at the school and district level.

6. *Standardization*. One of the strongest trends in the next decade will be a general move toward standardization, a move inevitable in any highly technical society. With respect to equipment and materials, several forces are at work that, potentially, could force standardization. The first of these is the so-called "state plan" by which many of the federal educational pro-grams dispense money to the states. In its most simplified form, the state submits a plan for a program, for example, dispensing funds under Title III of the National Defense Education Act. Once this plan has been ap-proved, the state, in effect, sits in control of the disbursement of the funds to its local and regional units. All the state has to do is to require standards for equipment in its plan and standardization becomes a reality. With the so-called "Compact of States" within the immediate future, providing for efficient communication between the states on educational matters, stand-ardization could soon become national.

A second force, which has been discussed for many years but has never been released, is possible if the major cities of the United States were to combine in order to write common specifications for equipment and mate-rials. Such cities represent a large share of the market and now contribute to the chaos in educational equipment standards by requiring annoying and, most often, useless differences in specifications. This raises the price per unit on such items as projectors, complicates bidding procedures, and localizes purchases. Economic considerations may force the end of this practice and a move to joint bidding or even centralized purchasing. Such a possibility is only a possibility and it is more likely that the provincial practices of the educational bureaucracies of such mammoth systems as New York and Los Angeles will remain at the level of their archaic city building codes. Of the two forces, the state force, even including central-ized purchasing (a procedure opposed vigorously by the audiovisual in-dustry), will probably prevail, and equipment and, to a certain extent, mate-rials standardization will occur during the decade under discussion.[18]

---

[18]Materials standards are equally, if not more, chaotic than equipment standards.

At a broader level, greater standardization will be forced on the present quasi-autonomous school system than now exists through such factors as the increasing influence of federal educational programs, even if no direct control is sought or applied; the inevitable cooperation of states and regions in educational matters; the introduction of whole systems of instruction; the possibility of a national assessment program; the reduction in the number of educational materials suppliers; the reduction in the total number of school districts in the United States; the general increase of communication; the prevalence of large-scale industrial thinking as it moves into the public sector; and the move toward computer data control.

7. *The National Assessment (Testing) Program.* Earlier in this paper it was mentioned that testing had been developed over the past several decades into a formidable subtechnology within the broader field of educational technology. The influence (some call it tyranny) of the New York Regents Examinations upon the curriculum of the schools in New York State has been commented upon for many years. Recently, this type of influence has extended throughout the nation with the examinations of the College Entrance Examination Board, the National Merit Scholarship Examinations, the Graduate Record Examination, etc. Attention here is devoted to examinations that affect the curriculum of the schools and not to other forms of testing, such as psychological and attitude.

The next decade will see the institution of some form of national educational assessment program. The word "assessment" is used advisedly because the sponsors of the idea (essentially the new educational establishment) are proposing to combine standard achievements testing techniques with sampling techniques similar to those used in public opinion polls to assess "how well the schools are doing." Such a program would not be achievement testing in the accepted sense. A good discussion of the pros and cons of this issue may be found in a recent *Phi Delta Kappan* (Hand, 1965; Tyler, 1965; "The Assessment Debate . . . ," 1965).

Once such an assessment program is underway, it will become another powerful force for standardization. The technical capability (test construction, sampling techniques, computerized statistics) already exists. Current moves underway to make this capability operational by taking the necessary political, social and economic steps will no doubt be successful. The claims already being made that such a program will not force a certain amount of standardization are rejected by the writer. That is not the issue anyway. The questions that remain to be answered are what kind of standardization and whether or not the standardization so created will be good or bad.

8. *Trends in Administration.* Many trends in the field of school and college administration could be singled out for projection. For example, at the brick-and-mortar level, school buildings will continue to improve and be made more compatible with the existing and developing instructional technology. Of all possible predications in the field of administration, four are

One of the earliest problems with teaching machines, for example, was that there were no standards for programs—either mechanical or educational. An abortive effort of a group of professional organizations to set up standards produced nothing.

selected for comment.

The first, discussed at length in Mr. Bushnell's paper, will only be mentioned briefly. Data-processing equipment and computers will become common tools for the school administrator in handling many routine problems; better decisions will be possible because of the immediate availability of better data. Centralized data-gathering centers will appear, probably as regional centers within the several states. Later, and inevitably, these will be joined into some kind of a national network.

Secondly, the most important function to be developed during the next decade will be a logistics of instruction. Everything within the new instructional technology—systems, complex use of materials, sophisticated equipment, new patterns of organization and buildings—requires formidable logistical support. The whole system will break down without it. Such support involves planning based on precise objectives and data, materials flow, equipment maintenance and replacement, backup manpower to the teacher, etc. Such thinking at present is almost completely foreign to school administrators at all levels except in a primitive form that provides sufficient pencils and sweeping compound for the year. Logistical thinking has rarely been applied to instruction; first, therefore, a theory of instructional logistics will have to be created. The pressures of the developing instructional technology will force it into being within the decade.

The third projection is somewhat broader than the first two but is related to them. Essentially, it is that organization patterns will move into larger and larger units for administrative purposes and that control will become more and more a part of what is sometimes called a corporate structure. In other words, the educational bureaucracy will enlarge, with an effort to make the parts (teachers, subadministrators, etc.) interchangeable.

The precise pattern of the enlarging units may take several forms during the decade ahead. The main point is that there will be an increase in size which, of course, will in turn increase the distance between the top levels and the point of contact with students. All state systems of education will tend to become real systems instead of the semisystems now existing. The regional laboratories, the Compact of States, and other such developments mentioned above, when combined with state plans, school district consolidation, and urban growth are all forces operating in this situation. It is hard to see anything but a diminution of local control of schools in the next 10 years. This diminution will tend to accelerate toward the end of the decade.

Finally, it is probable that one or more new private school systems, national in scope, will be started during this same period. These systems will be developed by the new industries moving into the educational field and will feature highly standardized, relatively fully automated, high-quality education designed essentially for upper- and upper-middle-class clientele. Such a system or systems will be accompanied by, but probably not related to, similar systems of private vocational training centers. This movement, too, will be gaining impetus by the end of the decade.

The school administrator required by these and many will quite clearly be a skilled professional manager, not the Latin scholar or part-time chemist-administrator, as desirable as such characteristics seem to many people; and he will not be so much the community-oriented, faithful service club member so highly valued in some school administration circles today. Subject matter scholars, through the technique of curriculum development projects, have learned to short-circuit school boards and administrators in matters of curriculum content. Increasingly, as methodology becomes more precise within educational technology, the same effect will be achieved by psychologists and educational engineers; thus, the issue of the ideal subject matter expert or liberal arts generalist qua school administrator will become completely dead, remaining to be mourned in the columns of literary magazines. The concept of the successful community-oriented administrator will also die, although a little more slowly, and the mourning will be heard at the annual steak fry.

9. *The Research, Development, Dissemination, and Adoption Syndrome and the Resistance to Innovation.* As the process of research, development, dissemination, and adoption begins to operate with force (see Part II), the traditional resistance of the educational system to change will crumble at an increasing rate during the next 10 years. Some enthusiasts for educational technology of any variety have consistently underestimated the power of the resistance of the system in the past; in the future, however, the timespan between idea and practice will increasingly be shortened. The current timespan between the development of a new process and its adoption by a substantial majority of units of the system has been estimated at about 35 years. During the next 10 years this timespan will be reduced to about one-quarter of that length, or from eight to 10 years.

10. *General Developments in Educational Psychology and Methodology.* Barring unexpected breakthroughs in understanding the physics and chemistry of the central nervous system which are, of course, possible, the situation in educational psychology as described above will not change much during the decade. Cognitive structure, inquiry, structure of knowledge, creativity, and student-response manipulation will still be key concepts both for pure and applied research and for the development and testing of instrumentation and materials. Increasingly, a dialog may be expected to develop between the cognitive structure-inquiry school and the operant conditioning school; a hard core of each will hold firm, but borders will become increasingly friendly. Experiments attempting to turn one system into another, as, for example, that recently reported by Schrag and Holland (1965), will increase.

In the meantime, newer viewpoints, particularly those associated with cybernetic principles and information theory, will begin to gain momentum. A recent book by Smith and Smith (1966) may be a bellwether. Research patterns using these developing theories will begin to intrigue younger psychologists. The decade will end with some newer points of view having enough adherents to threaten what will then be "old hat" psychology.

Because the psychological situation will remain unsettled, the educational engineers will build bridges of learning on an empirical basis. The decade will see a great commitment to the empirical approach in the production of instructional materials and hardware. Materials and processes will be tested increasingly on suitable populations and will be revised until they work.

The engineers, of course, will use whatever can be used from the studies of the pure psychologists; further, more research will be based on realistic student populations rather than small laboratory situations from which it is difficult to generalize.

The combination of the empirical approach and increased pure and applied research throughout the whole field of education and educational psychology represents one force that will turn a current form of thinking completely around so that it will point in the opposite direction.

The second force in operation relates to the national commitment to education as *the* uplifting force in our national life. For example, developing learning programs for such things as Operation Headstart and the Job Corps training centers are very difficult technical problems; they require, for their solution, large doses of educational technology considered broadly —methods, processes, machines, organization, skilled specialist manpower. Pious claims to the contrary, a knowledge of subject matter alone will be of little or no help in dealing with a 16-year-old illiterate from the Kentucky hills.

These two forces will combine into a pressure that will result in the rediscovery of educational methodology. From about 1950 to 1960 it was extremely fashionable to decry educational methodology as useless, as a fake medicine sold by charlatans, as something certainly not needed in the process of instruction. Those that grudgingly conceded that methodology did exist equated it with "tricks of the trade"—something that could be picked up overnight on an apprentice basis.

An illustrative example relates to the problems which led to the creation of Operation Headstart, a huge national operation designed to provide missing background for young, deprived children so that they might be ready to learn in school. It is now ironically forgotten that many critical books and articles appearing from 1950 to 1960 claimed that there was no such thing as "readiness" for learning and that professional educators, in maintaining that there was, were perpetrating a fraud on the American public. These books were read by the upper-level economic group and the articles appeared in the "best" magazines. This general downgrading of methodological concepts which resulted in its own mythology, is being attacked by events and will result in a complete destruction of this posture. Further, since control of subject matter is now secure through the technology of instructional systems, the entire dialog between method and subject matter will be wiped out.

11. *The Buildup of the New Educational Industry.* As was indicated earlier in this memorandum, one trend of the last decade was the emergence of the science-based industries into the educational scene. The result of this in the next decade will be a pattern of power struggle, mergers, acquisitions, and new combinations, such as joint ventures between such industries, universities and nonprofit corporations of the "think-tank" variety.

It is hard to see how the old-line publishing firms and audiovisual producers and suppliers can retain their current organization, appearance and way of doing business. The next 10 years will see enormous changes in the educational business, which will generally be in the direction of larger, more diversified enterprises that will absorb many of the smaller companies and will force others out of business. The time of the lone salesman working out of his house for the small company and calling on a friendly territory will run out in this decade.

The existence of this larger educational industry will gradually force a change in bidding and purchasing procedures, and, hence, will influence the formation of the logistics of instruction. Such an industry will also permit more communication between the training segment (military, industrial, etc.) and the pure educational segment (schools, colleges, etc.) because materials and equipment will be supplied to both. If the pattern continues in the direction it seems to be going at the present time, these industrial giants will, through contracts, also be operating educational enterprises. The current operation of some of the Job Corps training centers by industrial groups and nonprofit corporations will be followed by the development of a general contracting capability under which such companies will provide instructional materials and services, build buildings, process data, catalog books, and even hire or provide teachers and administrators for schools and colleges. Some of them may well develop their own school systems.

References

Anderson, Charnel. *History of Instructional Technology, I: Technology in American Education, 1650-1900,* Occasional Paper No. 1, Washington, D.C.: NEA, Technological Development Project, 1961.

"The Assessment Debate at the White House Conference," *Phi Delta Kappan,* Vol. 47, No. 1 (September 1965), pp. 17-18.

Ausubel, David P. *The Psychology of Meaningful Verbal Learning: An Introduction to School Learning.* New York: Grune & Stratton, 1963.

Biological Sciences Curriculum Study. *High School Biology, Special Materials: Teacher's Manual (Revised Edition).* Boulder, Colo: Biological Sciences Curriculum Study at the University of Colorado, 1964.

Bruner, Jerome S. *The Process of Education.* Cambridge, Mass.: Harvard University Press, 1965.

Bruner, Jerome. *on Knowing: essays for the left hand.* Cambridge, Mass.: Harvard University Press, the Belknap Press, 1962.

Campion, Lee E. and Clarice Y. Kelley. "Studies in the Growth of Instructional Technology, II: A Directory of Closed-Circuit Television Installations in American Education with a Pattern of Growth," Occasional Paper No. 10. Washington, D.C.: Department of Audiovisual Instruction, NEA, 1963.

Case, H. W. and A. Roe. "Basic Properties of an Automated Teaching System," abstract in *AV Communication Review,* Vol. 13, No. 4 (Winter 1965), p. 453.

Ellul, Jacques. *The Technological Society.* Translated by John Wilkinson. New York: Alfred A. Knopf, 1964.

Finn, James D. "Automation and Education: 3. Technology and the Instructional Process," *AV Communication Review,* Vol. 8, No. 1, Winter 1960, pp. 5-26.

Finn, James D. "Instructional Technology," *The Bulletin of the National Association of Secondary School Principals,* Vol. 47, No. 283 (May

1963), pp. 99-103.

Finn, James D. and Donald G. Perrin. *Teaching Machines and Programed Learning, 1962: A Survey of the Industry.* Occasional Paper No. 3. Washington, D.C.: NEA, Technological Development Project, 1962.

Finn, James D., Donald G. Perrin, and Lee E. Campion. "Studies in the Growth of Instructional Technology, I: Audiovisual Instrumentation for Instruction in the Public Schools, 1930-60—A Basis for Take-Off," Occasional Paper No. 6. Washington, D.C.: Department of Audiovisual Instruction, NEA, 1962.

Galbraith, John Kenneth. *The Affluent Society.* Boston: Houghton Mifflin Company, 1958.

Glaser, Robert, William W. Ramage, and Joseph I. Lipson. *The Interface Between Student and Subject Matter.* Pittsburgh: Learning Research and Development Center, University of Pittsburgh, 1964.

Guilford, J. P. and P. R. Merrifield. "The Structure of Intellect Model: Its Use and Implications," Los Angeles: University of Southern California Psychological Laboratory (Report #24), 1960.

Hand, Harold C. "National Assessment Viewed as the Camel's Nose," *Phi Delta Kappan,* Vol. 47, No. 1 (September 1965), pp. 8-13.

Harcourt, Brace and Company; Harper & Brothers; Henry Holt and Company; Houghton Mifflin Company; The Macmillan Company; Scholastic Magazines; Scott, Foresman and Company. *A Report to Educators on Teaching Films Survey.* 1948.

Hebb, D. O. *The Organization of Behavior. A Neuropsychological Theory.* New York: John Wiley & Sons, Inc. 1949.

Hilgard, Ernest R. *Theories of Learning.* (Second Edition). New York: Appleton-Century-Crofts, 1956. Third Edition, 1966.

Hoffmann, Banesh. *The Tyranny of Testing.* Riverside, N.J.: The Crowell-Collier Press, 1962.

James, H. Thomas. "Problems in Administration and Finance When National Goals Become Primary," *Phi Delta Kappan,* Vol. 47, No. 4, December 1965, pp. 184-187.

Kramer, Samuel Noah. *History Begins at Sumer.* Garden City, N.Y.: Doubleday & Co., Inc. 1959.

Leverenz, Humboldt W. and Malcolm G. Townsley. "The Design of Instructional Equipment: Two Views," Occasional Paper No. 8. Washington, D.C.: Department of Audiovisual Instruction, NEA, 1962.

Licklider, J. C. R. "Preliminary Experiments in Computer-Aided Teaching," in John E. Coulson (Ed.), *Programed Learning and Computer-Based Instruction: Proceedings of the Conference on Application of Digital Computers to Automated Instruction.* New York: John Wiley and Sons, Inc., 1962, pp. 217-239.

Lumsdaine, A. A. "Educational Technology, Programed Learning, and Instructional Science," in *Theories of Learning and Instruction,* The Sixty-third Yearbook of the National Society for the Study of Education, Part I. Chicago : The University of Chicago Press, 1964, pp. 371-401.

Miles, John R. and Charles R. Spain. *Audio-Visual Aids in the Armed Services: Implications for American Education.* Washington, D.C.: American Council on Education, 1947.

Miller, George A., Eugene Galanter, and Karl H. Pribram. *Plans and the Structure of Behavior.* New York: Henry Holt and Company, 1960.

Price, Derek J. de Solla. *Little Science, Big Science.* New York: Columbia University Press, 1963.

Redding, M. Frank (with additional material by Roger H. Smith). *Revolution in the Textbook Publishing Industry.* Occasional Paper No. 9. Washington, D.C.: NEA, Technological Development Project, 1963.

A Review of Current Programs. Watertown, Mass.: Educational Services, Inc., 1965.

Schrag, Philip G. and James G. Holland. "Programing Motion Pictures: The Conversion of a PSSC Film into a Program," *AV Communication Review,* Vol. 13, No. 4 (Winter 1965), pp. 418-422.

Silvern, Leonard. *Systems Engineering in the Educational Environment.* Hawthorne, Calif.: Northrop Corp., 1963.

Skinner, B. F. "Teaching Machines," *Science,* Vol. 128, No. 3330 (Oct. 24, 1958), pp. 969-977.

Skinner, B. F. "Why Teachers Fail," *Saturday Review,* Oct. 16, 1965, p. 80. (a)

Skinner, B. F. "Reflections on a Decade of Teaching Machines," in *Teaching Machines and Programed Learning, II: Data and Directions.* Washington, D.C.: Department of Audiovisual Instruction, NEA, 1965, pp. 5-20. (b)

Smith, Karl U. and Margaret Foltz Smith. *Cybernetic Principles of Learning and Educational Design.* New York: Holt, Rinehart and Winston, 1966.

Tyler, Ralph W. "Assessing the Progress of Education," *Phi Delta Kappan,* Vol. 47, No. 1 (September 1965), pp. 13-16.

Dialog in Search of Relevance

I would like to share with you today some thoughts or muses on the topic "Dialog in Search of Relevance—Balance and Emphasis." Dialog is very much an "in" word in educational circles. Dialog is really "old hat" to those of us who are in the media field. We have always had dialog with representatives from industry.

We have moved from the small businessman to the giant corporation—the "combines" which some of you are so worried about. So far nothing much has happened. There have been more failures than successes (among companies who are entering the educational market for the first time). Many companies are getting in and out quickly, some of them several times.

It has been the tendency of industry to collect behavioral scientists of the Skinnerian type and concentrate industry's efforts along this line. I think educators, in general, are puzzled by the move from the camera store to IBM. This change will be very remarkable when it occurs and my thesis is that we need some direction in where to go.

The key words to me are not the history and significance of the educational dialog, but rather its nature and direction. Direction is what industry needs and direction is what education should give. And I believe that the spokesmen for the entire field of education should be from our own field of educational technology.

The direction must be relevant for our time. To answer the question of what is relevant, I spoke by telephone with Dr. Arthur Pearl, School of Education, The University of Oregon, a man who I believe is destined to be the number one educational philosopher in the United States. Here are some of Dr. Pearl's comments which I recorded during our telephone conversation:

Education to be relevant has to meet four criteria which are crucial for living in our modern, technologically advanced and complicated world.

1. First, education must be relevant to the world of work. Everyone going to school should be getting a wide range of vocational choices in how to earn a living. Students should not be prepared

From a speech given at the Lake Okoboji Leadership Conference, 1967. This is a summation excerpted by Norman Felsenthal, the conference recorder, from the tape recording of Finn's address.

for *a* job but for a choice of jobs.

2. Education must be relevant to the political structure. A complicated society requires more understanding in legislative, executive and judicial decision making. This means student government has to be real to serve a real function.

3. A relevant education requires much more intellectual inquiry; people have to be turned on intellectually. It's not enough for people to do what a computer does better—code, store and retrieve facts. These people must be able to fret and to conceptualize; they have to be able to analyze; they have to be turned on to literature, art, music, science, and math. They have to become cultural carriers and they have to appreciate the various different cultures that a complicated society has.

4. Interpersonal relations must be a part of a relevant education. People must acquire the competence to live with their neighbors in a more crowded society.

Unless education deals with the central issues of our time, the program is worthless. Education that is noncontroversial is meaningless. A scholar must do more than acquire a vocabulary; he must be involved in action.

There is one point in education that must be nonnegotiable. That point involves the Bill of Rights, the rights of an individual in a free society. Many schools work for behavior modification—to get children to act alike and to look and dress alike. School administrators are very anti-Bill of Rights in this sense. Their system is very central—shape up or ship out. Media people who work with administrators must become passionate advocates of democracy. With complacency, a person becomes the enemy of democracy.

This relevance which Arthur Pearl talks about can be achieved by changing our emphases. We have tended to emphasize order and system. I believe we must now move in the direction of freedom—freedom for everyone in the system.

John Kenneth Galbraith defines technology as the systematic application of scientific or other organized knowledge to practical tasks. The main characteristic of technology is the breaking down of tasks into detailed subdivisions so that organized knowledge may be put to work.

"Planning involves inevitably the control of human behavior. The denial that we do any planning has helped to conceal the fact that control exists, even from those who are controlled," says Galbraith.

Galbraith is speaking of economic planning, but I believe his comments apply across the board to our technological culture and to any large-scale application of instructional technology.

There is a deeper dialog involving deeper philosophical statements which we all need to hear. Unless we listen to what some of our bright young people are saying, to what the New Left is trying to expound, to what some artists are expressing, we, as educators, may fail this country and all the young people in it. This deeper dialog is between the industrial state—which exercises impersonal control over people, no matter to what degree they are right or wrong—and the spokesmen for men as human beings—for man in microcosm. An educator, it seems to me, does not have to adopt totally the view of one side or the other. This is what I mean by relevance and balance.

Two observations can be made about the education young people receive in our colleges today.

First, students end up hating books. They acquire this hatred systematically. You can't learn to enjoy reading if you have to pull something out of every book you pick up—if you have to produce a paper or a seminar report.

This explains the attraction of university undergraduates to Marshall McLuhan. McLuhan has a philosophy that is basically anti-book and it's very attractive to people who come in contact with books the way university students do. The reason there are underground books at a university is not that these books have more to say, but that students can read them without compulsion. You destroy a book by turning it into an assignment; you can destroy anything that way.

Second, university students are removed from men with ideas. Universities are inhospitable to a learned man—a man with a vision of unified knowledge. It is impossible for him to be in a university because he isn't a good enough specialist. The university student comes in contact with intellectual technicians who are uninspiring and dull. No student can have any respect for them.

A student cannot know what the life of the mind is like because he hates books and because he never comes in contact with men who have lived the life of the mind.

What are my suggestions to you about the direction of the education-industry dialog, and I emphasize that we, the educators, are the ones that need to give this dialog direction? I would like to suggest some margins because I believe the marginal media man is always between this and that.

As a past supporter of behaviorism and shaping, I do not believe we should throw out the baby with the bath; but somehow we have got to get over on the human free side as well. We are sort of standing with one foot in both camps.

There are several things we need to do and I want to conclude my speech with five suggestions:

1. We must alter our theoretical framework which is now moving in the direction of behavioral shaping at too rapid a rate. We must slow down this trend, though I don't mean to imply that we should wipe it out.

2. We have to consider the intellectual disciplines and begin a dialog with the people in these disciplines in order to help them make their teaching at all levels more relevant. I do not believe it is the content that ever was irrelevant—some of the things in Greek history are relevant to our current period—but the disciplinarians do not understand this. And I believe you can get them to understand, if you can get them to give more attention to the human factors involved.

3. I believe we should insist that the products and efforts of industry

concentrate on the human being. In the next few years we need more to follow the lead of Carl Rogers and Abraham Maslow than we do B.F. Skinner and other behaviorists. With media and a different instructional design, we can move into the affective domain and be concerned with human beings.

4. We must also realize the need to effectively deal with power groups. I predict that we are going to have more conflict with teacher militancy; negotiations are going to hinge eventually on the application of educational technology—whether money is to be spent on a language laboratory or to raise teachers' salaries. We have to deal with other minority and power groups as well. These groups include women and students, two of the most down-trodden minorities of our times. We must fight against any form of censorship whatsoever. I would make this absolute. If you sit once again like you did in the early 1950s, we will go down the tubes, I think, for sure. We must move to general freedom, to openness, and be problem oriented in what we do and how we deal with people. We have got to encourage creativity, and this requires a different kind of medium from one designed for behavioral shaping.

5. Finally, I think we have to take a general ethical position. The question is very simple: Whose side are you going to be on? I don't mean to imply that industry is bad. And I claim no conspiracy theory. But isn't it time to give industry some direction? If no direction is given, it is inevitable that technology will take off on its own. We will have lost a battle and a war; a war that can be won very easily if you leave Okoboji with the determination to be on the side of human beings and the Bill of Rights no matter what.

Institutionalization of Evaluation

We live in an age of analysis. We also tend to synthesize and system-atize everything we analyze in order to solve problems. These processes of analysis, synthesis and systemization are some of the power tools of our high-order scientific-technological society, aided, of course, by such things as computers and punched cards which supply data about our bill-paying habits, our blood types, and our penchants for blondes, brunettes or red-heads.

At the moment, the concern of the study group to which this paper is addressed is evaluation; not mere or abstract evaluation, but evaluation of the successes, failures, feasibilities and non- (or un-) feasibilities of various Title III innovative projects set up in school systems throughout the United States. Funds for this effort are supplied by the United States Congress and the projects are administered by the U. S. Office of Education, moni-tored by the several state departments of education, and worried about by local school administrators.

This interest in evaluation, in my opinion, must be seen in a technical-social-political context within the entire educational enterprise. The age of analysis in which we live is generating an age of assessment in education. Thus, we have a campaign developing for a national assessment program (how well are the schools doing?); several states are also asking and an-swering the same question within their borders.[1] Other considerations (and pressures) aside, it would be no surprise, therefore, to see this evaluation *zeitgeist* penetrating the Title III program. How well are all of these Title III projects doing?

Further, as technological patterns of thinking and processing invade the previously primitive (from a technological point of view) educational cul-ture, it is inevitable that a drive for systemization should begin. For exam-ple, Hammond opens his paper on evaluation with the statement, "The need for a systematic approach to the evaluation of innovations has be-come one of education's most pressing problems" (Hammond, n.d.). Anal-ysis of a sophisticated variety must precede systemization, and such anal-

---

[1] There is the little matter of the very embarrassing performance of Los Angeles school children on reading tests, for instance.

Reprinted from *Evaluation and "PACE": A Study of Procedures and Effective-ness of Evaluation Sections in Approved PACE Projects, with Recommendations for Improvement,* February 29, 1968, pp. A-170-A-202.

yses can be found not only in Hammond but in Clark and Guba (n.d.); Guba (1967); and Shufflebeam (1967) to refer to some very recent examples. It should be noted that all of the work cited is exceptionally rigorous and highly sophisticated. It provides an excellent base from which to attack certain practical evaluation problems, not only for Title III projects but for any instructional process; this work is in the high technical tradition and, as such, is relatively new to professional education.[2]

Further, another sign of the analysis-synthesis-system approach to evaluation in education is the continual invention, development and refinement of instruments for use in evaluation processes. Test-makers are everywhere, inventing measuring devices ranging from pencil and paper tests to simulators.[3] Guba suggests many new measuring and feedback instruments are needed (Guba, 1967).

The general objective, then, seems to be in the direction of systemization of evaluation procedures. Sharp analyses, increasing and better instrumentation, process studies are all leading to Dr. Hammond's "systematic approach" to evaluation in general, and, if this study committee is any indication, to a systematic approach to evaluation for Title III projects. Such a movement to system should lead to a great deal of improvement.

I would like to point out, however, that this movement toward system in evaluation also may *not* lead to improvement. For, in order for an evaluation system to be applied across the country, it is necessary first to institutionalize it; this is to say that, unless other means are invented, the evaluation system must be initiated, monitored and controlled by a bureaucratic system. Institutionalization of the evaluation process could destroy the innovative possibilities of Title III.

There is nothing inevitable in this potential destruction of the innovative process by the technical organization[4] for evaluation. However, if the evaluation processes as institutionalized are not to be made into a missile system aimed at the heart of educational innovation, additional analysis and invention is absolutely necessary. The remainder of this paper will examine this problem and, in addition, report some observations on the evaluation provisions of a number of Title III proposals which were studied in some detail; hopefully, the problem examination and the proposal exam-

---

[2]It may be a sign of age, but the writer can remember when the word "evaluation" was used as an excuse in parts of the educational community to avoid rigorous research; evaluation meant that anything went—and, in many places today, it still does. The new tradition will obviously change things for the better on this point.

[3]This phenomena can be seen among technologically oriented graduate students. One of my students completed a study on the evaluation of visual material by photographing and then measuring the eye pupil size of the evaluator and comparing it with his stated evaluation. Another is going to measure pulse pressure in much the same way. In both cases, the instrumentation had to be developed. Both of these projects originated with the students themselves. It is no accident that Egon Guba did some very complex studies of television using an eye-movement television-film setup a few years ago.

[4]I should make explicit that I believe the evaluation process discussed above is a subtechnology within the broader concept of instructional technology.

ination can be tied together to develop some recommendations to close the paper.

Why Evaluation Anyway?

The basic question that needs to be asked to begin this analysis is: Why is evaluation important in the educational enterprise? There are at least five purposes or reasons that can be presented in answer to this question:

1. To add to the substantive knowledge of educational processes.
2. To provide information in order to adjust, discard or otherwise change the application of an on-going educational process.
3. To provide justification for a political-social-economic action relating to education.
4. To create a product (usually paper) which can move through educational bureaucratic systems and thus keep these systems operative.
5. To provide instruments which may be used to carry information on the success of the process to the educational community.

These five purposes do not necessarily operate in a discrete fashion: in other words, in any one situation several may appear in the form of a mix. It is fairly easy, however, to identify the emphasis in each case. The five purposes will be briefly discussed in the paragraphs that follow.

The distinction between the first and second has been noted by many of the recent analysts, such as Shufflebeam and Guba. The first—measurements conducted under carefully controlled conditions—theoretically provides material for the *corpus* of educational research; as such, the results should add to the substantive understanding of educational processes. And, as has been pointed out many times, such results are rarely directly applicable to the problems of the practitioner and are of little use to decision makers. It is possible, however, for such product-oriented research to come into existence as a fallout or byproduct of a much more comprehensive evaluation procedure. The use of such research techniques as the *only* means of approaching evaluation has been amply criticized in recent years.

The second purpose for evaluation is now thought to be the most important when examining ongoing innovative projects in education, such as those set up under Title III. Here the decision maker gets information on a feedback system which tells him how well the process is going, what changes need to be made, etc. Various models of this evaluation procedure have been proposed (Clark and Guba, n.d.; Guba, 1967; Shufflebeam, 1967).

This feedback evaluation system (if it may be called that), designed to aid decision makers dealing with practical educational problems, such as the operation of a Title III innovative center, has not yet been criticized to any extent due to its novelty for the field of education and the careful construction of the emerging theoretical models.

However, the feedback evaluation system is open to criticism. It assumes, at the outset, that the decision-making process in a given school, school system, or other educational entity is rational. It is not. The folklore of education is filled with examples of the school business manager selecting curricular materials, the high school dean of women throwing Salinger out of the library, and others too numerous to mention. Prior questions have to be asked whenever the feedback evaluation system is proposed. Who or what group is the decision maker? How does the power structure really work? What are the motivations? Unless these questions are answered and the rationality or irrationality of the particular system is analyzed, the beautiful, precise and rational models of the feedback evaluation system will not work—or, at the very least, work very imperfectly.

The third purpose for evaluation is the purpose of justification. In this case, a board of education, a state legislature, a committee of Congress, or numerous other bodies both public and private need information in order to take some action respecting education. This action may be in appropriating funds, hiring additional remedial reading teachers, purchasing a language laboratory, etc. Or, in the opposite case, it may be to fire the superintendent, set the building program back two years or reduce the audiovisual appropriation by 40 percent. These actions are *justified* by evaluation, whether formal or informal. In the case of the disposal of school personnel, the evaluation before action may be choleric and personal; increasingly, however, as statistics become everyday playthings of the mass media, justifications for political-economic-social educational action are couched in scientific garb, whether really scientific or not. We are all familiar with arguments which press in opposite directions for action by some public body based on the same evaluative report.

While it is obvious that public bodies with the appropriate authority over education have every right to evaluative information and, in fact, often need more than they get, it is equally true that the development of justification ought to be a secondary objective of educational evaluations carried out by professionals. This, of course, is a value judgment. It can be argued, for example, that the effort expended in developing a particular kind of justification evaluation in order to save a program known to be good is more important in the real world of politics that more technically adequate professional evaluation. The answer to this problem, it seems to me, is to pay attention to the two elements in the old cliché about the tail and the dog. An evaluation program set up only for justification purposes is unprofessional; a professional evaluation procedure on a program that is demonstrably good ought to develop sufficient data to justify its continuance or expansion.

The fourth purpose for evaluation recognizes the reality of the new industrial state—the corporate society. Such a state produces hierarchical bureaucracies (this phrase is, I suppose, redundant) in industry, government, labor unions, and volunteer organizations, as well as universities. Evaluative reports are, of course, necessary for the proper functioning of the enterprises which are the concern of these bureaucracies, particularly for the use of the technostructure, as Galbraith has called the decision-making groups in large industries.

The reader is reminded, however, that bureaucracies lead a life of their own that is somehow magically related to the flow of paper in and out of little wooden or wire baskets and conferences in conference rooms concerning the leapfrogging of this paper among the baskets. Paper, then, must be generated so that the system may lead its organic, inward life. Evaluation studies may be a large part of this pulsing circulatory system—the corpuscles, so to speak. It is emphasized that the relation of this particular form of corpuscular paper with the real, operational world may be nil or almost nil. In many cases that is not its purpose of existence.

It then follows that a careful distinction must be made between required evaluation which is necessary and has an effect on operations and decisions and that which only serves the life function of the bureaucracy itself. The first needs improvement; the second needs to disappear.

Finally, the evaluation process is undertaken to provide data on new developments in order that these data may be diffused throughout the educational community so that schools in distant places may understand and take advantage of the findings. This idea is a little tricky, as it could be held that the evaluation comes first and diffusion follows as a matter of course. In many cases this is, in fact, what happens. However, there are other cases in which the distinctions between evaluation process and diffusion instrument are not so clear. The generators of a good idea want to sell it. The evaluation can be the package. Obviously, such a package may not be the same as an evaluation package for researchers, decision makers, or bureaucrats.

These broad purposes for evaluation do not coincide very well, I fear, with the meticulously drawn detail in the charts of the experts. It may be that their only value is in delineating the perceptions of the responsible administrator on the firing line. Thus, for example, if a request for evaluation is seen as necessary for the functioning of bureaucratic life, it will be developed with that purpose in mind—a useful paper corpuscle designed for the bureaucratic arteries, not for real-time operations.

It may be more important, however, to examine the drive toward systematic evaluation in the Title III program from the point of view of these five broad categories of evaluation purposes. Evaluation of Title III for what? For diffusion, for checking and adjusting ongoing processes, for substantive knowledge, for justification (perhaps of the entire Title III program itself), or for improving the circulation of a bureaucracy? I believe that the mix of these purposes must be carefully measured before an intelligent judgment can be made concerning any agreed-upon evaluation procedure.

The Problem of Institutionalization

Leaving the question of purpose open for the moment, we can turn to what I believe is the heart of the matter, namely, the question of institutionalizing the entire Title III evaluation process.

If this is so, then the arguments introduced in the introductory material could stand further examination. It was stated that systematic evaluation was considered desirable; that this was part of the general drive toward analysis, synthesis and systemization within the educational culture; and that such systemization had its bad aspects as well as its good aspects.

Although I have nowhere seen the concept verbalized in a precise manner, it seems clear that we are being asked to provide guidelines for the institutionalization of evaluation for Title III projects. The reasoning seems to go something like this: (a) Present evaluation procedures are not good; they are spotty, at times sloppy and unscientific; many times they imitate the researcher's controlled experiment when they should be providing the decision maker with feedback information to correct the system, and a credibility gap exists on the diffusion front; (b) "hard data" must be developed for public bodies at all levels concerned with the Title III program; (c) the rather bad evaluation procedures now in use are neither generating hard data nor helping in decision making due to lack of knowledge, guidelines and skill among the operators of Title III projects; (d) by a thorough tightening up on the evaluation guidelines to be developed by experts, and by institutionalizing these guidelines with a system of information and controls, the evaluation procedure will be helped, public bodies will be made happy by the presence of hard data, and better decisions will be made in directing ongoing projects. This is, indeed, an enticing picture, and to raise questions about its fundamental premises seems to be akin to questioning the institution of motherhood, Sigmund Freud to the contrary.

However, I would like to question the entire concept of institutionalization of the Title III evaluation process and insert into the record a few arguments that might at least suggest institutionalization in a different form. It is granted at the outset that, with over 1,200 Title III projects on which a considerable sum of money is being spent and with the great need to develop viable educational innovations which can be adopted by the educational community, improvement in evaluation procedures is a necessity. Further, operations on the scale of Title III require quality controls which are only made possible by large-scale systematic evaluation procedures.

Granting all this, questions may still be raised and arguments considered. First, while the analyses of the experts—Guba, Shufflebeam, Stake, et al.— of the evaluation process are impressive and potentially fruitful, is it possible that they have, in fact, overanalyzed the process and, in doing so, slipped into the same trap that the conventional educational research man does when he attempts to apply controlled research techniques to evaluation processes operating under field conditions? Are these analyses, rather, important additions to our substantive knowledge and should they instead be used to generate more study of the process so that field applications would eventually develop? Have, in fact, these analyses departed from operational reality, at least in the sense that the practitioner would not know what to do with them? And, if one or more of these models were frozen into enforced guidelines, would this not result only in bureaucratic paper? I am not sure of the answers to these questions, but I feel that these possibilities deserve more consideration than they have been getting.

I have no question, however, on another point. The proposed models simply do not embrace all Title III projects. There is a tendency to forget that a portion of the Title III effort is designed to provide supplementary educational services to various geographic areas, and proposals have been submitted and projects funded for such service centers. Further, entire new educational program efforts ("A Six County Program in the Performing Arts") do not lend themselves too well at first to measures that are meaningful and always present difficult problems for evaluative information systems.

The service centers present the real challenge, however. There has been, I believe, a tendency on the part of U.S.O.E. to play down, if not ignore, the meaning and importance of the service center concept to units of the educational community. This is due, no doubt, to the decision which placed the emphasis on the innovative aspects of Title III rather than on the supplementary service center idea. It is still a fact, as noted above, that some of these centers have been approved and funded, and if there is to be systematic evaluation for Title III, these projects must be included.

In this connection, there are two problems. The first problem is somewhat technical. The difficulty lies in the fact that the sophisticated evaluation models do not exactly fit the problem of evaluating a service center—for example, a media center supplying media services to a group of school districts. These models, for all their claim to generality, tend to concentrate on innovation in the instructional process—curriculum, methodology, the mediation of instruction. It seems pretty obvious that when you set up a service center of some kind, the distance between the regional center and the student is highly attenuated from the point of view of evaluation—both as to time and distance. To expect the evaluation process as abstracted in these models to cut through from a regional center to a student in the fifth grade of George Washington School in one of six school districts, define the effect that the sudden acquisition of a film library had upon him in a year's time, and adjust the content or service of the library accordingly is also to expect that the films will be delivered to the school via flying saucers piloted by little green men.

I should hasten to add that the *principles* inherent in the models can be, in many cases, applied to the evaluation of service center operation. The problem is, that if the institutionalization of the evaluation process continues to proceed and harden along the lines it is apparently proceeding, harrassed administrators will be asked to evaluate a service center operation by standards that ought to be applied to the evaluation of a new approach to phonics in the teaching of reading. This simply would not make sense.

In addition, the media world is not without a certain sophistication in the evaluation of service center operations—and these procedures relate to the principles enunciated by Guba, as one model maker, but not to the tactics that seem to be implied. To cite a homely example, if film keeps coming back into the library from a given district all chewed up, the center director then has a practical measure readily at hand which requires further investigation immediately. He must find the answers to such questions as: How are the films projected? By students, teachers or both? If teachers, are they doing it properly or do they need some training? If the human factor is not the problem what about maintenance of the equipment? If the provisions for maintenance and control are all right, what about the performance of the Acme Repair Company on the projector service contract? etc. etc. Once these questions are answered, then changes can be made. This procedure is in line with the principles of Guba, Shufflebeam, et al., but not the suggested tactics that seem to flow from them. Problems such as this one will only become difficult if evaluation is institutionalized. Under such hardening of the categories[5] the evaluative universe is interpreted to be the instruc-

---

[5] A phrase picked up from Edgar Dale many years ago.

tional process and the responsible administrator is forced to proceed accordingly.

The second problem defies the models. An examination of proposals for Title III centers (mainly media service centers), both funded and unfunded, shows immediately that the funds are very badly needed to supply materials, equipment and services that are sadly lacking in the districts to be served. NDEA funds, articles about media, and fears about commercial domination to the contrary, the plain fact is that many, many schools in this country do not have enough of anything to do the job required of them. Under Title III they get *some* money for equipment, materials and services. It is like giving a drink of water to a man who has spent three days on the desert without it. How are you going to evaluate that? By the test of survival? What is survival in the educational setting? What these few centers funded by Title III mean is that all of the schools involved are experiencing an increase in their technological base. I submit it is only after this base has been functioning to the point where it requires *additional* technology does evaluation become meaningful. In the beginning, anything is better than nothing. There are, of course, still evaluation questions generally pointed toward improving operations and attaining efficiency. Other models, however, are needed for this.

In a sense, the broad educational programs mentioned above fall into the same category as the service center. In some of the proposals I examined, for example, broad programs in the performing and plastic arts and the humanities were proposed for regions which had absolutely nothing of this kind but the prints sold at Woolworth's, the local piano teacher, and the county pioneer pageant at the fair each fall. In one rather large area, for example, there was no school except a religious high school where a student could get instruction in the playing of any stringed instrument. Again, the desert-water analogy holds. Some things are obvious. Music, art or the theater brought into a community make things better, period. It seems, in a way, ridiculous to measure or count such efforts; members of Congress should be happy with the invasion of the arts as a happening; experimenters or journalists might have to wait a few years for experimentation or diffusion. It is granted that the best possible operation is needed, but a narrow institutionalization of the evaluation process for Title III projects will not provide that better operation.

If the evaluation theory we are apparently following does not exactly fit the service center and large program projects, the incongruence must show somewhere. It does in reference to objectives, an important point to notice when thinking about evaluation. Almost every reference to objectives (the achievement of which are to be measured) refers to *behavioral* objectives or some variant thereof. In addition, performance tests, criterion tests, etc. are easily picked-up bywords in discourse on evaluation. It is as if the jargon of programed instruction has suddenly become the *lingua franca* of all educational evaluation—or, for that matter, of all education.

Now, some of my best friends are behavioral objectives, but I would not want my media service center to marry one. Seriously, objectives are one thing and behavioral objectives are another. Behavioral objectives are a microcosm, to be entered into when students are directly related to content, processes, media, or people in the classroom. To apply them to large programs embracing all of the arts throughout a wide region, a library service center for a county school system, or a data processing installation represents a beautiful confusion of form with substance—setting up the conditions of operation for an educational Parkinson's Law. Even smaller sectors of the educational enterprise directly related to instruction may not need objectives stated in behavioral terms.

I wish to make it very clear that I am not attacking behavioral objectives as such. They can be made to accomplish spectacular things with certain instructional processes and are legitimate targets for evaluation. On the other hand, sometimes we will need *system* objectives, which are not the same thing at all. By stretching a point, it might be said that an evaluator might want to measure (and change) *the behavior of a system* (such as a library, a full-scale curriculum operation, or something else), but I do not believe that this type of objective was exactly what B. F. Skinner had in mind (or Ralph Tyler many years earlier).

There is, of course, nothing in the models with which we have been dealing that requires behavioral objectives; and it is also a truism that evaluators can't evaluate for any purpose without objectives. The fact remains that all the discourse about evaluation is conducted *as if* there were no other types of objectives in the educational universe, even when the discussants unconsciously know better. Thus the incongruence between the theory (or theories) of evaluation under analysis and the real-time world of operation can be shown to be a possibility.

This exploratory discussion relating to some specific problems of institutionalizing Title III evaluative processes can now be brought into focus at the philosophical level. To review, there is apparently great concern as to the quality of existing evaluation, there is a desire to produce "hard data" for persuasive purposes, there is a need for accurate information as to progress and to adjust for improvement, and there is the necessity of diffusing information on successful practices—success being determined by competent evaluation. Further, the size of the Title III effort (over 1,200 units) and its wide distribution geographically with enormous differences in the resources and abilities of the educational units involved, all press for standardization (at an acceptable level of competence) of evaluation procedures.

There is, however, a deeper drive involved in this effort—or, at least, I believe it to be so. The industrial state is the corporate, bureaucratic state. The imperatives of technology, we are reminded by many observers such as Galbraith, have replaced ideology in much of our culture. Technology requires large-scale organization, orderly processes, group planning, and, where possible, it seems to me, a kind of neatness in the system that one might associate with a computer installation or a "clean room" in an electronics factory.

In a fundamental sense, the entire Title III effort at educational innovation is a move toward bringing the educational enterprise into the modern industrial state. Title III *is* educational technology. This may sound strange to those educators who define educational technology as a term synonymous with language laboratories, computers, or television. I would remind them of Galbraith's definition of technology, although many similar definitions might be cited. Galbraith said, "Technology means the systematic application of scientific or other organized knowledge to practical tasks" (Galbraith, 1967, p. 12). He goes on to point out that the main characteristic of technology is the breaking down of tasks into detailed subdivisions so that organized knowledge may be put to work, and that this analytical procedure "is not confined to, nor has it any special relevance to mechanical processes" (Galbraith, 1967, p. 13). I submit that the very selection of evaluation as a field to analyze in connection with the Title III program is evidence of this movement toward technology.

In any of the units of a society of high technology, such as the United States, very extensive planning is necessary. Since the units are very large, for the most part (Galbraith's 500 "mature corporations," large government, etc.), the planning affects and controls millions. *A systemization of the Title III evaluation process is a form of planning.* This concept leads us to the crux of the argument. Galbraith has noted, for example, that " . . . planning involves, inevitably, the control of human behavior. The denial that we do any planning has helped to conceal the fact of such control even from those who are controlled" (Galbraith, 1967, p. 23). He was speaking of economic planning, but I believe this concept to be totally generalizable in our technological culture.

Planning involves the creation and management of systems; systems require, or at least imply, bureaucratic control. Hence, unless, as indicated in the earlier portion of this paper, additional means are invented to fit the peculiarities of the institution with which we are dealing—the American educational enterprise—the development of a systematic, technically competent evaluation process for Title III will result in bureaucratic control that I believe would mean the end of the dream that Title III would bring needed innovation to American education.

Such a prospect is difficult enough, but further complexities must be examined. Galbraith has pointed out that, with high technology and large organization, as in the mature industry, the planning and management processes are in the hands of fluid groups of experts, each bringing complex information into the group processes where decisions are made. The fluid groups he calls the technostructure. We come now to the rub. *American education, as a sector of the political economy, is very primitive from a technological point of view and has practically no technostructure.*

If the concept of a lack of a technostructure in American education is accepted, it is possible to explain many things.[6] Galbraith does this in another context when explaining why socialist countries have had "the most uniformly dismal experiment of countries seeking economic development" (Galbraith, 1967, p. 101). Speaking of India and Ceylon, he goes on to say that, in these countries,

   . . . if the minister is to be questioned, he must have knowledge. He

---

[6]Consider the inability of the old-line staff administrative patterns to handle aspects of the new educational technology (hardware and materials logistics, etc.); consider the problems of the ghetto from this point of view; etc.

cannot plead that he is uninformed without admitting to being a non-entity. . . . Technical personnel are less experienced than in the older countries. Organization is less mature. These lead to error, and suggest to parlimentarians and civil servants the need for careful review of decisions by higher and presumably more competent authority. Poverty . . . calls for further review. And rigid personnel and civil service rules, the established British answer to primitive administrative capacity, extend into the public firm and prevent the easy constitution and reconstitution of groups with information relevant to changing problems. (Galbraith, 1967, pp. 101-102)

It seems to me that it is easy enough to transfer this Galbraithian concept to the American educational system. To begin with, there is no large-scale organization in the technological sense and, as noted, no techno-structure. If development is to occur, it becomes obvious to those responsible—in our case, U.S.O.E. and our study committee—that review and control of decisions and operations relating to evaluation are absolutely necessary when dealing with such "primitive administrative capacity," and it *is* primitive from this point of view.

At this point, however, it is necessary to exorcize a ghost. I am suggesting that national bureaucratic control or even systemization of evaluation seems to be necessary under the circumstances, but I am further suggesting that this may be unwise (the reasons for this will be discussed below). It then might follow that all I am interested in is a reduction in the size of the bureaucracy and the removal of the controls to the state level. Nothing could be further from the truth. Bureaucratic control from the state level will merely extend the "primitive administrative capacity" from the school district upward. I venture to say that no state department in the United States has an adequate technostructure or is about to get one—and this includes New York and California, both of which have been praised in many quarters. I believe that institutionalizing Title III evaluation processes under state departments of education will concentrate many undesirable elements of such a system. My arguments against institutionalization in the form that seems to be implied by events must be seen in this light. Once the arguments are considered, it may be possible to suggest a better solution.

To return to the main theme, given the assumption of planning and systematic (translate bureaucratic) control of Title III evaluation procedures, certain undesirable effects seem to inevitably flow from many (but not all) such developments. High technical solutions to some problems require exactly such arrangements—getting to the moon, stamping out an epidemic, etc. In such situations large-scale technology and its peculiar requirements seem to fit fairly well and the people involved are relatively comfortable. However, all large-scale applications of technology (systematic organization) do not fit—particularly where they impinge in certain ways on human beings (recall Galbraith's sentence on control).

This lack of human fit of many of the technological developments in the United States—depersonalization of university life, smog and the automobile, social decisions made by the corporate structure over which those being decided about have no control—has given rise, in the last decade, to a heated dialog which has erupted on numerous occasions into violence.

It is a mistake, however—a serious mistake—to view the dialog only from the point of view of the violence or certain individual issues such as Vietnam, civil rights, or rent strikes. For anyone who cares to take the time to inquire, a much deeper dialog, a much deeper emerging philosophical statement are there to hear. "To hear" is used advisedly, for I believe unless we listen to what some of these bright young people are saying, to what the New Left is trying to expound, to what some artists are expressing, we, as educators, may fail this country and all of the young people in it.

What is this dialog? It is a dialog between high-order technological organization, the industrial state, impersonal controls over people and spokesmen, no matter to the degree that they are right or wrong, for men as human beings—for man in microcosm. An educator, it seems to me, does not necessarily have to adopt totally the view of one side or the other. Some educators, at least, ought to see the thousand dilemmas present in this confrontation and seek solutions which are, first, educative, and secondly, human without reducing our culture back to some primitive stage where we live in the hills in shacks. I believe we should seek in general what William Jovanovich saw in the future when he predicted "the emergence of a new kind of intellectualism which will reconcile content with style, social purpose with personal sensibility" (Jovanovich, 1967, p. 59).

Assuming that I am right in understanding that the effort to "improve" the evaluation processes associated with Title III will move in the direction of national systemization and control (or, worse, state systemization and control), the criticisms of these spokesmen for the defense of man as man have relevance. They should be seriously thought about, for within the intimate environment in which each man lives, they attack systemization and control with a vengeance.

Let us begin with one of the best known spokesmen for this point of view, Paul Goodman. Recently he was asked to address the National Security Industrial Association and took the opportunity to berate this industrial-military technology group. At one point he said:

> Your thinking is never to simplify and retrench, but always to devise new equipment to alleviate the mess that you have helped to make with your previous equipment. (Goodman, 1967, p. 16)

And, then he went on:

> Your systems analyses of social problems always tend toward standardization, centralization, and bureaucratic control, *although these are not necessary in the method.* [Italics mine] (Goodman, 1967, p. 16)

Finally, he stated a principle or theme that reappears time and time again in this literature:

> In a society that is cluttered, overcentralized, and overadministered, we should aim at simplification, decentralization, and decontrol. (Goodman, 1967, p. 17)

A great deal of this new literature is being created by young people. A whole issue of the *American Scholar* was recently devoted to writing by people under 30. In it, Michael Rossman made an effort to explain the deep philosophical base of the so-called National Student Movement. In doing this, he expressed much about their concern with man as individual man and even explained (and this is a little hard for an older person to understand) "participatory democracy" both as philosophy and as tactic. Three concepts appear in much of this literature, and they appear in Rossman. They are: Engagement, Encounter and Involvement. The concern is with humans relating to humans—with true *encounter*. Rossman puts it this way:

> ... the present Old Left among us ... aims at the mass; at the racial, economic or occupational population. But the unit in terms of which the Movement conceives change tends to be the small group.

> ... The way to influence large groups is by local example, rather than global persuasion.

> ... *direct personal involvement* is the Movement's human backbone.

> ... In saying that people must be involved in the decisions that shape their lives, the emphasis is on *involved*.

> ... political dialogue must be cast in a different vocabulary than that possible with the comfortable separation of the Changer and the Changed.[7] (Rossman, 1967, pp. 595-596)

Harper's recently published a symposium consisting of a series of dialogs between well-known older commentators on the national scene and a panel of young people similar to those appearing in the *American Scholar*. In this case, it was, at times, hard to distinguish between the older and younger viewpoints. Again, the theme reappeared time and again. The issue was impersonalism, bureaucratic control versus general encounter, and the human condition. It is tempting to go on quoting a great deal because the material seems so relevant, but I shall try to restrain myself. Paul Potter, one of the "older" members of the panel, said:

> ... there is a growing belief that the only force really shaping the future is the force of unleashed technology controlled by giant, impersonal bureaucracies. ... economic planners ... cluck truculently about the "great leveling force of technological development" that will in time assimilate all revolutions and all cultural diversities into one grand machine-civilization. (Potter, 1967, p. 49)

In commenting on Potter's article, young Robert Gross said:

> ... we have to end the domination of this society by the large, rigid bureaucracies which pay little attention to the needs of the people they are intended to serve. (Gross, 1967, p. 50)

And Alfred Kazin commented:

> ... the more immediate and abundant our technical power, the more we lose the naive, spontaneous imagination. (Kazin, 1967, p. 51)

Other parts of the text refer to "students who are demanding flexibility and personal relevance," "non-rational ways of getting at knowledge,"

---

[7]It should be noted that, for the professional educator, Rossman has come provocative things to say about teaching and learning in higher institutions and proposes some interesting reforms.

"taking strength from the free private life." There is also, however, another thread which suggests that something better might be made of this "technocratic totalitarianism," as Potter called it, and he went on to say:

The technology and the bureaucracy can be mastered and put to work to create for everyone what we've begun to have a taste of.... (Potter, 1967, p. 50)

A Suggested Accommodation

The argument has come full circle, and the potentials of accommodation are there if they can be identified. It seems to me important to suggest a new approach to the problem of institutionalizing evaluation for Title III projects which would accommodate need for and technique of consistent, high-quality evaluation procedures with human, local needs and differences in projects and concepts. Personally, I feel that many of the critics of technical bureaucratic control cited above *offer* little as replacement for this control with a sort of leaderless "participatory democracy" which, in a technical sense, will not even achieve their own objectives. And yet, much of what they have to say is important.

We have been concerned with models of high-quality evaluation procedures; with purposes of evaluation; with implied arrangements to insert controls in the system so that legitimate purposes may invariably be supported with technical competence. All of the elements of a bureaucratic system are there—whether in the eventual rough and tumble of administrative or legislative politics this control is placed at the federal or state level. Of course, such control could apparently be nonenforced by guidelines or some other system which in fact would quickly encrust into a straitjacket. On the other hand, controls are needed so that competent, useful evaluation may take place. This, to repeat, is the problem of accommodation.

And I hold that it is a solvable problem and that the possible solution, as Boyd Bode used to be fond of saying, "lies at hand." Many of the elements are present in Project EPIC of Tucson, Arizona (Hammond, n.d.). Project EPIC is, essentially, a sort of local evaluation service center funded by Title III funds and assisting local school agencies within the area it services.

Given this idea as a start, it is possible to make a series of recommendations that can, I believe, achieve the sought-after accommodation between the need for evaluation and the human variation which inevitably occurs at the end of the line. Such an accommodation will not be as neat as a "clean room" in an electronics factory; on the other hand, it will not be so messy as to be useless; in fact, it might have enough variation in it to release creative energy—which was the general idea of Title III in the first place.

Recommendations

1. Title III funds be used to set up a series of regional evaluation centers throughout the United States designed to provide training and assistance to local educational agencies.
2. The function of these centers be to provide advice, training and services and, particularly, to diffuse the general idea of the importance, usefulness and nature of a high-quality evaluation system.

3. It be understood that the evaluation centers are only persuasive and helpful in nature and that, if an educational agency chooses not to respond, it be allowed to without penalty—actual or implied.

4. These centers also engage in a certain amount of applied and field research with the purpose of developing viable and variable evaluation procedures which can embrace all types of evaluation needs and purposes.

5. A back-up national board be set up to assist the centers and the U.S. O.E. and Congress. This board would have the following functions:

a. Locate and rotate manpower between the centers. Much of this manpower could be one-year leave-of-absence type; other slots could be filled with qualified graduate students on an intern basis.

b. Act as the assembling agency for results which ought to be diffused and as the communication agency between the centers. As such it should act as both the stimulus and the conscience for the centers.

c. Engage in broad-scope research and development studies in the field of evaluation.

d. Provide an information source for all government agencies—local, state, federal.

e. Relate to and diffuse information to the educational community about other national, private evaluative efforts, such as the National Assessment Program, etc.

Under no circumstances should this board be thought of as a control mechanism in the bureaucratic sense.

If these recommendations are analyzed for the purpose for which they were made—to create a system which would achieve the objectives of necessary high-quality evaluation procedures for local, human purposes without inserting another bureaucracy into the system—details of operation and administration should become reasonably clear. The human being at the end of the line—administrator, teacher or media specialist—can have his opportunity for involvement and encounter. And it is highly likely that we can raise the quality of evaluation immensely.

References

Clark, David L. and Egon G. Guba. *An Examination of Potential Change Roles in Education.* Bloomington, Ind.: The National Institute for the Study of Educational Change (mimeo), July 1967. undated.

Galbraith, John Kenneth. *The New Industrial State.* Boston: Houghton Mifflin Company, 1967.

Goodman, Paul. "A Causerie at the Military-Industrial," *New York Review of Books,* Volume 9, No. 9 (November 23, 1967).

Gross, Robert A. "To Mr. Potter," *Harper's Magazine,* Vol. 235, No. 1409, October 1967, pp. 50-51.

Guba, Egon G. *The Basis for Educational Improvement.* Bloomington, Ind.: The National Institute for the Study of Educational Change (mimeo), July 1967.

Hammond, Robert L. *Evaluation at the Local Level.* Tucson, Ariz.; Project EPIC (mimeo), undated.

Jovanovich, William. "My Illusions and Yours," *Harper's Magazine,* Vol.

235, No. 1409, October 1967, pp. 56-60.

Kazin, Alfred. "Art on Trial," *Harper's Magazine*, Vol. 235, No. 1409, October 1967, pp. 51-55.

Potter, Paul. "The Future is Not Inevitable," *Harper's Magazine*, Vol. 235, No. 1409, October 1967, pp. 47-50.

Rossman, Michael. "The Movement and Educational Reform," *The American Scholar*, Vol. 36, No. 4, Autumn 1967.

Shufflebeam, Daniel L. *The Use and Abuse of Evaluation in Title III.* Columbus, Ohio: The Ohio State University Evaluation Center (mimeo), July 1967.

Section V

Bibliography

Bibliography

1940

Finn, James D. "Among Ourselves—Notes from and by the Department of Visual Instruction of the National Education Association," a monthly column in *Educational Screen*. November 1940, p. 382; December 1940, pp. 422-423.

1941

——————. "Adequate Training for a Director of Audio-Visual Education," *Education*, Vol. 61, No. 6 (February 1941), pp. 337-343.
——————. "Among Ourselves—Notes from and by the Department of Visual Instruction of the National Education Association," a monthly column in *Educational Screen*. January 1941, p. 21; February 1941, pp. 69-71; March 1941, pp. 110-119; April 1941, pp. 153-161; May 1941, pp. 204-207; June 1941, pp. 245-249, September 1941, pp. 293-294; October 1941, p. 340; November 1941, p. 391; December 1941, p. 430.

1942

——————. "Among Ourselves—Notes from and by the Department of Visual Instruction of the National Education Association," a monthly column in *Educational Screen*. January 1942, pp. 23-25; February 1942, pp. 66-68; March 1942, p. 95, April 1942, p. 144; May 1942, pp. 188-190; June 1942, pp 232-233.

1943

——————. "Film Distribution," (part of a symposium on the Training Film Activities of the U.S. Army), *Journal of the Society of Motion Picture Engineers*, Vol. 41 (September 1943), pp. 251-254.

1948

——————. Review of *Radio, Motion Picture and Reading Interests: A Study of High School Pupils*, by Alice P. Sterner. *Educational Screen*, Vol 27, No. 4 (April 1948), p. 188.
——————. Review of *Audiovisual Aids to Instruction*, by William Exton Jr. *Educational Screen*, Vol. 27, No. 6 (June 1948), p. 283.
——————. Review of *Education for What is Real*, by Earl C. Kelley. *Educational Screen*, Vol. 27, No. 7 (September 1948), pp. 340-342.

Bibliography

1949

_____, with Edgar Dale and Charles F. Hoban Jr. "Research on Audio-Visual Materials." In N.B. Henry (Ed.), *The Forty-eighth Yearbook of the National Society for the Study of Education: Audio-Visual Materials of Instruction.* Chicago: The University of Chicago Press, 1949, pp. 253-293.

1950

_____. "A Study of Military Audio-Visual Programs," *Abstracts of Doctoral Dissertations,* No. 60. Ohio State University Press, 1950, pp. 103-109.
Reprinted in: McClusky, F. Dean and James S. Kinder. *The Audio-Visual Reader.* Dubuque, Iowa: Wm. C. Brown Co., 1956, pp. 350-353.
_____, with Edgar Dale. "Audio-Visual Materials," *Encyclopedia of Educational Research.* New York: The Macmillan Co., 1950, pp. 84-97.
_____, with Fred F. Harcleroad and William Allen. *The Tachistoscope: Some Uses in Southern California.* Bulletin of the Research Committee of the Audio-Visual Education Association of California, Southern Section. Los Angeles: Audio-Visual Education Association of California, 1950.

1951

_____. "Audio-Visual Building Coordinators in the Rocky Mountain Region." In Fred Harcleroad and William Allen (Eds.), *Audio-Visual Administration.* Dubuque, Iowa: Wm. C. Brown Co., 1951, pp. 58-59.
_____. "Audio-Visual Aids." In Irving Melbo, et al., *Report of the Survey of the Monrovia-Duarte Union High School District.* Los Angeles: Published by the author, 1951, pp. 161-163.
_____. "The Social Studies." In Irving Melbo, et al., *Report of the Survey of the Monrovia-Duarte Union High School District.* Los Angeles: Published by the author, 1951, pp. 142-144.

1953

_____, with Myron S. Olson. *Secondary Education—A Resource Syllabus.* Second Edition. Los Angeles: College Bookstore, 1953.
_____, with Robert O. Hall and Mendel Sherman. *Report of the Study of the Audio-Visual Program of the Whittier Union High School District.* Whittier, California: Published by the author, May 1953.
_____. "Professionalizing the Audio-Visual Field," *Audio-Visual Communication Review,* Vol. 1, No. 1 (Winter 1953), pp. 6-17.
_____. "Television and Education: A Review of Research," *Audio-Visual Communication Review,* Vol. 1, No. 2 (Spring 1953), pp. 106-126.
First draft published in: *Brochure of Background Materials: Governor's Conference on Educational Television.* California State Department of Education, November 1952.

Bibliography

Reprinted in: McClusky, F. Dean and James S. Kinder, *The Audio-Visual Reader*. Dubuque, Iowa: Wm. C. Brown, Co., 1954, pp. 335-346.

—————————. "Why Teaching Tools?", *Teaching Tools*, Vol. 1, No. 1 (Fall 1953), p. 5.

1954

—————————, with Robert O. Hall. "The Audio-Visual Program." In Melbo, Irving, et al. *Report of the Survey of the Phoenix Union High Schools and Phoenix College System*. Los Angeles: Published by the author, 1954, pp. 301-307.

—————————, with Fred F. Harcleroad. "The Selection and Education of Audio-Visual Personnel." Chapter 6, *The School Administrator and His Audio-Visual Program*, First Yearbook of the Department of Audiovisual Instruction. Washington, D.C.: Department of Audiovisual Instruction, NEA, 1954, pp. 121-142.

—————————, with Fred F. Harcleroad. "The Training of the Non-Theatrical Film Worker in the University," *Journal of the University Film Producers Association*, Vol. 7, No. 2 (Winter 1954), pp. 4-8.

—————————. "Direction in Audio-Visual Communication Research," *Audio-Visual Communication Review*, Vol. II, No. 2 (Spring 1954), pp. 83-102.

—————————. "Creativity in Teaching?", *Teaching Tools*, Vol. 1, No. 2 (Winter 1954), p. 55.

—————————. "Shape of Thing to Come," *Teaching Tools*, Vol. 1, No. 3 (Spring 1954), p. 99.

—————————. "Research and the Teacher," *Teaching Tools*, Vol. 1, No. 4 (Summer 1954), p. 143.

—————————. "The Caretaker's Daughter," *Teaching Tools*, Vol. 2, No. 1 (Fall 1954), p. 3.

1955

—————————. "Audio-Visual Education," *The American People's Encyclopedia*, 1955.

—————————. "An Open Letter to the *Saturday Review*," *Film and A-V World*, March 1955, pp. 131-135.

—————————. "A Look at the Future of AV Communication," *Audio-Visual Communication Review*, Vol. 3, No. 4 (Fall 1955), pp. 244-256.

—————————. "Teachers, Comic Books, and Teaching Tools," *Teaching Tools*, Vol. 2, No. 2 (Winter 1955), p. 47.

—————————. "The Sound and the Fury of Rudolf Flesch," *Teaching Tools*, Vol. 2, No. 3 (Spring 1955), p. 91.

—————————. "A Lesson in Grammar for Arthur Bestor," *Teaching Tools*, Vol. 2, No. 4 (Fall 1955), p. 127+.

1956

—————————. "Needed: A New Concept of Communication in General Education," *Current Issues in Higher Education*. Washington, D.C.: Association for Higher Education, NEA, 1956, pp. 224-231.

—————————. "The Testing of Public Relations: Audio-Visual Materi-

als," *PR, The Quarterly Review of Public Relations*, Vol. 1, No. 5 (October 1956).
_____. "The Teacher's Handy Pocket Guide to the Jargon of the Critics," *Teaching Tools*, Vol. 3, No. 1 (Winter 1956), pp. 3+.
_____. "What is Educational Efficiency?", *Teaching Tools*, Vol. 3, No. 3 (Summer 1956), pp. 113-114.
Reprinted in: *Education*, Vol. 77, No. 5 (January 1957), pp. 262-265.
_____. "AV Development and the Concept of Systems," *Teaching Tools*, Vol. 3, No. 4 (Fall 1956), pp. 163-164.

1957

_____. *The Audio-Visual Equipment Manual.* New York: Henry Holt (Dryden).
_____, with Robert O. Hall and William C. Himstreet. *The Littlest Giant.* Los Angeles: University of Southern California Press, 1957.
_____. "Automation and Education: 1. General Aspects," *Audio-Visual Communication Review*, Vol. 5, No. 1 (Winter 1957), pp. 343-360.
_____. "Automation and Education: 2. Automatizing the Classroom—Background of the Effort," *Audio-Visual Communication Review*, Vol. 5, No. 2 (Spring 1957) pp. 451-467.
_____. "Teacher Productivity," *Teaching Tools*, Vol. 4, No. 1 (Winter 1957), pp. 7-9.
_____. "A-V Development and the Responsibility of the Public," *Teaching Tools*, Vol. 4, No. 2 (Spring 1957), pp. 63-64+.
_____. "The Forgotten Mission," *Teaching Tools*, Vol. 4, No. 4 (Fall 1957), pp. 146-147.

1958

_____. "NCFA Educational Program," *Proceedings: Forty-fourth National Consumer Finance Association Convention.* Washington, D.C.: NCFA, 1958, pp. 48-50.
_____, with John C. Schwartz Jr. and Stanley Brown (Eds.). *Reports on Select California Research Projects to Serve as Guide Lines for Those Concerned with America's Education.* Los Angeles: Audio-Visual Education Association of California Research Bulletin, 1958.
_____. "Educational Relations of the Consumer Finance Industry," *Pennsylvania Consumer Finance News*, Vol. 8, No. 3 (September 1958), pp. 12-16.
_____. "The Good Guys and the Bad Guys," *Phi Delta Kappan*, Vol. 40, No. 1, October 1958, pp. 2-4+.
_____. "Professor Adler and Technology," *Journal of Higher Education*, Vol. 29, No. 8 (November 1958), pp. 455-458.
_____. "Where Are the 'Means and Appliances?' ", *Teaching Tools*, Vol. 5, No. 1 (Winter 1958), pp. 6-7.
_____. "Merit Rating for School Boards? Some Suggestions," *Teaching Tools*, Vol. 5, No. 3 (Summer 1958), pp. 108-109.

Bibliography

_____. "The Middle of the Line," *Teaching Tools*, Vol. 5, No. 4 (Fall 1958), p. 160.

1959

_____, with Leonard H. Bathurst and Frank H. Oetting. *Reports on Select California Audio-Visual Research Projects to Serve as Guidelines for Those Concerned with America's Education*. Los Angeles: Audio-Visual Education Association of California, 1959.

_____. "Some Notes for an Essay on Griswold and Reading," *Audio-Visual Communication Review*, Vol. 7, No. 2 (Spring 1959), pp. 111-121.

_____. "From Slate to Automation," *Audio-Visual Instruction*, Vol. 4, No. 3 (March 1959), pp. 84-85+.

_____. "A-V 864 vs A.D. 1970," *Teaching Tools*, Vol. 6, No. 1 (Winter 1959), pp. 4-5.

1960

_____. "Directions for Theory in Audio-Visual Communication," *The New Media in Education*, Western Regional Conference on Educational Media Research, Sacramento, California, April 20-22, 1960, edited by Jack V. Edling. Sacramento: Sacramento State College, 1960, pp. 54-61.

_____. "Reflections on a Roller Coaster," *AVISCO Newsletter*. Garden City, N.Y.: Long Island Audiovisual Council, 1960, pp. 30-38.

_____. "Automation and Education: 3. Technology and the Instructional Process," *AV Communication Review*, Vol. 8, No. 1 (Winter 1960), pp. 5-26.
Reprinted in the following publications:
1. Lumsdaine, A. A. and Robert Glaser (Eds.). *Teaching Machines and Programmed Learning: A Source Book*. Washington, D.C.: Department of Audiovisual Instruction, NEA, 1960. (Shortened version.)
2. *Phi Delta Kappan*, Vol. 41, No. 9 (June 1960), pp. 371-378. (Shortened version, entitled "Technology and the Instructional Process.")
3. *AV Communication Review*, Vol. 9, No. 5, Supplement 2 (September-October 1960), pp. 84-94. (Same version as *Phi Delta Kappan*, but entitled, "A New Theory for Instructional Technology.")
4. Smith, Wendell I., and J. William Moore (Eds.). *Programmed Learning: Theory and Research*. Selected Readings. New York: D. Van Nostrand Co., 1962, pp. 3-17. (Same version as No. 3.)
5. Russel, Roger W. (Ed.). *Frontiers in Psychology*. New York: Scott, Foresman and Company, 1964, pp. 59-66. (Same version as No. 3.)
6. Gross, Ronald and Judith Murphy (Eds.). *Frontiers of Education*. New York: Harcourt Brace, 1964. (Original version.)

_____. "Major Development in Audio-Visual Education in 1959," *Film World*, Vol. 16, No. 1 (January 1960), pp. 16-17.
Reprinted in· *Faculty Memo*, Audio-Visual Instructional Materials Center, Department of Education, University of Chicago.

Bibliography

_____. "Assignment DAVI Personnel: 1970," *Educational Screen and Audiovisual Guide*, (August 1960), pp. 430-431.

_____. "Technological Innovation in Education," *Audiovisual Instruction*, Vol. 5, No. 7 (September 1960), pp. 222-228.

_____. "Teacher Understanding—A Key to the New Technology," *CTA Journal*, September 1960, p. 5.

_____. "Teaching Machines: Auto-Instructional Devices for the Teacher," *NEA Journal*, Vol. 49, No. 8, November 1960, pp. 41-44. (Reprinted in two books of readings.)

1961

_____. "New Techniques of Teaching for the Sixties," *Teacher Education: Direction for the Sixties*, American Association of Colleges for Teacher Education, National Education Association, 1961, pp. 31-42.

_____, with Elinor Richardson. "The Principal Faces the New Technology," *National Elementary Principal*, Vol. 40, No. 4 (January 1961), pp. 18-22.
Reprinted in: *Graflex Audiovisual Digest*, second edition. Rochester, N.Y.: Graflex, Inc., 1963. (Shortened version, entitled, "The New Technology.")

_____. "Educational Technology—A New Force," *ALA Bulletin*, February 1961, pp. 118-121.

_____. "The Tradition in the Iron Mask," *Audiovisual Instruction*, Vol. 6, No. 6 (June 1961), pp. 238-243.

_____. "Technology and the Teacher: Perspective on Technology: I: Does Pokeberry Juice Have to Go?", *Professional Growth for Teachers*, Vol. 7, No. 4, 1961.

_____. "Technology and the Teacher: For the Principal—Growth in the Middle," *Professional Growth for Principals*, Vol. 7, No. 4, 1961.

1962

_____. "8mm Sound Film in Instructional Systems," *8mm Sound Film and Education*, edited by Louis Forsdale. New York: Columbia University, 1962, pp. 142-144.

_____. "The Instructional Materials Section—An Evaluation," from *An Analysis of the Department of Educational Services*. University Park, Penn.: Pennsylvania State University, 1962.

_____. "Instructional Technology and the Credential Requirement." September 1962. (Unpublished.)

_____, with Donald G. Perrin. *Teaching Machines and Programed Learning: 1962: A Survey of the Industry*. National Education Association, Technological Development Project, Occasional Paper No. 3. Los Angeles: University of Southern California, School of Education, January 1962.

_____, with Donald G. Perrin and Lee E. Campion. *Studies in the Growth of Instructional Technology: I: Audiovisual Instrumentation for Instruction in the Public Schools, 1930-1960—A Basis for*

Bibliography

Take-Off. National Education Association, Technological Development Project, Occasional Paper No. 6. Los Angeles: University of Southern California, School of Education, August 1962.

_____. "Technology's Challenge: The Sweeping Compound Theory of School Administration Has Got to Go," *Professional Growth for Administrators*, Vol. 7, No. 4, 1962, p. 2.

_____. "Technology: Revolution in the Classroom," *The Challenge of Change.* New York: The American Textbook Publishers Institute, 1962, pp. 13-17.

_____. "Technology—Potential for Educational Revolution?", *Tomorrow's Teaching.* Frontiers of Science Foundation Symposium, Oklahoma City, Oklahoma, January 13-14, 1962. Reprinted with some changes in: *Improving Instruction Through Audio-Visual Media.* California State Department of Education, 1963.

_____, with Donald G. Perrin. *Bibliography on New Media and Instructional Technology.* Washington, D.C.: Technological Development Project, National Education Association, February 1962.

_____, with Joan Rosengren. "8mm Sound Film: A Full Dress Conference at TC," *Audiovisual Instruction*, Vol. 7, No. 2, February 1962, pp. 90-93.

_____. "Review of Educational Research," Vol. 32, No. 2 (April 1962), special issue of "Instructional Materials: Educational Media and Technology," prepared by the Committee on Instructional Materials: Educational Media and Technology, James D. Finn, Co-Chairman.

_____. "A Walk on the Altered Side," *Phi Delta Kappan*, Vol. 44, No. 1, October 1962, pp. 29-34.

_____. "The Engines Are Ready for the Thrust," *Audiovisual Instruction*, Vol. 7, No. 7, September 1962, pp. 438-440.

_____. "Take-Off to Revolution," *The American Behavioral Scientist*, Vol. 6, No. 3 (November 1962), pp. 12-15.

_____. "Footnote to an Editorial," *Educational Screen and Audiovisual Guide*, December 1962, p. 718.

1963

_____. "Technology and Educational Development," *AVISCO Newsletter.* Garden City, N.Y.: Long Island Audiovisual Council, 1963.

_____, with Donald G. Perrin. *Instrumentation for Instruction, 1955-1970.* National Education Association, Technological Development Project. Los Angeles: University of Southern California, School of Education, 1963. (Unpublished final report.)

_____. "Teaching Machines," Review of *Teaching Machines* by Benjamin Fine, *The Journal of Teacher Education*, Vol. 14, No. 1 (March 1963), p. 105.

_____. "Instructional Technology," *The Bulletin of the National Association of Secondary School Principals*, Vol. 47, No. 283 (May 1963), pp. 99-103.

_____. Guidelines for the Assessment of the Unique Educa-

Bibliography

tional Potentials of the Various Media," *Educational Media in Transition.* A report to the U.S. Office of Education. New York: Educational Media Council, October 1963.

_____. "Machines Revisited," *Programed Instruction,* Vol. 3, No. 1 (October 1963), p. 2.

_____. "A Royal Society for the Improvement of Media Knowledge." Some remarks on the status and future of the Educational Media Council. Based on an extemporaneous speech delivered to the Council at Washington, D.C., October 8, 1963. (Unpublished.)

_____. "A Review of the Keating Report," *American School and University,* December 1963, pp. 34-35.

1964

_____. "The Franks Had the Right Idea," *NEA Journal,* Vol. 53, No. 4, April 1964, pp. 24-27.

_____. "A Revolutionary Season," *Phi Delta Kappan,* April 1964, pp. 348-354.

_____, with Boyd M. Bolvin and Donald G. Perrin. *A Selective Bibliography on New Media and Instructional Technology.* Staff Paper Number 1, Instructional Technology and Media Project, School of Education. Los Angeles: University of Southern California, April 1964.

_____, with Boyd M. Bolvin. *An Information Indexing, Storage, and Retrieval System for Documents in the Field of Instructional Technology.* Staff Paper Number 2, Instructional Technology and Media Project, School of Education. Los Angeles: University of Southern California, June 1964.

1965

_____. "Conventional Wisdom and Vocational Reality." Presented at the Fourth Annual Conference on Higher Education, San Francisco, California, May 7-8, 1965. (Unpublished.)

_____. "New Uses for Old Clichés." Commencement address delivered at Los Angeles Valley College, June 18, 1965. (Unpublished.)

_____. "An Uncomfortable Rejoinder. Comments on W. H. Ferry's proposition concerning vocational education in junior colleges." Prepared for the Conference on Technical Training in Junior Colleges, Center for the Study of Democratic Institutions, Santa Barbara, California, June 25, 1965. (Unpublished.)

_____. "Comments and Recommendations Concerning the Design of Von Kleinschmidt Hall for the Viewpoint of Applying Instructional Technology." 1965. (Unpublished.)

_____. "Instructional Technology," *Audiovisual Instruction,* Vol. 10, No. 3, March 1965, pp. 192-194.

_____. "The Marginal Media Man, Part I: The Great Paradox," *Audiovisual Instruction,* Vol. 10, No. 10, December 1965, pp. 762-765.

Bibliography

1966

_____. "The Emerging Technology of Education," in *Educational Implications of Technological Change*, Appendix Volume IV, Technology and the American Economy. Prepared for the National Commission on Technology, Automation, and Economic Progress. Washington, D.C.: U.S. Government Printing Office, February 1966, pp. 33-52.

_____. "Educational Technology, Innovation, and Title III," prepared for PACE (Projects to Advance Creativity in Education), 1966.

_____. *The New Educational Media and the Harvard Graduate School of Education: Some Considerations Relating to a Program*. A Memorandum Addressed to the Faculty, Graduate School of Education, Harvard University. Prepared at the request of Professor David Purpel by James D. Finn with the assistance of Robert O. Hall. 1966.

_____. "A Possible Model for Considering the Use of Media in Higher Education." An edited extract from *Educational Media and Harvard University*, A memorandum prepared at the request of Professor David Purpel, Graduate School of Education, by James D. Finn, with the assistance of Robert O. Hall and Bruce Humphrey. Los Angeles: University of Southern California, Fall 1966.
Published in *AV Communication Review*, Vol. 15, No. 2, Summer 1967, under the same title, pp. 153-157.

_____. "The Technological Revolution in Education." *New Media and Changing Educational Patterns*. Sacramento, Calif.: State Department of Education, 1966, pp. 21-22.

1967

_____, with Diana Caput. *Interim Report on the Search Bibliography and Document Collection*. U.S. Public Health Service, Medical Information Project, Research Memorandum Number 1. Los Angeles: University of Southern California, School of Education, School of Medicine, October 1967.

_____, with Royd Weintraub. *An Analysis of Audio-Visual Machines for Individual Program Presentation*. U.S. Public Health Service, Medical Information Project, Research Memorandum Number 2. Los Angeles: University of Southern California, May 1967.

_____, with Stephen Abrahamson and Diana Caput. *Strategy and Tactics for Program Preparation*. U.S. Public Health Service, Medical Information Project, Research Memorandum Number 3. Los Angeles: University of Southern California, May 1967.

_____. "Dialog in Search of Relevance." Speech presented at Lake Okoboji Leadership Conference, 1967.

_____. *A Study of the Concentration of Educational Media Resources to Assist in Certain Education Programs of National Concern. Part II: Educational Media and Vocational Education*. Prepared for a contract with the U.S. Office of Education by the Educational Media Council. May 1967.

Bibliography

_____. "Personnel in Educational Media/Technology: Some Comments on the State of the Profession: I. The Public Schools." Included in the final report of a Title III national study by the National Academy of Scholars, 1967.

_____. "Personnel in Education Media/Technology: Some Comments on the State of the Profession. II. Higher Education." Included in the final report of a Title III national study by the National Academy of Scholars, 1967.

1968

_____. "Institutionalization of Evaluation," in *Evaluation and "PACE": A Study of Procedures and Effectiveness of Evaluation Sections in Approved PACE Projects, with Recommendations for Improvement*, February 29, 1968, pp. A-170-A-202.

Reprinted in: *Educational Technology*, December 1969, pp. 14-23.

_____. "What is the Business of Educational Technology? Some Immodest Comments on Mr. Muller's Paper." Prepared for a Conference sponsored by Designing Education for the Future, An Eight-State Project. July 1968. (To be published.)